AF352809

Polymer Processing

Polymer Processing

Modeling and Correlations Finalized to Tailoring the Plastic Part Morphology and Properties

Special Issue Editors

Giuseppe Titomanlio
Vito Speranza

MDPI • Basel • Beijing • Wuhan • Barcelona • Belgrade • Manchester • Tokyo • Cluj • Tianjin

Special Issue Editors
Giuseppe Titomanlio
University of Salerno
Italy

Vito Speranza
University of Salerno
Italy

Editorial Office
MDPI
St. Alban-Anlage 66
4052 Basel, Switzerland

This is a reprint of articles from the Special Issue published online in the open access journal *Materials* (ISSN 1996-1944) (available at: https://www.mdpi.com/journal/materials/special_issues/Polymer_Processing).

For citation purposes, cite each article independently as indicated on the article page online and as indicated below:

LastName, A.A.; LastName, B.B.; LastName, C.C. Article Title. *Journal Name* **Year**, *Article Number*, Page Range.

ISBN 978-3-03928-754-3 (Pbk)
ISBN 978-3-03928-755-0 (PDF)

Contents

About the Special Issue Editors

Giuseppe Titomanlio graduated in chemical engineering at University of Naples. Soon after graduation, he moved to the University of Palermo, first as an assistant professor, then as an associate professor after, and since 1986 as full professor of transport phenomena. In 1991, he moved to the University of Salerno. From 2004 to 2010, there he was Dean of Chemical Engineering. From 2001 to 2004, he was President of GRICU (the Italian association of university researchers in chemical engineering areas). His research has focused on the properties and processing of polymers. In particular, his research includes rheology and properties of solid polymers, and melts, analysis, and simulation of phenomena related to polymer processing and, in particular, to injection molding, problems of general interest to the processing of polymeric materials such as the crystallization kinetics at high cooling rates and high-pressure and flow-induced crystallization. He also focuses on the use of polymers for biomedical and pharmaceutical applications. The results of these activities resulted in about 200 papers published in international scientific journals and numerous communications at international conferences. He is often invited as a lecturer at conferences on polymer processing and properties all around the world. Over the years, Prof. Titomanlio has coordinated research units in several research projects of local, national, and European relevance. He is active in the Polymer Processing Society (PPS), of which he was a member from the very beginning of the Society. He was organizer and chairman of the Regional Conference of the PPS held in Palermo in September 1991 and of the 24th International Conference of the PPS, held in Salerno in June 2008; he was President of the PPS for the biennium 2012–2014.

Vito Speranza was born in Avellino in 1965. In 1992, he graduated in electronical engineering at the University of Salerno defending a thesis on optimal interpolation of electromagnetic fields over a cylinder. He received a Ph.D. degree in April 1999 defending a thesis entitled "Analysis and Simulation of the Injection Molding Process". From 1999 to 2002, he was a research fellow at the University of Salerno, Dept. of Chemical Engineering, in modeling and simulation of processing of polymeric materials. As of September 2000, he was teaching assistant for various courses (Transport Phenomena, Principles of Chemical Engineering, and Fluid Dynamics) in Chemical Engineering. Since March 2002, he has been research engineer in the technical staff at the University of Salerno, Dept. of Industrial Engineering (former Dept. of Chemical Engineering). His main research interests focus on analysis and simulation of injection molding of thermoplastics, structure development in polymer processing, and morphology and polymer surface analysis. He has been involved in several national and international research programs and in several collaborations with companies and research centers. He is author of one patent, 70 publications in international peer-reviewed scientific journals, 2 book chapters, and about 80 publications of proceedings at international conferences.

Preface to "Polymer Processing"

During the transformation of polymeric materials into final usable objects, polymers (usually viscoelastic fluids) undergo complex histories of deformation and temperature distributions. The morphology of each element of the final object is determined by the deformation and temperature history of the particle, which, at the end of a complex evolution, solidified in that position. Crystallization kinetics, when active, is deeply influenced by the molecular stretch acquired by the effect of the flow. The distribution of final morphology determines the properties of the final object and these can undergo remarkable changes (more than one order of magnitude) via the effect of morphology variations. These observations have been shared for some time within the scientific community; however, only in special cases is understanding clear or are phenomena described occuring along chain of processing–morphology–properties. On the basis of these considerations, the objective of this Special Issue was the collection of progression or reviews clarifying the relationships among processing conditions, as well as the resulting morphology and properties of the final objects. This objective included both experimental correlations and modelling both in relation to any polymer processing operation and to any polymer of technological interest. This Special Issue includes 11 research articles and 1 review. Hashimoto et al. experimentally studied ultra-high molecular weight polyethylene (UHMWPE) films stretched under different operating conditions by adopting different temperatures and stretching speeds, and with different operational configurations by adopting both uniaxial and biaxial stretching modes, in order to evaluate the stretching effect on the film final structure. Zhu et al. adopted a Lagrangian approach to obtain numerical results about the mixing mechanism and performance of a novel four-screw extruder. Sun et al. developed a 3D numerical simulation of reactive extrusion processes with the aim of better understanding the effect of operational and geometric parameters on both mixing and reaction processes in the preparation of PP/TiO2 nanocomposites. Speranza et al. accurately analyzed morphology via atomic force microscopy (AFM) and discussed, in relation to the operating conditions, the morphologies developed along the cross-sections of moldings, obtained by adopting a system able to rapidly change the cavity surface temperature during the process. The processes adopted to obtain those samples were numerically simulated to apply a criterion for the achievement of fibrillar morphology based on histories of molecular stretching and mechanical work. The results of the criterion were found to be consistent with the morphology distributions along the cross-section of each sample by Liparoti et al. The replication of micro- and nano-features on PLA molded samples, obtained by injection-molding tests performed with a modulated cavity surface temperature during the process, was analyzed by Liparoti et al. Li et al. constructed an adaptive optimization method to reduce stress and deformation exerted on a polymer stent produced with the micro-injection molding process. Nagato et al. investigated the replication of microlens arrays on polymethylmethacrylate (PMMA) films produced using a laser-assisted thermal imprinting process (LATI) under locally different processing conditions with a pressure and laser power heating mold surface. Hao et al. provided an innovative preparation method of insulation pressboards adopted in converter transformers; the method is based on the coating of the pressboards with a polytetrafluoroethylene (PTFE) functional film by radio frequency magnetron sputtering. The method was found to be effective in enhancing both the electrical insulation and the oil insulation of the prepared pressboards. Hamidi et al. investigated the processability and properties of silk-reinforced composites, produced with vacuum-assisted resin transfer molding (VARTM). Ruan predicted the effect of flow and temperature on the spherulitical and shish-kebab structures by adopting a simulation model based on a multiscale approach. Finally, Gonzalez-Gutierrez et al. reviewed the material extrusion additive manufacturing (MEAM) techniques: they focused the review on the techniques that adopt polymers filled with high contents of ceramic and metallic powders.

Giuseppe Titomanlio, Vito Speranza
Special Issue Editors

 *materials*

Editorial

Polymer Processing: Modeling and Correlations Finalized to Tailoring Plastic Part Morphology and Properties

Giuseppe Titomanlio and Vito Speranza *

Department of Industrial Engineering, University of Salerno, via Giovanni Paolo II, 132, 84084 Fisciano, Italy; gtitomanlio@unisa.it
* Correspondence: vsperanza@unisa.it; Tel.:+39-089-96-4145

Received: 11 March 2019; Accepted: 9 April 2019; Published: 14 April 2019

Abstract: The analysis of polymer processing operations requires the description of simultaneous transient momentum and heat transfer down to material solidification. The aim of the analysis is to improve and, hopefully, optimize the final properties that are determined by the final morphology of the part. In this special issue, consisting of 1 review and 11 research articles detailing several polymer processing operations, experimental and numerical analyses have been conducted in order to identify and describe the main relevant phenomena, that affect the product morphologies and properties.

Keywords: injection molding; laser-assisted thermal imprinting; extrusion; additive manufacturing; microinjection molding; numerical simulation; morphology; film stretching; composite laminates

---

During the transformation of polymeric materials into final usable objects, the polymers (usually viscoelastic fluids) undergo very complex histories of deformation and temperature distributions. The morphology of each element of the final object is determined by deformation and temperature history of the particle which, at the end of a complex evolution, solidified in that position. It is well known that also crystallization kinetics, when active, is deeply influenced by the molecular stretch acquired by effect of the flow. On the other hand, the distribution of final morphology determines the properties of the final object and these can undergo remarkable changes (more than one order of magnitude) by effect of morphology variations.

These observations have been shared for some time within the scientific community, however, only in special cases, clear understanding or description of phenomena taking place along the chain processing-morphology-properties have been identified and reported. On the basis of these considerations the objective of this special issue was the collection of progress or reviews clarifying relationships among processing conditions, as well as the resulting morphology and properties of the final objects. This objective includes both experimental correlations and modelling both in relation to any polymer processing operation and also to any polymer of technological interest.

This special issue includes 11 research articles and 1 review. Nagato et al. investigated the replication of microlens arrays on polymethylmethacrylate (PMMA) films produced by using a laser-assisted thermal imprinting process (LATI) under different processing conditions of pressure and laser power heating mold surface locally [1]. Sun et al. developed a 3D numerical simulation of reactive extrusion processes with the aim of better understanding the effect of operational and geometric parameters on both mixing and reaction processes in the preparation of PP/TiO$_2$ nanocomposites [2]. Speranza et al. accurately analyzed morphology via atomic force microscopy (AFM) and discussed, in relation to the operating conditions, the morphologies developed along the cross sections of moldings, obtained by adopting a system able to rapidly change the cavity surface temperature during the process [3]; the processes adopted to obtain those samples were numerically simulated in order to apply a criterion for the achievement of fibrillar morphology based on histories of molecular stretching and mechanical work; the results of the criterion were found to be consistent with the morphology distributions along the cross section of each sample by Liparoti et al. [4]. Li et al. proposed an adaptive optimization method in order to reduce stress and deformation exerted on a polymer stent obtained with the micro-injection molding process [5]. Hashimoto et al. carried on an experimental study on ultra-high molecular weight polyethylene (UHMWPE) films stretched under different operating conditions, namely adopting different temperatures and stretching speeds, and with different operational configurations, by adopting both uniaxial and biaxial stretching modes, in order to evaluate the stretching effect on the film final structure [6]. Zhu et al adopted a Lagrangian approach to obtain numerical results about the mixing mechanism and performance of a novel four-screw extruder [7]. Hamidi et al. investigated processability and properties of silk reinforced composites, obtained with vacuum assisted resin transfer molding (VARTM) [8]. Liparoti et al., analyzed the replication of micro-features and nano-features on PLA molded samples, obtained by injection-molding tests performed with a modulated cavity surface temperature during the process [9]. Hao et al. provided an innovative preparation method of the insulation pressboards adopted in converter transformers; the method is based on the coating of the pressboards with a polytetrafluoroethylene (PTFE) functional film by radio frequency magnetron sputtering; the method was found effective in enhancing both the electrical insulation and the oil insulation of the prepared pressboards [10]. Ruan predicted the effect of flow and temperature on the spherulitical and shish-kebab structures by adopting a simulation model based on a multiscale approach [11]. At last, Gonzalez-Gutierrez et al. prepared a review on the material extrusion additive manufacturing (MEAM) techniques: in particular, they focused the review on the techniques that adopt polymers filled with high contents of ceramic and metallic powders [12].

Funding: This research received no external funding.

Conflicts of Interest: The authors declare no conflict of interest.

References

1. Nagato, K.; Yajima, Y.; Nakao, M. Laser-Assisted Thermal Imprinting of Microlens Arrays—Effects of Pressing Pressure and Pattern Size. *Materials* **2019**, *12*, 675. [CrossRef] [PubMed]
2. Sun, D.; Zhu, X.; Gao, M. 3D Numerical Simulation of Reactive Extrusion Processes for Preparing PP/TiO$_2$ Nanocomposites in a Corotating Twin Screw Extruder. *Materials* **2019**, *12*, 671. [CrossRef] [PubMed]
3. Speranza, V.; Liparoti, S.; Pantani, R.; Titomanlio, G. Hierarchical Structure of iPP During Injection Molding Process with Fast Mold Temperature Evolution. *Materials* **2019**, *12*, 424. [CrossRef] [PubMed]
4. Liparoti, S.; Speranza, V.; Pantani, R.; Titomanlio, G. Process Induced Morphology Development of Isotactic Polypropylene on the Basis of Molecular Stretch and Mechanical Work Evolutions. *Materials* **2019**, *12*, 505. [CrossRef] [PubMed]
5. Li, H.; Liu, K.; Zhao, D.; Wang, M.; Li, Q.; Hou, J. Multi-Objective Optimizations for Microinjection Molding Process Parameters of Biodegradable Polymer Stent. *Materials* **2018**, *11*, 2322. [CrossRef] [PubMed]
6. Hashimoto, Y.; Nishitsuji, S.; Kurose, T.; Ito, H. Structural Formation of UHMWPE Film Tracked by Real-Time Retardation Measurements during Uniaxial/Biaxial Stretching. *Materials* **2018**, *11*, 2292. [CrossRef] [PubMed]
7. Zhu, X.Z.; Tong, Y.; Hu, Y.X. Chaotic Manifold Analysis of Four-Screw Extruders Based on Lagrangian Coherent Structures. *Materials* **2018**, *11*, 2272. [CrossRef] [PubMed]
8. Hamidi, Y.K.; Yalcinkaya, M.A.; Guloglu, G.E.; Pishvar, M.; Amirkhosravi, M.; Altan, M.C. Silk as a Natural Reinforcement: Processing and Properties of Silk/Epoxy Composite Laminates. *Materials* **2018**, *11*, 2135. [CrossRef] [PubMed]
9. Liparoti, S.; Speranza, V.; Pantani, R. Replication of Micro- and Nanofeatures in Injection Molding of Two PLA Grades with Rapid Surface-Temperature Modulation. *Materials* **2018**, *11*, 1442. [CrossRef] [PubMed]
10. Hao, J.; Liu, C.; Li, Y.; Liao, R.; Liao, Q.; Tang, C. Preparation Nano-Structure Polytetrafluoroethylene (PTFE) Functional Film on the Cellulose Insulation Polymer and Its Effect on the Breakdown Voltage and Hydrophobicity Properties. *Materials* **2018**, *11*, 851. [CrossRef] [PubMed]

11. Ruan, C. "Skin-Core-Skin" Structure of Polymer Crystallization Investigated by Multiscale Simulation. *Materials* **2018**, *11*, 610. [CrossRef] [PubMed]
12. Gonzalez-Gutierrez, J.; Cano, S.; Schuschnigg, S.; Kukla, C.; Sapkota, J.; Holzer, C. Additive Manufacturing of Metallic and Ceramic Components by the Material Extrusion of Highly-Filled Polymers: A Review and Future Perspectives. *Materials* **2018**, *11*, 840. [CrossRef] [PubMed]

Article

Structural Formation of UHMWPE Film Tracked by Real-Time Retardation Measurements during Uniaxial/Biaxial Stretching

Yoshinori Hashimoto [1,2,*], Shotaro Nishitsuji [1], Takashi Kurose [1] and Hiroshi Ito [1]

[1] Department of Organic Materials Science, Graduate School of Organic Materials Science, Yamagata University, 4-3-16 Jonan, Yonezawa, Yamagata 992-8510, Japan; nishitsuji@yz.yamagata-u.ac.jp (S.N.); takashi.kurose@yz.yamagata-u.ac.jp (T.K.); ihiroshi@yz.yamagata-u.ac.jp (H.I.)

[2] Toyo Seiki Seisaku-syo, Ltd., 4-23, Ukima 5-chome, Kita-ku, Tokyo 115-0051, Japan

* Correspondence: y-hashimoto@toyoseiki.co.jp

Received: 23 October 2018; Accepted: 10 November 2018; Published: 15 November 2018

Abstract: This work reports on an experimental study of the stretching of ultra-high molecular weight polyethylene (UHMWPE) film in various uniaxial/biaxial stretching modes at various temperatures and stretching speeds. We examined the stress-birefringence relationship as a stress-optical rule (SOR) under uniaxial stretching and evaluated the stress-optical coefficient (SOC). Wide-angle X-ray diffraction (WAXD) measurements were applied to evaluate the contribution to birefringence of the crystalline and amorphous phases and to characterize stretching modes. In simultaneous biaxial stretching, the melting temperature (T_m) proved critical to structural formation. We applied thermal analysis techniques and tensile testing to evaluate higher order structures after each stretching mode.

Keywords: uniaxial/biaxial stretching; retardation; birefringence; molecular orientation; stress-optical rule

1. Introduction

Long-chain structures of semi-crystalline polymers have been used to create films with advanced properties. An understanding of the structural evolution of polymer film during stretching is essential to establishing ideal processing conditions, including temperature, stretching speed, draw ratio and stretching mode. Stretching of polymer film changes the crystallinity and the structure of crystalline polymers; these changes are more complex than for amorphous polymers.

Film stretching speed in industrial process nowadays reaches 400 m/min for polymers such as polypropylene (PP) and polyethylene terephthalate (PET) which can provide high production rate. However, it is an important task to make it compatible with the required properties of stretched film. Fast stretching speed generates sudden structural formation which narrows the suitable processing window. Insight into different types of stretching modes helps in controlling processing window and understanding the nature of the film. In laboratory research, on-line structure evaluations during film stretching have been studied to understand higher order structures in addition to off-line evaluation.

Recent studies indicate that UHMWPE film has been focused as an application in the field of lithium ion battery separator used in computer and communication applications [1,2]. However, while there are many reports on polymers such as PP and PET, still few reports exist on fast stretching of UHMWPE film. In UHMWPE film processing, solvent is usually added to resolve the difficulty of processing due to very high viscosity. Structural evaluations of biaxially stretched UHMWPE film without solvent have been studied by Uehara et al. [3]. Properties of UHMWPE and its relationship with stretched films in various conditions were also studied [4–6]. Stretching speeds conducted in reported studies were of the order of 0.1 m/min. Our motivation is to investigate structural formation of UHMWPE film under wider range of stretching speed and various stretching modes by on-line retardation measurement method.

Evaluations of birefringence, wide angle X-ray diffraction (WAXD), infrared spectroscopy and other measures are used to investigate higher order structures. Birefringence, which arises from the polarity of polymer molecules and molecular orientation, is an optical property defined by difference in the refractive index in orthogonal directions and indicates average orientation, including both the crystalline and amorphous phases. Investigations of the linear relationships at the onset of stretching between birefringence and stress, known as stress-optical rule (SOR), helps to illuminate the process of structural formation. Hassan et al. [7–9] discuss on-line measurements of birefringence and the development of higher order structures in investigations of mechano-optical behavior of PET films during stretching, reporting that SOR continues to apply up to the onset of strain hardening. Simultaneous measurements of stress and birefringence had been conducted by Kotaka et al. [10] in an elongational flow on low-density polyethylene blended with UHMWPE, confirming that stress-optical coefficient (SOC) was independent either of strain rate, stress, molecular weight and its distribution, and showed agreement with the value in reported studies [11,12]. Ryu et al. [13] have shown that SOR does not hold at low temperature close to glass transition temperature (T_g) or at high strain rates due to the contributions of glassy stress. In our previous study [14], we investigated SOC of cyclic olefin copolymer films stretched at temperatures above T_g, SOC values were found to approach the photo-elastic coefficient (PEC) under fast stretching speed.

In uniaxial stretching modes, the free-width and constrained-width stretching are known. Free-width stretching occurs as the width of the film shrinks and has been studied as a fundamental stretching mode. Relationship between draw-ratio and strengthening in mechanical properties was investigated by Gao et al. [15], the orientation behavior drawn to different draw-ratios was studied using deformation models by Bandeira et al. [16]. In contrast to free-width stretching, constrained-width stretching maintains constant width by clamping the sides of the film. This can be regarded as a type of biaxial stretching, since the polymer chains tend to orient toward the transverse direction (TD). Meng et al. [17] studied the effect of constraint in the TD, that it can provide broader stretching window as compared to free-width stretching. Ward et al. [18] investigated the deformation behavior in both free-width and constrained-width uniaxial drawings and showed it can be described by the deformation of a molecular network.

Biaxial stretching modes includes simultaneous and sequential biaxial stretching and are effective methods to obtain films with enhanced properties [19]. In general, simultaneous biaxial stretching makes it possible to achieve isotropic properties for the machine direction (MD) and for the TD. However, it requires complex stretching equipment and generally leads to limited production rates. In crystalline polymers, biaxial stretching is performed by breaking the crystalline structure and tie molecules. Thus, polymers of low crystallinity have a wide processing window. In sequential biaxial stretching, the initial stretching along the MD results in alignment parallel to the MD, resulting in strain-induced crystallization. With the second stretching along the TD, the elongation of the amorphous phase dominates compared to the crystalline phase. Types of orientation by different stretching modes were defined and classified by Heffelfinger et al. [20], which has been commonly referred in understanding of stretching modes.

Investigations of structure development under various stretching modes have been studied on various polymers, revealing the attainable properties of films. Ajji et al. [21] investigated the biaxial stretchability, developed structure, molecular orientation and shrinkage of linear low-density polyethylene (LDPE). Cakmak et al. [22] investigated the effects of stretching mode on the crystalline texture of polylactic acid (PLA) films. Under simultaneous biaxial stretching, PLA films showed in-plane isotropy with poor crystalline order. Under sequential biaxial stretching, development of oriented crystallization was confirmed at first stretching in the MD. Second stretching in the TD showed destruction of the former crystalline structure in the MD, establishing a second growth of oriented but poorly ordered crystalline structure. Kojima et al. [23] investigated biaxial oriented films of PE produced by blow extrusion. In biaxial stretching of film, there have been significant patent [19], concerning biaxial oriented polyethylene film with improved optical and sealability properties.

In the current study, we used the newly developed stretcher combined with a retardation-measuring high-speed camera able to track retardation as on-line measurements under fast stretching speed up to 60 m/min, a speed comparable to industrial process. The effects of temperature, stretching speed, and draw-ratio (DR) on the structural development of UHMWPE film under different stretching modes were studied. Retardation and birefringence of the stretched films were measured using polarizing microscope as off-line measurements. WAXD measurements were carried out to evaluate crystallinity and orientation, the contribution of stretching mode to birefringence is discussed. Thermal analyses and tensile tests were performed to assess the structure of the stretched UHMWPE films.

2. Materials and Methods

2.1. Materials

We examined a UHMWPE film (Saxin NEWLIGHT® #13W [24]) in this study. This film has M_w of 5.5 million g/mol. Raw UHMWPE particles are compressed and skived to a thickness of 130 μm. The material has a white semi-transparent appearance. The surface is very smooth. The density of the material is 0.94 g/cm^3; the melting temperature (T_m) is 137 °C. The degree of polymerization is about 10 times that of high-density polyethylene (HDPE). The film features extremely long linear chain structure and maintains its form beyond T_m, permitting stretching across a wider stretching temperature window. Differential Scanning Calorimetry (DSC) scans identified a crystallinity of 59%.

2.2. Stretching Process

Figure 1 and Table 1 show DSC scans of the sample film and stretching window from this study. Figure 2 shows the film stretcher used, combined with high-speed camera. The sample film was moved by a pneumatic device to the clamping position in the chamber. The chamber was controlled to settle to equilibrium temperature, with a heater blower located underneath the clamping position. The preheating time before stretching was 3 min. Stretching temperatures ranged within the 100–140 °C range. Stretching modes included uniaxial free/constrained width stretching and simultaneous/sequential biaxial stretching. True stress was calculated by dividing engineering stress by instantaneous sectional area. Note that certain additional stretching conditions were tested to confirm the validity of our discussion.

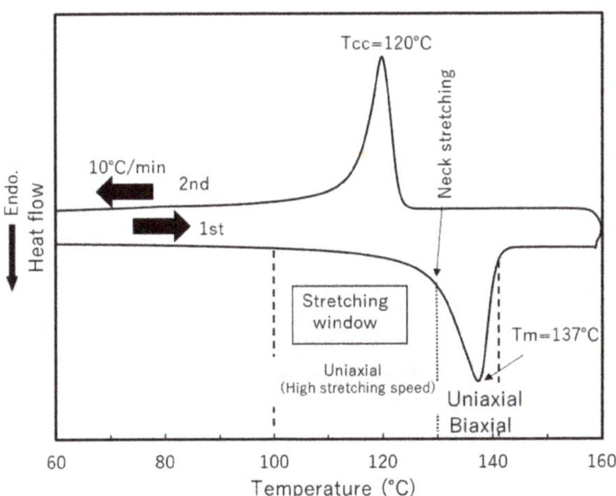

Figure 1. DSC measurement results for sample film showing stretching window in the experiment.

Table 1. Stretching mode, temperature and achieved draw-ratio in the experiment; stretching speeds are 1, 10, and 60 m/min for uniaxial mode, 1 m/min for biaxial mode.

Stretching Mode	Temperature (°C)			
	100	**120**	**130**	**140**
Uniaxial free-width	7×1	7×1	7×1	6×1
Uniaxial constrained-width	4×1	4×1	4×1	6×1
Simultaneous biaxial	-	Break	4×4	6×6
Sequential biaxial	-	Break	Break	6×6

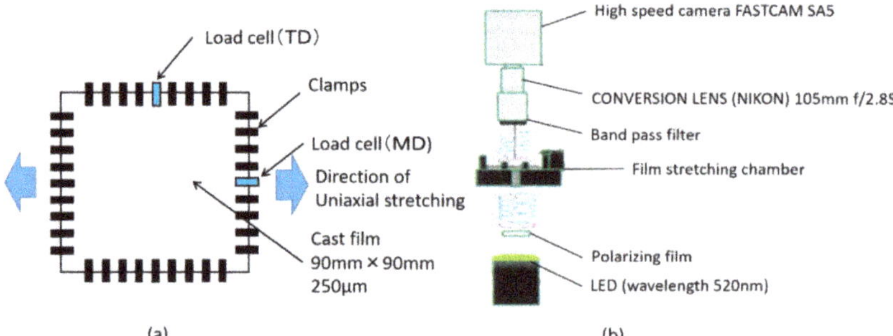

Figure 2. Schematic diagram of (a) film stretcher and (b) high-speed camera used for real time retardation measurements.

2.3. Retardation and Birefringence Measurements

We arranged the film stretcher by attaching a high-speed camera (Photron Limited, Tokyo, Japan, FASTCAM SA5) enabling to track retardation of the film during stretching at high temporal resolution. The camera is provided with micro-scale polarizing elements and a pixel reading circuit. Monochromatic wave is emitted through polarizing film toward the film stretching chamber, where birefringence of the film produces polarized light. Result was obtained as convolution of retardation within the quarter of the wavelength range, therefore the degree of convolution was tracked and multiplied by 130 nm. While retardation shows distribution of a certain degree, we focused on the center of the film and an average value along about 1 cm width was evaluated.

For off-line measurements, polarizing microscope (Olympus Corporation, Tokyo, Japan, BX51-P) was used. Thick Berek compensator was installed to measure retardation of suitable range. The relationship between birefringence and retardation is:

$$\Delta n = |n_{MD} - n_{TD}| = R/d \tag{1}$$

where d is thickness and n_{MD} and n_{TD} are refractive indices along MD and TD, respectively. In crystalline polymer, Δn was assumed to obey Stein's additivity law [25]:

$$\Delta n = X_c f_c \Delta n_c + (1 - X_c) f_a \Delta n_a \tag{2}$$

where first term on the right-hand side is contribution from crystalline phase, composed of crystallinity X_c. degree of orientation f_c and intrinsic birefringence of crystalline phase Δn_c. Second term is contribution from amorphous phase, where ratio of amorphous phase is derived from $(1 - X_c)$. Here, form birefringence was neglected and $\Delta n_c = 0.058$ was used [25]. These values were evaluated by WAXD analysis. Stress-optical constant (SOC) is given by:

$$SOC = \Delta n / \sigma \tag{3}$$

where σ is the true stress.

2.4. Wide-Angle X-ray Diffraction(WAXD)

Wide-angle X-ray diffraction (WAXD) measurements of the obtained stretched films were performed using X-ray diffraction instrument (Rigaku Corporation, Tokyo, Japan, Rigaku Micro). The Cu-Kα measurement (wavelength = 1.54 Å) radiation from an anode operating at 40 kV and 30 mA was used to detect the crystalline phase. Stretched films were cut and stacked to a thickness of about 0.3 mm in order to obtain sufficient accuracy. The orientation of crystalline phase was calculated via applying the azimuthal scan through the diffraction angle with respect to the selected plane and the ratio of half-width of diffraction peak was used for calculation. 1D-WAXD profiles were obtained from circularly integrated intensities of 2D-WAXD image acquired. Subsequently, decomposing the peaks of 1D-WAXD profiles into crystalline and amorphous phase as:

$$X_c = \sum A_c / \left(\sum A_c + \sum A_a \right) \tag{4}$$

where A_c and A_a are the fitted areas of crystalline and amorphous phase, respectively.

2.5. Light Transmittance

We measured light transmittance dispersions of the stretched films using customized light transmittance measuring equipment (Lambda Vision Inc., Kanagawa, Japan) within range of 350–1050 nm wavelength, which correspond to wavelength range of visible light. Software (Lambda Vision Inc., Kanagawa, Japan, ColorLabIV-LCD Ver.5.51) was used for analysis.

2.6. Thermal Analysis

2.6.1. Differential Scanning Calorimetry (DSC)

We evaluated the crystallinity and the melting temperature (T_m) of UHMWPE by DSC (TA Instruments, New Castle, DE, USA, Q200) at 30–160 °C at a heating rate of 10 °C/min.

2.6.2. Dynamic Mechanical Analysis (DMA)

Dynamic mechanical analyzer (TA Instruments, New Castle, DE, USA, RSAIII) was used to measure the viscoelasticity of the stretched films. Heating rate was 2 °C/min in 40–160 °C range.

2.6.3. Thermal Mechanical Analysis (TMA)

Thermal mechanical analyzer (TA Instruments, New Castle, DE, USA, Q400 TMA) was used. The film was cut into strip and clamped in the temperature controlling chamber. 0.02 N force was applied to the film during the measurement.

2.7. Mechanical Property Measurements

Dumbbell specimen was punched out from the stretched films by the dumbbell cutter DIN 53504-S3. Tensile testing machine (Toyo Seiki Seisaku-syo, Ltd., Tokyo, Japan, Strograph VG) was used to perform engineering stress-strain measurements with the strain rate of 50 mm/min.

3. Results and Discussion

3.1. Retardation and Birefringence Measurements

Figure 3 shows retardation distribution image of uniaxial film stretching. Scale for the images is 9 cm by 9 cm. We can see that in free-width uniaxial stretching, retardation is fairly uniform. Whereas in constrained-width stretching, retardation reflects stress concentration at clamping point, and pleat like deformation along TD can be observed. The effect of constraint become apparent in the following results.

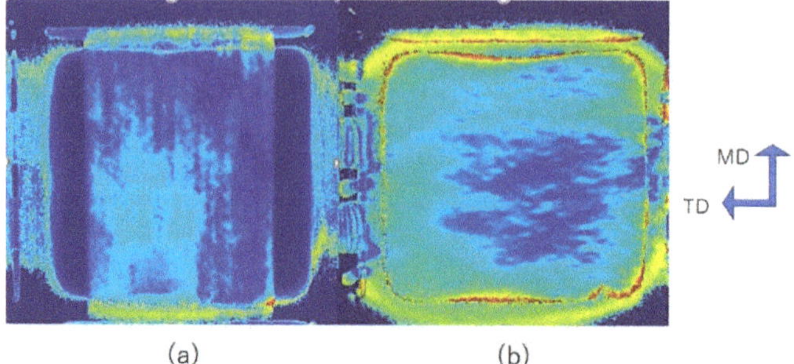

(a)   (b)

Figure 3. Retardation image of uniaxial: (**a**) free-width (**b**) constrained-width stretching along the MD at DR = 4 with stretching speed of 1 m/min.

Figure 4 shows the stress-strain curves and retardation behavior for uniaxial free-width stretching at 100 °C and 120 °C. The retardation results denote values relative to the preheated state, at which absolute retardation changes from null. It should also be noted that after DR = 2, decrease in retardation was difficult to track by on-line measurement, as retardation integration assumes monotonical behavior, thus we omitted beyond DR = 2 from the graphs. However, we observed relaxation after the end of stretching at each DR and confirmed that no significant relaxation was occurring. We can see that retardation reaches a maximum value by DR = 2, then decreases slowly. From the starting point of DR = 1, dependence on stretching speed becomes apparent at 120 °C. In the previous study of amorphous polymer [8], the relationship between stretching temperature and speed is such that lower temperatures and faster stretching produce comparable behavior. Here we can confirm same nature of dependence on temperature and stretching speed. However, variation range in SOC is narrower in stretching at 100 °C. This may be because polymer crystalline molecules at lower temperatures are more rigid, impeding structural change even at fast deformation rates. Stretching at 130 °C exhibited neck formation until DR = 4; the thickness of the stretched portion remained constant. Stress at 130 °C was too low, and the emergence of wrinkles on the surface of the films presented problems; thus, here we discuss only temperatures below this point.

Figure 5 shows the relationship between birefringence and true stress with free-width uniaxial stretching. We see that higher temperatures move the SOC closer to the value for the melt state. The SOC for high-density linear polyethylene melted at T = 423 K was reported to be 2.35 GPa^{-1} by Janeschitz [12], 1.3 GPa^{-1} was reported by Koyama et al. [11] for low-density polyethylene at molten state. Our results above give SOC = 0.11–1.0 GPa^{-1}, which is comparable to accepted research.

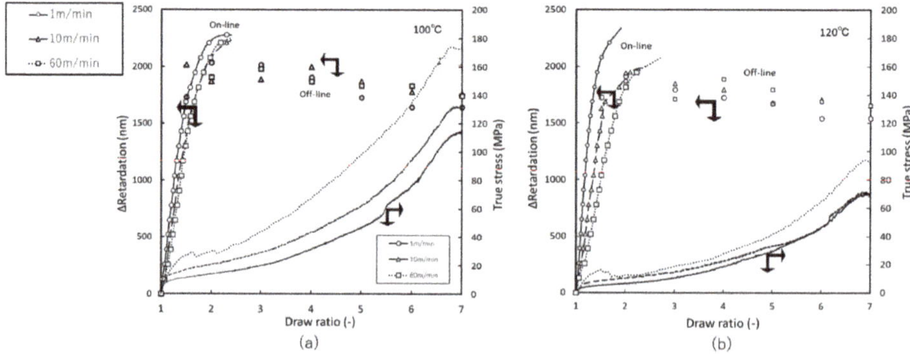

Figure 4. True stress and retardation vs. draw ratio of free-width uniaxial stretching at: (**a**) 100 °C (**b**) 120 °C.

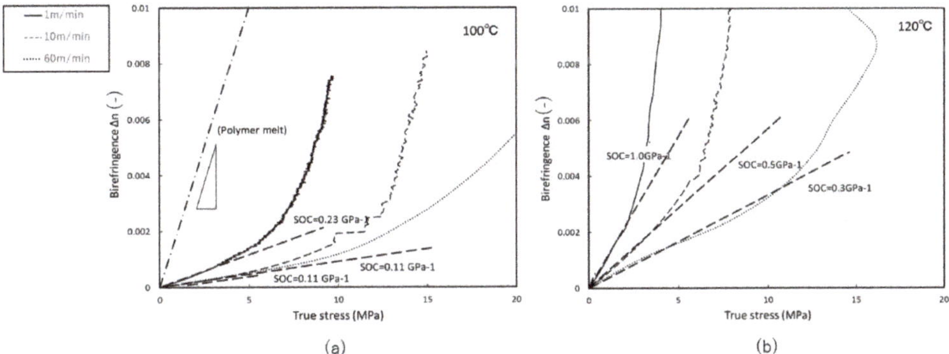

Figure 5. Birefringence vs. true stress of uniaxial free-width stretching at: (**a**) 100 °C (**b**) 120 °C.

Figures 6 and 7 present the same argument for uniaxial constrained-width stretching. Constraints along TD resulted in lower stress compared to uniaxial free-width stretching. The dependence of birefringence on stretching speed becomes more apparent compared to Figure 4. In uniaxial constrained-width stretching, the SOC was found to be in the range of 0.05–1.6 GPa^{-1}, a broader range than for uniaxial free-width stretching. This suggests that the stretching window for constrained-width stretching allows more control of the structural formation.

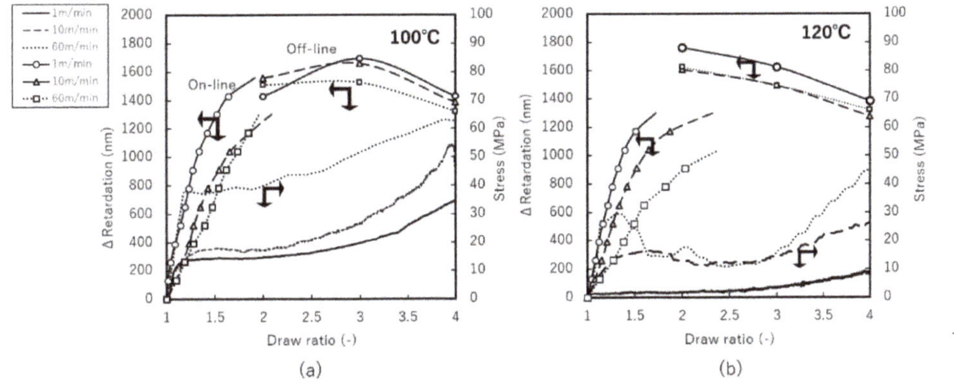

Figure 6. True stress and retardation vs. draw ratio of uniaxial constrained-width stretching at: (**a**) 100 °C (**b**) 120 °C.

Figure 8 shows the relationship between the stress-draw ratio and retardation image at each DR. Scale for the images is 9 cm by 9 cm. At 130 °C, preheated film exhibits an unstable retardation distribution, which remains unchanged even after changing preheating times. As stretching begins, we observe deformation characterized by significant distortion until DR = 2.5 × 2.5. This behavior is similar to neck forming deformation in uniaxial stretching at 130 °C, in which stretched and non-stretched regions are intermingled in-plane. By DR = 3 × 3, stretching is completed and results in a uniform isotropic film with gloss and clarity. Compared to stretching at 140 °C, stress declines when temperatures are above T_m. The boundaries of the raw UHMWPE powder expand as DR increases. In the early stages until DR = 2 × 2, we observe anisotropic retardation, followed by in-plane isotropic optical behavior.

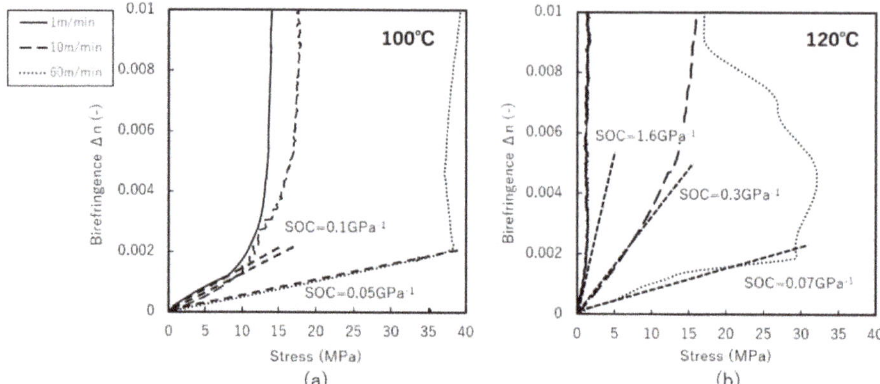

Figure 7. Birefringence vs. true stress of uniaxial constrained-width stretching at: (**a**) 100 °C (**b**) 120 °C.

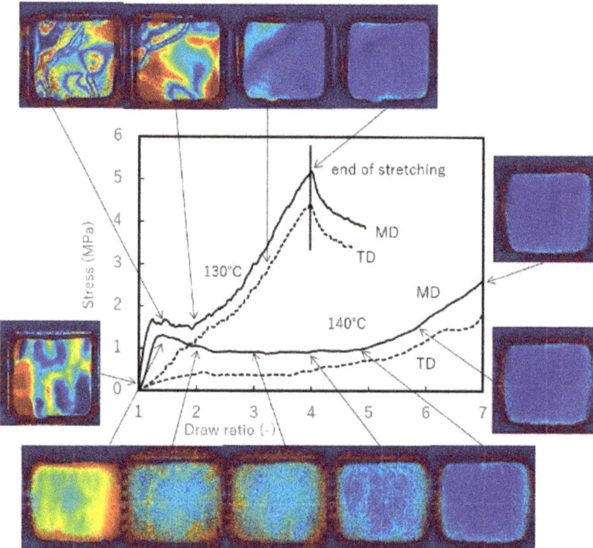

Figure 8. Stress vs. draw ratio of simultaneous biaxially stretched films at 130 °C and 140 °C with stretching speed of 1 m/min.

Observations by optical micrograph provide more insight. Figure 9 shows optical micrographs of biaxially stretched films. In biaxially stretched films, traces of the boundaries of the raw particles are visible as well. Between the original boundaries, we see a fibrillar oriented form. However, as mentioned earlier, their optical properties and thus their morphology are quite different, depending on stretching temperature. As will be shown in WAXD analysis, stretching at 130 °C lowers crystallinity and results in higher transparency. At temperatures above 140 °C, fibrillar portion gains more mobility, resulting in a greater remaining non-stretched region as stress decreases. Tiny holes begin appearing in the fibrillar region at low DR; thus, stretching at higher temperature only degrades the quality of the film and narrows the stretching window. It is noted that in uniaxially stretched films, the boundaries of the raw UHMWPE powder are observed at intervals of about 100–130 μm, elongated along the stretching axis and exhibiting affine deformation.

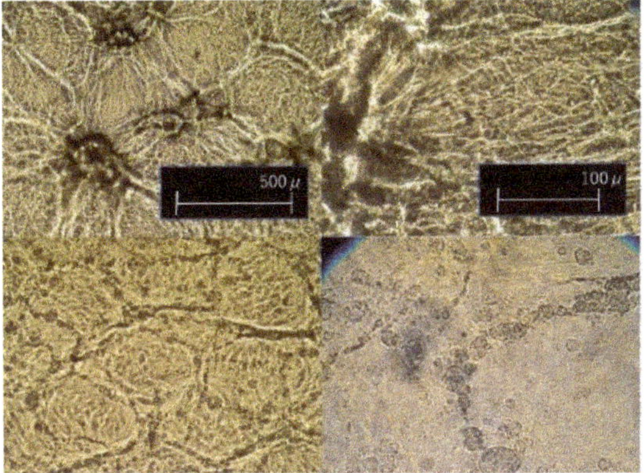

Figure 9. Optical micrographs of simultaneously biaxially stretched films at stretching speed of 1 m/min, DR = 4 × 4. (**upper**) 140 °C (**lower**) 130 °C.

3.2. Wide-Angle X-ray Diffraction (WAXD)

Figure 10 shows birefringence and the contributions from the crystalline and amorphous phase evaluated by WAXD. Degree of orientation and crystallinity were evaluated by Equation (2) to give the contribution from both crystalline and amorphous phase. Off-line birefringence measurements are plotted here, taking into account the shrinkage after stretching as the real DR; thus, they differ slightly from intended DR. In the case of uniaxial free-width stretching, we see that the contribution from the crystalline phase increase steeply up to DR = 2, then remains more or less constant. This is attributed to the offsetting contributions of increasing degree of orientation and decreasing crystallinity obtained by WAXD. The amorphous contribution plotted here, the remaining birefringence component, shows a growing contribution against DR, suggesting that the amorphous orientation increases monotonically.

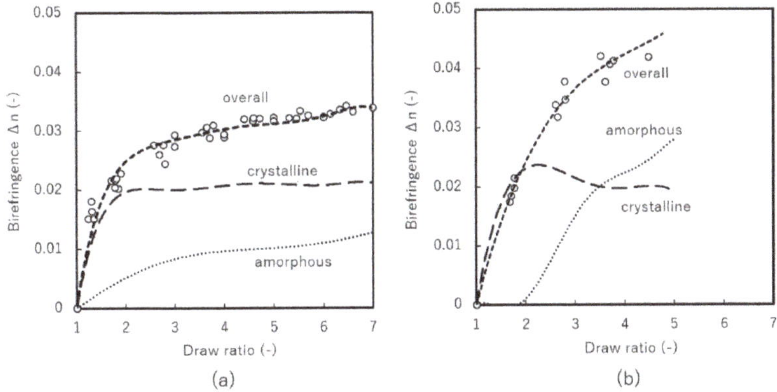

Figure 10. Birefringence vs. draw-ratio of uniaxial: (**a**) free-width (**b**) constrained-width stretched films at 100 °C and 120 °C with stretching speed of 1 m/min.

Uniaxial constrained-width stretching shows different behavior. Birefringence appears not to approach the limit value under stretchable DR. Additionally, contributions from the crystalline phase gradually decrease, indicating that amorphous contributions become more significant. At the early stage of stretching until DR = 2, the amorphous contribution is negligible. This suggests that constraints along TD delay amorphous rearrangement. In both uniaxial stretching modes, the orientation of the amorphous phase is estimated to be significantly lower than that of the crystalline phase, considering that intrinsic birefringence of amorphous phase is estimated to be around 0.2 [26,27]. Note that this discussion assumes the absence of any significant relaxation in birefringence as observed by high-speed camera.

Figure 11 compares the crystallinity of uniaxial free-width stretched film at different temperatures as evaluated by WAXD. It shows that higher temperatures and faster stretching speed tend to increase the crystallinity attained. The tendency of polymer chains to align is expected to increase at higher temperature, which would agree with our result.

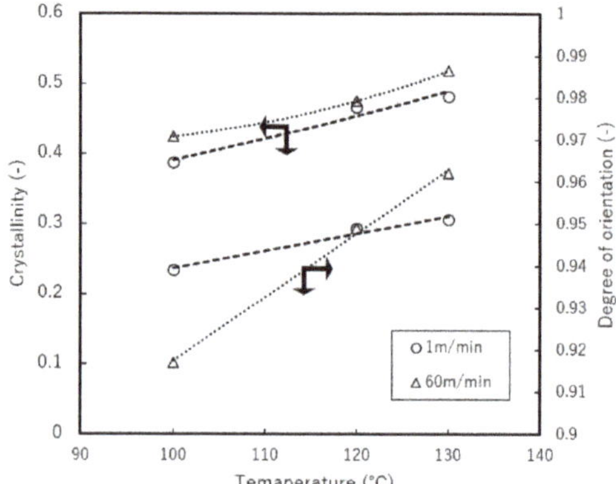

Figure 11. Crystallinity and orientation dependence on stretching temperature for uniaxial free-width stretched films by DR = 7.

Figure 12 shows 2D WAXD patterns for simultaneous biaxial stretching. We see that the diffraction pattern is isotropic, with constant intensity along azimuthal angle. The crystallinity of the sample film as evaluated by WAXD is 44%. The crystallinity measured by DSC was 59%, but since crystallinity measurement of stretched film has difficulty due to thermal shrinkage, we refer to the measurement by WAXD. Stretching at 130 °C shows that the diffraction peak from (200) plane becomes negligible, while the contribution from the amorphous phase dominates. The crystallinity is evaluated to be 23%. The crystallinity of the film stretched at 140 °C is 43%, nearly equal to the sample film. To emphasize the differences between these films, we measured light transmittance. Film stretched at 130 °C exhibited transmittance of 47%, whereas for film stretched at 140 °C was 20%, a finding consistence with our results of crystallinity.

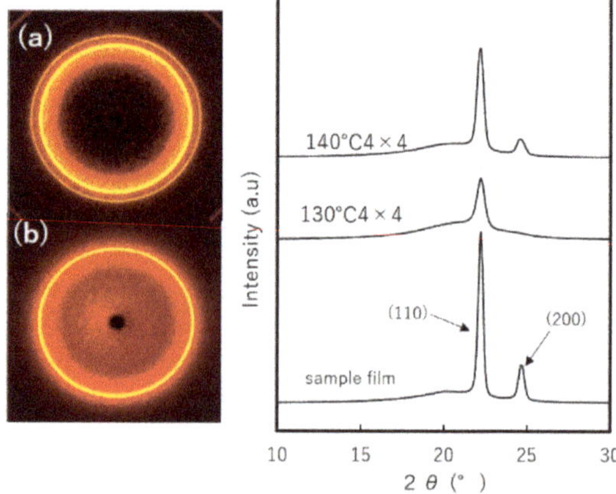

Figure 12. WAXD patterns of simultaneously biaxial stretched films 4 × 4 at: (**a**) 140 °C (**b**) 130 °C.

3.3. Thermal Analysis

3.3.1. Dynamic Mechanical Analysis (DMA)

Figure 13 shows the results for storage modulus E' and loss modulus E'', damping factor tan δ measurements. The storage moduli of the stretched films are more dependent on temperature than the film sample. The modulus of the sample film at T_m, decreases significantly as the crystalline structure breaks down. Takayanagi model is known to evaluate modulus of semi-crystalline polymer which treats the material as composite of two-phase; crystalline and amorphous [28]. Difference in degree of crystallinity of the stretched films, may suggest different modulus behavior with temperature. However, stretched films exhibited similar linear changes on E'. Stable behaviors beyond T_m suggest that they maintain their form at high temperature. On loss modulus, film stretched at 130 °C with lower crystallinity decreased monotonically which reflects amorphous nature; at 140 °C, it remained stable with temperature. These contribute to the very broad curve for tan δ, with its peak shifting toward lower temperature. Inconsistency with simple physical model suggests taking into account the influence of orientation and morphology for more detailed analysis.

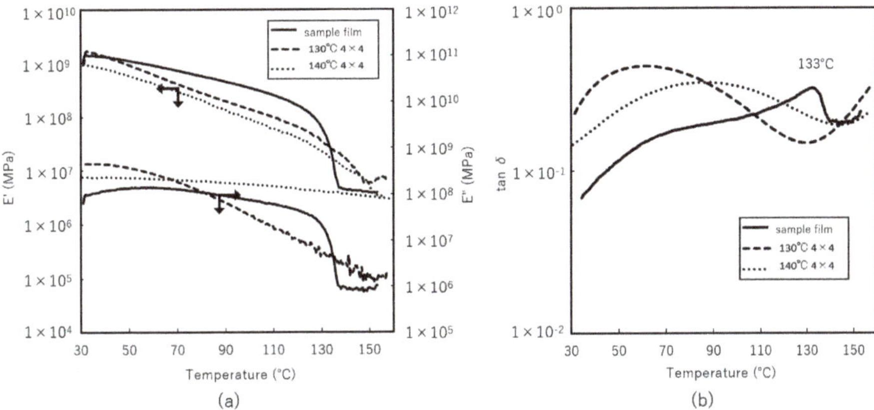

Figure 13. Dynamical mechanical analysis of simultaneous biaxially stretched films. (**a**) Storage and loss modulus (**b**) damping factor tan δ.

3.3.2. Thermal Mechanical Analysis (TMA)

Figure 14 shows dimensional changes of uniaxially and biaxially stretched films with temperature. Uniaxially stretched films begin to shrink toward thermodynamically stable state. Comparing the results to Figure 11 shows that shrinkage behaviors correspond to degree of crystallinity and orientation. Fast stretching results in slightly higher crystallinity and greater stability with temperature changes. Biaxially stretched films initially exhibited positive thermal expansion at lower temperature, followed by dramatic shrinking. At low temperature, in-plane orientation and entanglement of polymer chains can interrupt shrinkage in one direction compared to uniaxial stretched films. Above certain temperatures, the mobility of the polymer chains, particularly in the amorphous phase, contributes to shrinkage, as observed with film stretched at 130 °C, at which crystallinity declines and early shrinkage appears. Requirements for lithium-ion battery separators include thermal shrinkage less than 5% after 60 min at 90 °C [1], biaxial stretching can be an effective stretching mode to satisfy such demands.

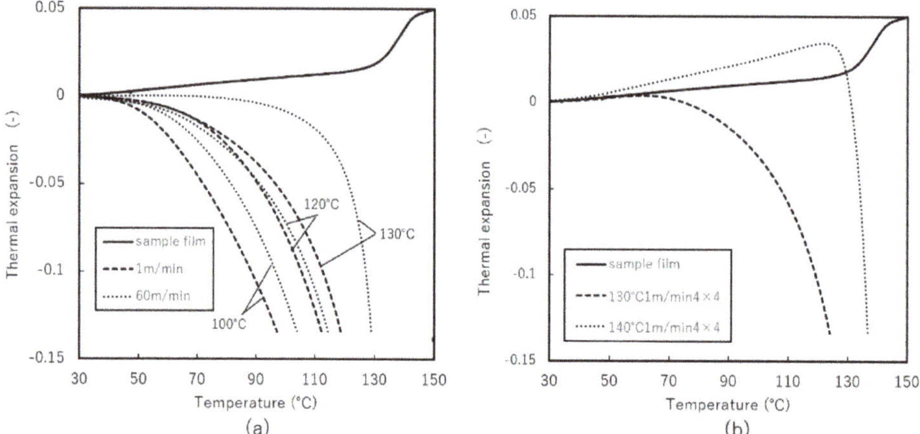

Figure 14. Thermal mechanical analysis of: (**a**) uniaxially free-width (**b**) simultaneous biaxially stretched films.

3.4. Tensile Tests

Figure 15 shows tensile test results for uniaxial constrained-width and simultaneous biaxially stretched films. The yield strength of the film sample was 30 Mpa; the strain-to-fracture ratio was about 4.5. Along the MD, strength increases in proportion to the DR of the film, while the strain-to-fracture ratio declines in inverse proportion to DR. Along the TD, the stress-strain curve exhibited neck formation until a strain of about 3–4. Stress at this stage is independent of the DR of the film. Strain to fracture increases with the DR of the film. For biaxially stretched films, breaking strength and strain increase significantly in films stretched at 130 °C. The greater ductility and strength can be attributed to the in-plane orientation of the amorphous phase (Figure 12). In sequential biaxial stretching, mechanical properties along the MD and TD are reversed between DR = 4 × 4 and 6 × 6. This coincided with the point at which stresses along the MD and TD overlap, suggesting that the stress-strain curve is an important predictor of isotropy. Biaxial stretching at 140 °C enabled to stretch up to high DR, but the strength of stretched films was not enhanced, as only DR larger than 4 × 4 could improve its strength.

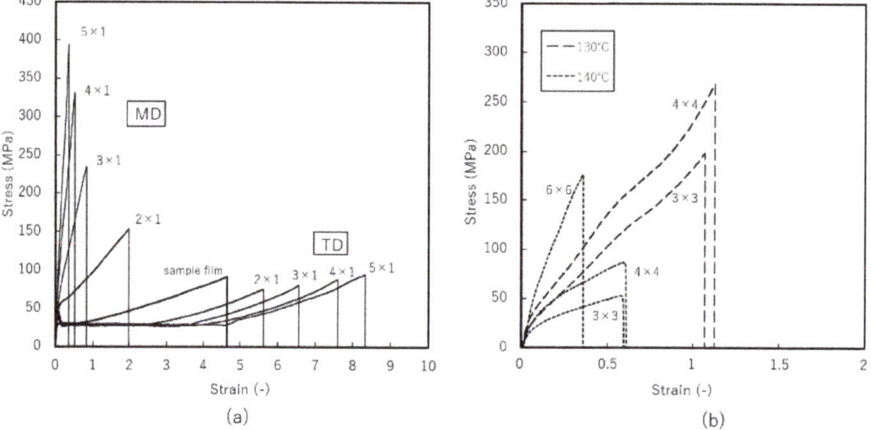

Figure 15. Tensile test results of: (**a**) uniaxial constrained-width stretched films at 120 °C (**b**) simultaneous biaxially stretched films at 140 °C with stretching speed of 1 m/min.

4. Conclusions

We observed various film-stretching modes by retardation-measuring high-speed camera. Evaluations identified SOC in the 0.05–1.6 GPa^{-1} range. Uniaxial constrained-width stretching resulted in a broader range of SOC at low to high stretching speeds. Birefringence analysis showed at the early stage, constrained in the TD constrains orientation of the amorphous phase. WAXD and thermal analysis showed high crystallinity under high temperature at high stretching speeds. In biaxial stretching, a suitable stretching window occurred only near T_m. Biaxial stretching below T_m showed significant changes in crystallinity and enhanced mechanical properties comparable to uniaxial stretching. Good thermal stability of biaxially stretched films were confirmed assuming energy storage application. Structural changes observed via retardation images captured with a retardation camera were consistent with results from optical micrographs. Our novel system tracking fast stretching process and in-plane retardation distribution is expected to extend laboratory scale research to industrial processes, emerging from trial and error processing to tailor-made processing.

Author Contributions: Conceptualization, Y.H. and H.I.; Methodology, Y.H., S.N., T.K., F.A.; Writing-Original Draft Preparation, Y.H.; Supervision, S.N., T.K. and H.I.

Funding: This research received no external funding.

Conflicts of Interest: The authors declare no conflict of interest.

References

1. Deimede, V.; Elmasides, C. Separators for Lithium-Ion Batteriese: A Review on the Production Processes and Recent Developments. *Energy Technol.* **2015**, *3*, 453–468. [CrossRef]
2. Gao, P.; Li, R. Nanoporous UHMWPE Membrane Separators for Safer and High-Power-Den sity Rechargeable Batteries. *Glob. Chall.* **2017**, *1*, 1700020. [CrossRef]
3. Uehara, H.; Tamura, T.; Hashidume, K.; Tanaka, H.; Yamanobe, T. Non-solvent processing for robust but thin membranes of ultra-high molecular weight polyethylene. *J. Mater. Chem. A* **2014**, *2*, 5252–5257. [CrossRef]
4. Uehara, H.; Kakiage, M.; Yamanobe, T.; Komoto, T.; Murakami, S. Transient crystallization during drawing from ultra-high molecular weight polyethylene melts having different entanglement characteristics. *Polymer* **2006**, *47*, 8053–8060. [CrossRef]
5. Uehara, H.; Kakiage, M.; Sekiya, M.; Yamanobe, T.; Komoto, T.; Sasaki, S.; Murakami, S. In situ SAXS analysis of extended-chain crystallization during melt-drawing of ultra-high molecular weight polyethylene. *Polymer* **2007**, *48*, 7385–7392. [CrossRef]
6. Uehara, H.; Tamura, T.; Kakiage, M.; Yamanobe, T. Nanowrinkled and nanoporous Polyethylene Membranes via Entanglement Arrangement Control. *Adv. Funct. Mater.* **2012**, *22*, 2048–2057. [CrossRef]
7. Cakmak, M.; Hassan, M.; Unsal, E.; Martins, C. A fast real time measurement system to track in and out of plane optical retardation/birefringence, true stress, and true strain during biaxial stretching of polymer films. *Rev. Sci. Instrum.* **2012**, *83*, 123901. [CrossRef] [PubMed]
8. Hassan, M.K.; Cakmak, M. Mechano optical behavior of polyethylene terephthalate films during simultaneous biaxial stretching: Real time measurements with an instrumented system. *Polymer* **2013**, *54*, 6463–6470. [CrossRef]
9. Hassan, M.K.; Cakmak, M. Mechanisms of structural organizational processes as revealed by real time mechano optical behavior of PET film during sequential biaxial stretching. *Polymer* **2014**, *55*, 5245–5254. [CrossRef]
10. Kotaka, T.; Kojima, A.; Okamoto, M. Elongational flow and birefringence of low density polyethylene and its blends with ultrahigh molecular weight polyethylene. *Polymer* **1998**, *39*, 2149–2153. [CrossRef]
11. Koyama, K.; Ishizuka, O. Birefringence of Polyethylene Melt in Transient Elongational Flow at Constant Strain Rate. *J. Polym. Sci. Part B Polym. Phys.* **1989**, *27*, 297–306. [CrossRef]
12. Janeschitz-Kriegl, H. *Polymer Melt Rheology and Flow Birefringence*, 1st ed.; Springer: Berlin, Germany, 1983; p. 113, ISBN-13 978-3-642-68824-9.
13. Ryu, D.S.; Inoue, T.; Osaki, K. A birefringence study of polymer crystallization in the process of elongation of films. *Polymer* **1998**, *39*, 2515–2520. [CrossRef]

14. Hashimoto, Y.; Ito, H. Structural Formation of Cyclic Olefin Copolymer (COC) Films as Revealed by Real Time Retardation Measurements during Fast Stretching Process. *Technologies* **2018**, *6*, 60. [CrossRef]

15. Mackley, M.R.; Gao, P.; Nicholson, T.M. Development of anisotropy in ultra-high molecular weight polyethylene. *Polymer* **1990**, *31*, 237–242. [CrossRef]

16. Bandeira, B.; Lewis, E.L.V.; Barton, D.C.; Ward, I.M. The degree of crystalline orientation as a function of draw ratio in semicrystalline polymers: A new model based on the geometry of the crystalline chain slip mechanism. *J. Mater. Sci.* **2016**, *51*, 228–235. [CrossRef]

17. Li, D.; Meng, L.; Xu, J.; Chen, X.; Tian, N.; Lin, Y.; Cui, K. Constrained and free uniaxial stretching induced crystallization of polyethylene film: A comparative study. *Polym. Test.* **2014**, *36*, 110–118. [CrossRef]

18. Ward, I.M.; Gordon, D.H.; Duckett, R.A. A study of uniaxial and constant-width drawing of poly(ethylene terephthalate). *Polymer* **1994**, *35*, 2554–2559. [CrossRef]

19. DeMeuse, M.T. *Biaxial Stretching of Film*, 1st ed.; Woodhead Publishing: Cambridge, UK, 2011; ISBN-13 978-1845696757.

20. Heffelfinger, C.J.; Burton, R.L. X-Ray Determination of the Crystallite Orientation Distributions of Polyethylene Terephthalate Films. *J. Polym. Sci.* **1960**, *47*, 289–306. [CrossRef]

21. Ajji, A.; Zhang, X.; Elkoun, S. Biaxial orientation in HDPE films: Comparison of infrared spectroscopy, X-ray pole figures and birefringence techniques. *Polymer* **2005**, *46*, 3838–3846. [CrossRef]

22. Cakmak, M.; Ou, X. Influence of biaxial stretching mode on the crystalline texture in polylactic acid films. *Polymer* **2008**, *49*, 5344–5352. [CrossRef]

23. Kojima, M.; Magil, J.H.; Lin, J.S.; Magonov, S.N. Structure and Morphology of Biaxially Oriented Films of Polyethylenes. *Chem. Mater.* **1997**, *9*, 1145–1153. [CrossRef]

24. Available online: https://www.saxin.jp/product/about-newlight/ (accessed on 9 November 2018).

25. Norris, F.H.; Stein, R.S. The X-Ray Diffraction, Birefringence, and Infrared Dichroism of Stretched Polyethylene. *J. Polym. Sci.* **1956**, *21*, 381–396. [CrossRef]

26. Hoff, M.; Pelzbauer, Z. Birefringence and orientation of polyethylene zone-drawn under various thermomechanical conditions. *Polymer* **1992**, *33*, 4158–4163. [CrossRef]

27. Rossignol, J.M.; Seguela, R.; Rietsch, F. Intrinsic Amorphous Birefringence of Semi-Crystalline Polyethylene from Combined Birefringence and Infrared Dichroism Measurements. *J. Polym. Sci. Part C Polym. Lett.* **1989**, *27*, 527–532. [CrossRef]

28. Takayanagi, M.; Imada, K.; Kajiyama, T. Mechanical Properties and Fine Structure of Drawn Polymers. *J. Polym. Sci. Part C Polym. Synposia* **1967**, *15*, 263–281. [CrossRef]

Article

Chaotic Manifold Analysis of Four-Screw Extruders Based on Lagrangian Coherent Structures

Xiang Zhe Zhu [1,*], **Ying Tong** [1] **and Yue Xin Hu** [2]

[1] School of Mechanical Engineering, Liaoning Shihua University, Fushun 113001, China; lshtongying@163.com
[2] School of Chemistry and Materials Science, Liaoning Shihua University, Fushun 113001, China; yxhu1981@163.com
* Correspondence: xzzhu@126.com; Tel.: +86-2456-8650-42

Received: 14 October 2018; Accepted: 9 November 2018; Published: 14 November 2018

Abstract: The four-screw extruder (FSE) is a novel equipment for polymer processing. In this paper, from a new viewpoint of Lagrangian coherent structures (LCS), two-dimensional fluid transport and chaotic mixing characteristics within three kinds of novel industrial FSEs are explored based on LCS to better understand the flow and mixing natures in the FSEs. Firstly, based on the finite-time invariant manifold theory, the finite-time Lyapunov exponent (FTLE) and LCS of FSEs are calculated by considering the different initial time. Hyperbolic LCSs from the FTLE maps are adopted to identify chaotic mixing manifolds in FSEs. Moreover, particle tracking and Poincaré sections are used to illustrate the different fluid motions in the above three isolated regions. Finally, the effects of relative rotating directions and layout of four screws on the chaotic manifolds in FESs are discussed in order to enhance local mixing performance. Furthermore, quantitative mixing measures, such as the segregation scale, logarithmic of stretching, and mean-time mixing efficiency are employed to compare the mixing efficiencies in three kinds of FSEs. The results show that the relative rotating directions and positions of four screws can change the chaotic manifolds and increase mixing performance in local poor mixing regions. FTLE and LCS analysis are helpful to better understand the chaotic mixing nature in the novel screw extruders.

Keywords: extrusion; four-screw extruder; finite-time Lyapunov exponents (FTLE); Poincaré section; chaotic manifold

---

1. Introduction

As a classic piece of mixing equipment, single-screw and twin-screw extruders are widely used in the polymer-processing industry [1]. With the development of the polymer industry, some novel screw mixing elements in screw extruders, such as the pin mixing section [2], pitched-tip kneading disk [3], and screw profile with slots [4], are devised to obtain highly efficient mixing and a fine final product. Recently, a novel multiple-screw extruder, the four-screw extruder, was developed and has attracted more interest due to its many advantages, such as longer residence-time distribution, high output, great shear rate, and mixing efficiency [5]. The four screws of the typical four-screw extruder constitute a square arrangement resulting in four intermeshing regions and one special central region. However, the traditional twin-screw extruder only has two intermeshing regions and without central region. Due to the complicated geometry, the flow and mixing mechanisms in the four-screw extruder are very complex in comparison with the typical twin-screw extruder. Therefore, it is necessary to study the effects of four-screw extruder's geometry, screw rotating manner on the flow, and the nature of the mixing in the four-screw extruder. More importantly, as a dynamic flow system, a fundamental understanding of the flow and mixing mechanisms in the extruders is helpful in controlling the desired flow for optimizing the geometrical structures and process conditions to guarantee the homogeneity of final product, which is also a primary aim of this article.

The key mixing mechanisms of laminar flow in screw extruders include dispersive and distributive mixing from a Eulerian viewpoint. In general, most of the studies have focused on the dispersive and distributive mixing to understand the mixing mechanisms in single- and twin-screw extruders [6–12]. Typically, Connelly et al. [10] adopted the segregation scale and cluster-distribution index to evaluate the distributive mixing in a 2D twin mixer. Moreover, the mixing index and shear stress were used to evaluate the dispersive mixing in the mixer. Domingues et al. [11] predicted the dispersive mixing performance from the extent deformation of the fluid drops, and mixing capability was quantified by the maximum line-stretching rate and the strain-rate-type identifier. Zhang et al. [12] adopted residence time distributions (RTD) to evaluate the distributive mixing in twin-screw mixers by particle tracking. Based on numerical and experimental investigations, many efforts were made to understand the relationship between the phenomena of the particle's global behavior and the mixing process [13–15]. However, it is difficult to reveal the potential relationship between the chaotic manifolds and fluid transport in the mixers using the traditional Eulerian method.

Many studies have proved that chaotic mixing is an effective method to enhance mixing efficiency in the laminar flow of polymer processing in order to conquer the nature of laminar flow [16–19]. The Poincaré section and Lyapunov exponent (LE) are often carried out as general parameters to describe the chaos in a dynamical system. Lee et al. [18] suggested exponent stretch rate based on LE as chaotic mixing measures in single-screw extruder. The hyperbolic fixed point and chaotic manifold were obtained from the Poincaré map in the perturbed system. Hwang et al. [19,20] studied the evolution of chaotic mixing in the chaos screw (CS) nonlinear dynamical model using a Poincaré map based on a fourth-order Runge–Kutta scheme. The changes from the homoclinic fixed point and elliptic rotations to the resonance bands or KAM tori were analyzed depending on the commensurability of frequency ratio of the corresponding orbits. Niu [21] used the Poincaré section to locate the KAM curves and quasiperiod areas. The chaotic behavior of a dynamic system was evaluated by the Lyapunov exponent. Wang et al. [22] calculated the measure of the Kolmogorov–Sinai entropy rate from LE to discover the homogeneity of the system.

Recently, as a new measurement of chaotic mixing, Lagrangian coherent structures (LCS) were proposed to identify the chaotic manifold [23–25] in a dynamic flow system. LCS provide an effective tool to capture the potential dynamical features of a flow system, which could be missed in a traditional Eulerian analysis based on the velocity or vorticity field. The ridges of the finite-time Lyapunov exponent (FTLE) present the most stretching and repelling structures, which are called LCS (It is noted that several studies reported that the ridges of the FTLEs are not LCSs, as they have nonzero flux across them [25]). Many researchers used the FTLE and LCS to study fluid mixing and the transport process in internal mixers [26–29] and identify the vortex pinch-off [30]. Santitissadeekorn et al. [26] used the LCS to investigate the transport behavior and mixing process in a batch mixer. Robinson et al. [27] extracted the manifold structures from the forward and backward FTLE with a rational integration time, and the mixing characteristics in different conditions were discussed by using the LCS. Moreover, they used the same method to identify and visualize the 3D manifold intersections in a helical ribbon mixer [28]. In addition, Conti and Badin proposed a new method based on Covariant Lyapunov Vectors to describe hyperbolic patterns in two-dimensional flows [29]. Because the ridges of FTLE are equal to the boundary of transportation, the LCS were also widely used in other subjects, such as the atmosphere [31], oceans [32,33], biology [34], and electromagnetism [35]. However, the studies of using the LCS to analyze time-varying flow in polymer processing considering the moving parts are relatively limited.

As a novel extrusion device, four-screw extruders share distinct fluid-transport and mixing mechanisms in comparison to traditional twin-screw extruders. This paper explores the chaotic mixing and fluid-transport characteristics within three kinds of novel industrial four-screw extruders by employing FTLE, LCSs, and Poincaré sections. We focus on the two-dimensional chaotic advection and mixing in four-screw extruders using computational fluid dynamics. The chaotic manifold structures of the four-screw extruders were visualized by extracting the hyperbolic LCSs from the FTLE maps.

Moreover, the Poincaré sections are used to assist in explore the mixing process in extruders and how material transport characteristics respond to changes in LCSs. Furthermore, quantitative mixing measures by means of the segregation scale, logarithmic of stretching. and mean-time mixing efficiency are employed to provide a rigorous method for comparison three kinds of four-screw extruders. The primary aim of this article is to better understand the nature and inherent of flow and mixing natures in the four-screw extruder.

2. Materials and Methods

2.1. Government Equations

In this study, 2D non-Newtonian and transient flow conditions are employed by using the finite-element method (FEM). The forms of continuity and momentum equations can be expressed as follows [36]:

$$\nabla \cdot \mathbf{v} = 0 \tag{1}$$

$$-\nabla p + \nabla \tau = \mathbf{v} \cdot \nabla \mathbf{v} \tag{2}$$

where p is the pressure, τ is the extrastress tensor, and $\mathbf{v}$ is the velocity vector.

The stress tensor is given as:

$$\tau = 2\eta(\dot{\gamma})\mathbf{D} \tag{3}$$

In which η is the viscosity, $\dot{\gamma}$ is the shear rate, and D is the rate-of-deformation tensor.

In this study, HDPE is processed in the four-screw extruders. It is appropriate to use the Carreau–Yasuda model to describe its rheological behavior as follows:

$$\eta = \eta_\infty + (\eta_0 - \eta_\infty)\left[1 + (\lambda\dot{\gamma})^a\right]^{\frac{n-1}{a}} \tag{4}$$

where η is the dynamic viscosity, η_∞ is the infinite shear viscosity, λ is a model-specific relaxation time, and $\lambda = \sqrt{2\Pi_D}$. Π_D is the second invariant of the rate of deformation tensor and n is the power-law index. In this study, the material parameters of the HDPE melts at 200 °C are as follows: η_∞ = 0 Pa·s, η_0 = 113088 Pa·s, λ = 3.11 s, n = 0.36, a = 0.26.

2.2. FTLE and LCS

The Lagrangian description is considered as a better method to understand the idiosyncrasies of steady fluid flow. The LCS was used to describe the coherent structures of two-dimensional turbulent flow that defined as manifolds upon the dynamics. These manifolds are useful to understand the results of material transport from experimental and numerical flow data [26], and especially to explain the underlying mixing reasons of the high-viscosity fluid flow resulting in laminar flow.

The two-dimension flow dynamical system of four-screw extruders, the fluid point trajectory satisfied as:

$$\begin{cases} \dot{x}(t) = v(x(t),t) \\ \quad x(t_0) = x_0 \end{cases} \tag{5}$$

where x is a fluid point trajectory on an arbitrary interval of time $[t_0, t]$. The solution of the dynamical system given in Equation (5) in a certain time as a flow map can be viewed as a flow map. It is denoted by $\varphi_{t_0}^t$ and satisfies as follows [37]:

$$\varphi_{t_0}^t : D \rightarrow D : x_0 \mapsto \varphi_{t_0}^t(x_0) = x(t; t_0, x_0) \tag{6}$$

Then, a finite-time version of Cauchy–Green deformation tensor by displacement grads tensor that form the trajectory $x(t)$ of the dynamical system is obtained by

$$C = \frac{d\varphi_{t_0}^{t_0+T}(x)}{dx}^* \frac{d\varphi_{t_0}^{t_0+T}(x)}{dx} \tag{7}$$

where M* denotes the adjoint of M. C is the Cauchy–Green deformation tensor. The maximum and minimum eigenvectors of the C in Equation (7) imply that there are compression and expansion along the trajectory, respectively; and the maximum stretching occurs in direction aligned with the eigenvector associated with the maximum eigenvalue of C.

So, the FTLE with a finite integral time T can be defined as:

$$\sigma_{t_0}^T(x) = \frac{1}{|T|} \ln \sqrt{\lambda_{\max}(C)} \tag{8}$$

In which, $\sigma_{t_0}^T(x)$ denotes the FTLE; T is associated to point $x{\in}D$ at time t_0; $\lambda_{\max}(C)$ is the maximum eigenvalue of C. The LCSs are approximately obtained by the ridge of the FTLE field at time t for the initial position [37] and represent the stable and unstable material lines in the unsteady fluid flow. It may clearly understand the mixing and transportation behaviors in the unsteady flow. So, the LCS is a useful tool to quantify mixing in various notions.

For a periodic flow described in this paper, the LCSs corresponds to the hyperbolic invariant manifolds. The geometry of the hyperbolic manifolds can be found by calculating the spatial distribution of the finite-time Lyapunov exponent (FTLE) [23–26]. A repelling LCS normally appear in this field as a maximum ridge in the forward time FTLE field. Similarly, attracting LCS produce a maximum ridge in the backward time FTLE field.

2.3. Configurations of Three Types of Four-Screw Extruders

Based on the typical FSE, we developed two other types of new configurations. Therefore, four-screw extruders mainly involve three types geometries according to the arranged forms and rotating directions of the four screws, as follows: (i) The basic system (BS) is the corotating square-arranged four-screw extruder [5], in which the four screws arrange in square and all corotate counterclockwise, as shown in Figure 1a. (ii) The counter-rotating system (CRS), called a counter-rotating square-arrayed four-screw extruder, has four screws arranged in a square and the two top screws rotate counterclockwise, but the two bottom screws rotate clockwise, as shown in Figure 1b. (iii) The screw equidistance distribute system (SEDS), called an equidistance arrayed four-screw extruder, has four screws arranged in a rhombus shape and all corotate counterclockwise, as shown in Figure 1c. Detailed configurations and model parameters of the three types of FSEs at cross-sections are shown in Figure 1 and Table 1.

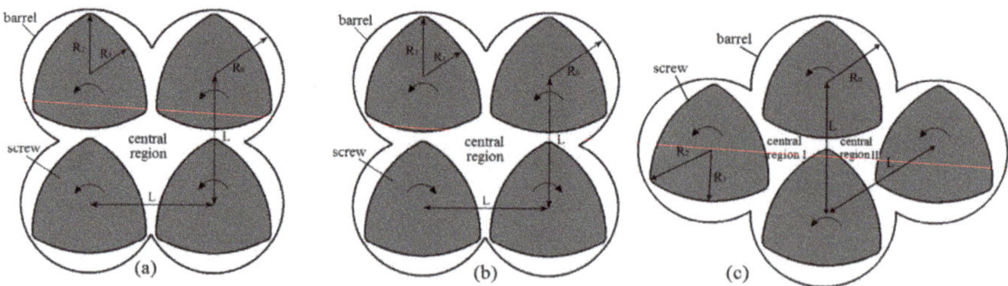

Figure 1. Three types of four-screw extruders at cross sections: (**a**) basic system (BS); (**b**) counter-rotating system (CRS); (**c**) screw equidistance distribute system (SEDS).

Table 1. Geometric parameters of four-screw extruders.

Parameter	Value
Barrel diameter R_0	18.5 mm
Screw root diameter R_1	13 mm
Screw tip diameter R_2	17 mm
Centerline distance of four screws L	33 mm
Screw clearance	3 mm
Clearance of screw and barrel	1.5 mm
Rotational speed of four screws	0.5 r/min
Leads of our screws	3

In comparison with typical twin-screw extruders, the three types of four-screw extruders have more intermeshing regions than twin-screw extruders, as shown in Figure 1. In addition, as can be seen from Figure 1a,b, the two models of corotating and counter-rotating FSEs, namely, the BS and CRS, both have four intermeshing regions and one central region where four screws cannot sweep the extrusion material. With the four-screw rotation, the area of the central region continually changes from big to small. However, in comparison with the BS and CRS, the SEDS has five intermeshing regions and two central regions with a relatively small area, as shown in Figure 1c.

2.4. FE Models and Computational Details

In this article, two-dimensional finite-element (FE) models without considering axial movement are employed to better understand the influences of geometric configurations on the flow-transportation and chaotic-mixing characteristics of four-screw extruders at cross sections. This is useful to achieve optimum shape design of cross section and improve mixing efficiency in practical applications. Three types of four-screw extruders were constructed using Solidworks software (2015, Dassault Systemes, Concord, MA, USA), as shown in Figure 1. Based on the mesh superposition technique (MST) [36], the FE models of the screws and barrel were established by using Gambit software (2.4.6, ANSYS Inc., Pittsburgh, PA, USA) without remeshing for periodical geometric changes. The quadrilateral and triangle elements were adopted to mesh the rotors and flow domain, respectively. In order to catch the small velocity changes in the small clearances between the rotors and walls and near the walls, four boundary layer grids in the FE model were employed, as shown in Figure 2. There is a total of 728,000 elements in the FE model of BS and CRS, respectively, and 637,800 elements in the FE model of SEDS. In order to ensure computational accuracy, more than 240 iterations were carried out per screw revolution. In our simulations, we chose a small-time step, $dt = 0.5$ s, which corresponds to 1.5° of screw rotation.

In this paper, two-dimensional flow fields in the mixer are calculated by using a commercial CFD code, ANSYS Polyflow, based on the generalized Newtonian approach. The convergence criterion was set to 1×10^{-4} in all the numerical cases. Moreover, based on the velocity fields in the extruders, the fluid-particle positions at time $t + T$ were located by using the fourth-order Runge–Kutta scheme. Then, spatial gradient $d\varphi_t x(t)/dx$ was used to determine the Cauchy–Green deformation tensor for each initial point, and the FTLE map in the mixer was obtained at time t of each grid point from Equation (8).

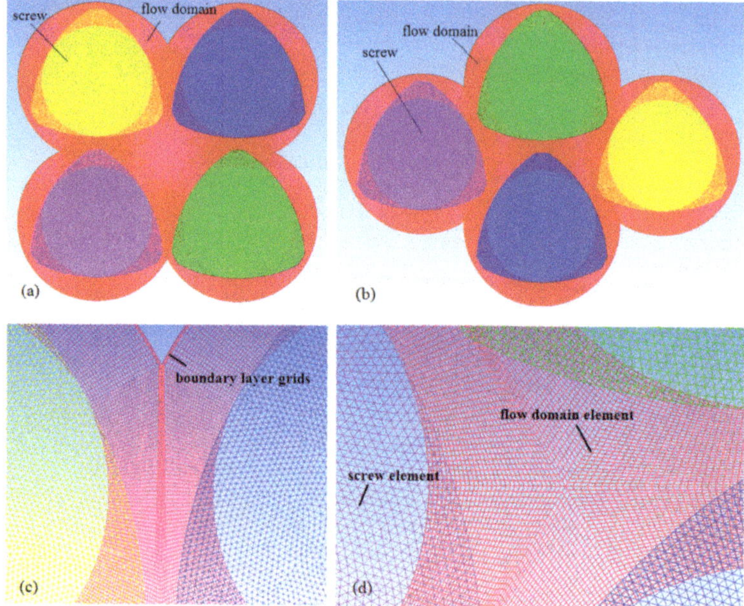

Figure 2. Finite-element models of the four-screw extruders in cross sections: (**a**) BS or CRS; (**b**) SEDS; (**c**) enlarged view of the gap region; (**d**) enlarged view of central region.

2.5. Grid Independence and Time-Step Validations

The effects of cell numbers in FE models on the numerical results is employed to validate the grid-independent [38]. Three kinds of FE models at different mesh interval sizes consist of 20,400 cells, 72,800 cells and 110,800 cells, respectively. The effects of grid elements on x-directional velocities of the red detected line across the screw channel are shown in Figure 3a. From this figure, it can be seen that the magnitudes of *x*-directional velocities in three kinds of FE models are almost same. Considering the computational cost, the FE model consisting of 72,800 cells was selected to accurately study the mixing mechanisms in the four-screw extruders.

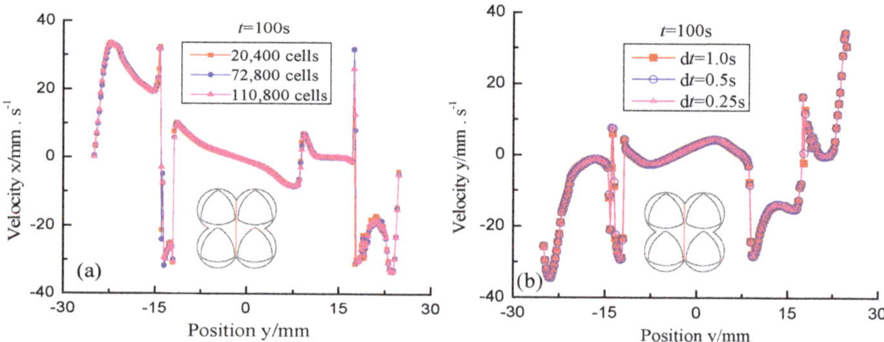

Figure 3. Effect of grid density and time step on the fluid velocity of the detected line: (**a**) Grid independence test; (**b**) time-step validation.

To illustrate the effect of time steps on prediction, the red detected line was selected in the flow domain and was advected for 100 s using three different time steps, namely, d*t* = 1.0 s, d*t* = 0.5 s, and dt = 0.25 s, to detect velocity changing of the detected line in y direction, as shown in Figure 3b. This shows the trajectory agreement in detail. Time step dt = 0.5 s was chosen in our simulations considering the cost of computing.

3. Results and Discussion

3.1. Chaotic Manifolds in the Basic System

The Lagrangian method has high sensitivity to initial conditions. The spatial map of FTLE describes the dynamical evolution of each particle over interval time T. When the initial condition is stable, in principle, LCS positions are uniquely determined by their end positions. From a Lagrangian perspective quantifying a fluid-transport process, it is important to define the original condition and the integration time. After many calculations with different integration times, rotating revolution T is chosen as integration time for all simulations.

The forward- and backward-time FTLE maps in the BS with integration time T = +a revolution and different initial time are shown in Figures 4 and 5, respectively. It is clear from Figures 4 and 5 that the maximal ridges in the FTLE maps show the location and change rule of repelling and attracting LCSs, and each screw is enclosed by the LCS. LCS form a closed diamond in the central region of the flow domain. Those fluid particles trapped in the central region are unable to mix with the rest of the flow domain. In addition, for all configurations of FTLE maps, LCSs slightly develop with the increase of initial time corresponding to different BS geometries. This leads to the similar diamond LCS structure in the central region, implying similar dynamic characteristics.

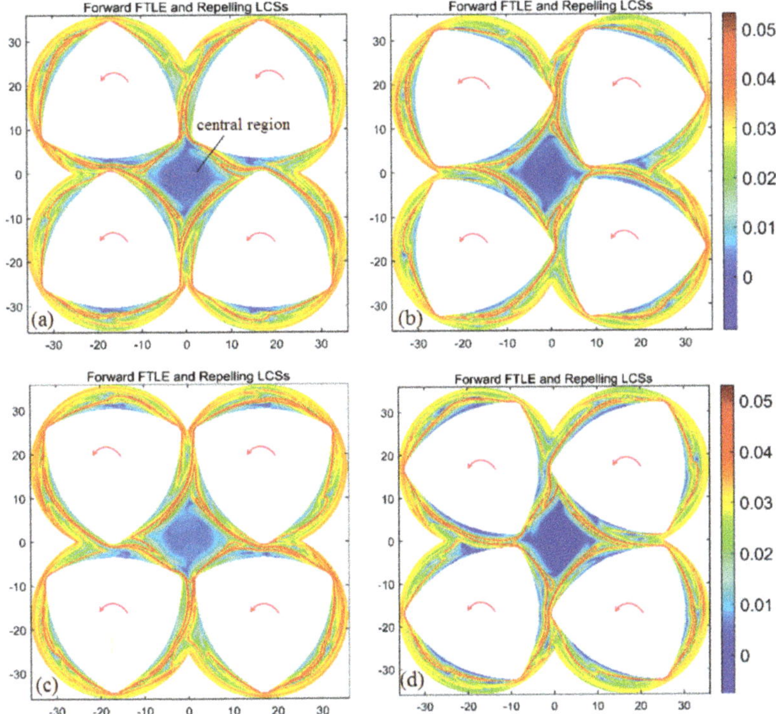

Figure 4. Forward-time finite-time Lyapunov exponent (FTLE) maps of the BS with integration time T = + a revolution and different initial time: (**a**) t_0 = 0 s; (**b**) t_0 = 10 s; (**c**) t_0 = 20 s; (**d**) t_0 = 30 s.

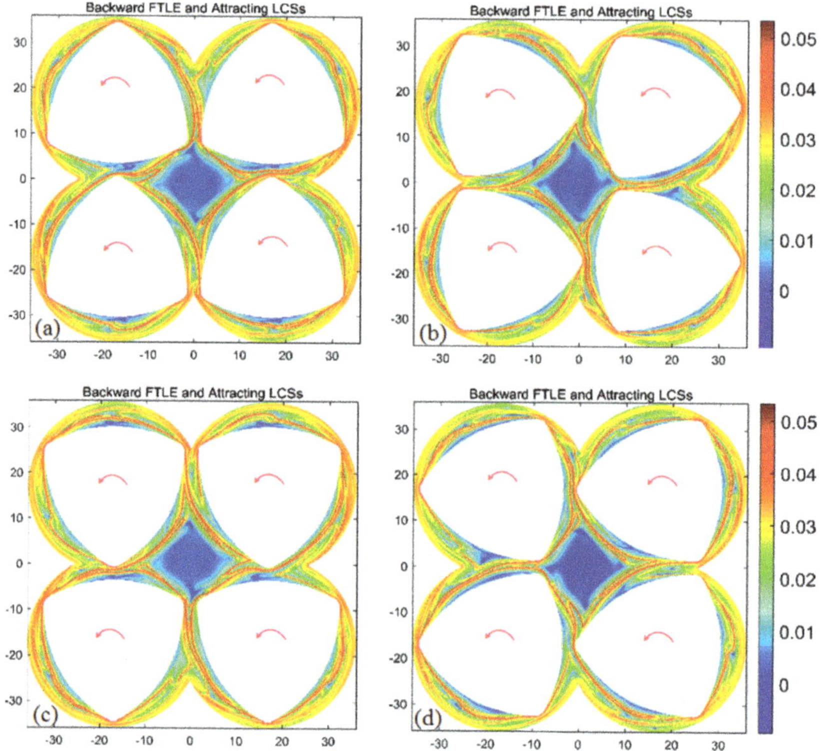

Figure 5. Backward-time FTLE maps of the BS with integration time $T = -a$ revolution and different initial time: (**a**) $t_0 = 0$ s; (**b**) $t_0 = 10$ s; (**c**) $t_0 = 20$ s; (**d**) $t_0 = 30$ s.

To locate the position of the hyperbolic fixed point in the flow system, the main repelling and attracting LCSs, namely, hyperbolic LCSs, in the BS with initial time $t_0 = 0$ s and $t_0 = 10$ s were redrawn, as shown in the Figure 6. It is clear from Figure 6 that there are four intersection points of hyperbolic LCSs, which are called hyperbolic periodic points. These intersections of manifolds indicate the presence of chaotic orbits. Moreover, hyperbolic periodic points are the dominant sources of mixing in the flow system. The fluid groups approaching the hyperbolic point are tangentially stretched away from the stable manifold and are folded along the length of the unstable manifold. Therefore, these flow regions that exist the hyperbolic periodic points have better mixing efficiency.

In the central region of the BS, the repelling and attracting LCSs tangle each other and emerge closed diamond of LCS structures, whose vertex is near the hyperbolic points. When a pair of hyperbolic LCSs come closer to being coincident, the tangle becomes a material transport barrier. The regions surrounded by the tangle only allow a small flux of fluid across the boundary. Therefore, we consider the tangle as the boundary that separates the physical flow domain into three portions with different mixing characteristics, as shown in Figure 6a. The first portion, which is called the outer mixing region, is located between the barrel wall and the repelling LCS. The second portion is called the inner mixing region and is located between the repelling LCS and the screw wall. The third portion encircled by the diamond LCSs is called the central region. The coincident hyperbolic LCS is the material transport boundary. A pair of hyperbolic LCSs are connected to adjacent intersection points to form a lobe structure set, which is the bridge of the material transport between adjacent regions with different mixing characteristics. In the outer mixing region, the stretching of fluid particles is better than that in the inner mixing region, and the mixing of fluid in the central region is the worst. So, the FTLE as a novel measure has potential advantages to quantity-mixing efficiency.

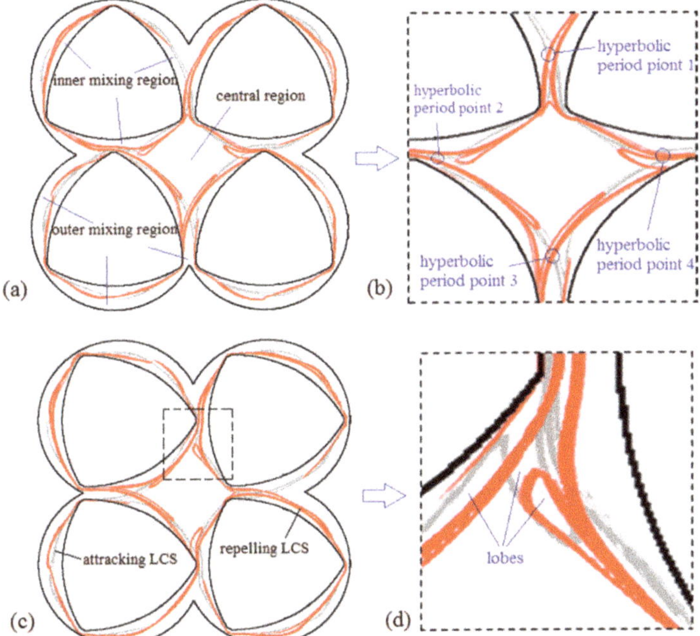

Figure 6. Repelling and attracting Lagrangian coherent structures (LCSs) redrawn from the FTLE maps with different initial time: (**a**) hyperbolic LCSs with initial time $t_0 = 0$ s; (**b**) enlarged view of central region; (**c**) hyperbolic LCSs with initial time $t_0 = 10$ s; (**d**) enlarged view of intermeshing region.

To further understand the relative good or poor mixing characteristics in the three portions identified by the LCSs, particle tracking was used to illustrate the different fluid motions in the BS. Initially, we set free 3000 particles in a rectangle region located on the axis of the flow domain, which would pass through two hyperbolic periodic points, as shown in Figure 7a. Particle evolution over five revolutions in the BS is described in Figure 7b–f. It is noted that the simulations over five revolutions is sufficient to obtain a developed flow.

From the above analysis, we know that there are four hyperbolic periodic points at the center of the domain. Nearby, in every hyperbolic periodic fixed point, the particles would move apart faster. With the four screw rotations, the particle groups near hyperbolic periodic points 1 and 3 move along the unstable manifold into the left and right chambers, respectively, as shown in Figure 7b. The tangle also acts as the fluid-transport barrier; the attracting LCS indicate the mixing behaviors of tracer particles. In Figure 7c, the particles in the chamber (outer mixing region) are encircled by the barrel wall and hyperbolic LCSs, and only move with the screw rotation. Because those particles gradually move away from the hyperbolic periodic points, they are weakly stretched along the screw rotation. After a revolution, shown in Figure 7d, the particle tracers arrived at horizontal hyperbolic points 2 and 4. With the rotation of four screws, the particles affected by the horizontal hyperbolic points were further divided in opposite directions and entered different mixing characteristic regions. Then, the mixing process continued. From Figure 7f, it can be seen that hyperbolic LCSs overlap each other to form a series of lobe structures in the intermeshing region. Particles trapped in the outer mixing region identified by the LCSs employ material change with the inner mixing region by means of the lobe structures. The particle transports have perfect agreement with the boundary characteristics of the hyperbolic LCSs.

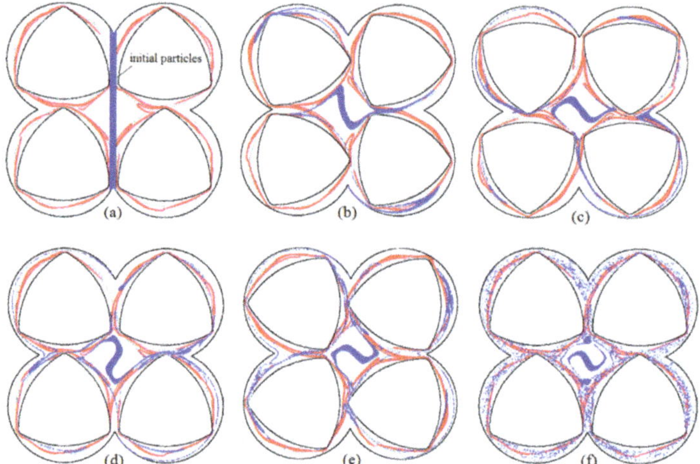

Figure 7. Particle-tracking results in the BS merging hyperbolic LCSs with different time: (**a**) $t = 0$ revolution; (**b**) $t = 5/12$ revolution; (**c**) $t = 10/12$ revolution; (**d**) $t = 1$ revolution; (**e**) $t = 15/12$ revolutions; (**f**) $t = 5$ revolutions.

It is noted that the movement of those particles in the central region is limited by the transport boundary of LCSs. The closed-diamond LCSs mean that the tracers are substantially isolated and could not efficiently mix with the rest of the domain. Therefore, the dead zone in the central region formed by the closed LCSs must be considered to optimize the geometric deign and enhance the mixing efficiency of the BS. In the next section, in order to decrease or remove the dead zone, we study the effect of screw rotational direction and the arrayed manner of the four screws on the chaotic manifolds in four-screw extruders.

3.2. Chaotic Manifolds in CRS and SEDS

Based on the base model of FSE, changing the direction of screw rotation is a relatively simple method to change the chaotic manifolds of a four-screw dynamic system. For the CRS, the two upper screws rotate counterclockwise, while the two screws below rotate clockwise. Figure 8 shows the backward-time FTLE maps in the CRS with a different initial time, where maximal ridges can correspond to the attracting LCS/unstable manifold. It can be seen from Figure 6 that the unstable manifolds in the CRS are complex in comparison with the BS. In the central region of the CRS, the closed diamond of the LCS structure disappears, and an open architecture of LCS structure appears. At the same time, many kinks appear in the CRS, implying the perfect folding action. It is known that the rate of fluid transport across the tangle of a manifold pair is proportional to the area of the lobes [27]. The kink structures in the CRS benefit from increasing the area of lobes, which is important to improve the fluid transport on both sides of the tangles. By changing the rotational direction of the screws, the closed fluid-transport boundary is destroyed to enhance local mixing efficiency in the central region and global mixing efficiency in the flow domain.

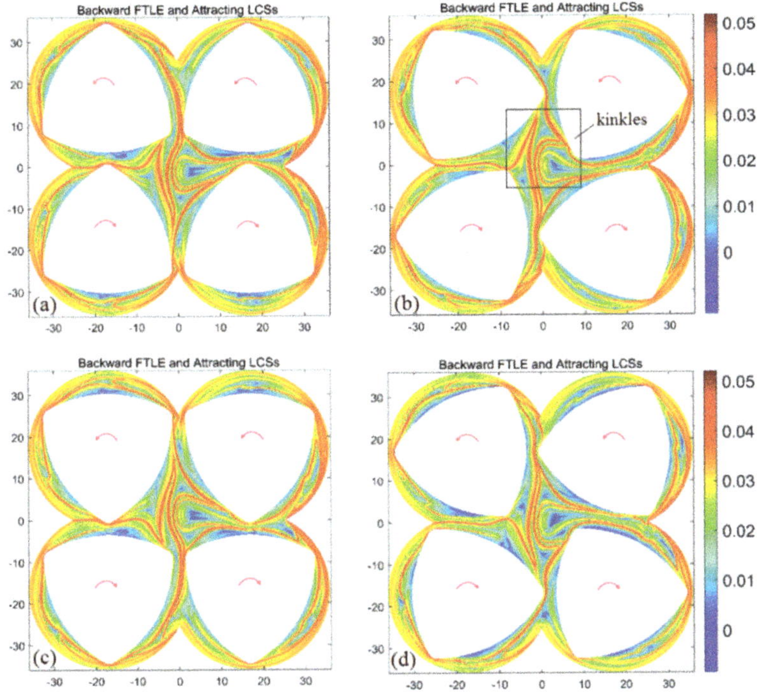

Figure 8. Backward-time FTLE maps in a CRS with integration time T = a revolution and different initial time: (**a**) $t_0 = 0$ s; (**b**) $t_0 = 10$ s; (**c**) $t_0 = 20$ s; (**d**) $t_0 = 30$ s.

Based on the base model of BS, the relative positions of the four screws are varied to construct the SEDS model in order to obtain different flow domain from BS. For the SEDS, we explored the relationship between the four screw positions and the hyperbolic LCSs of the flow system, which are responsible for determining mixing performance.

Figure 9 shows the forward-time FTLE maps in the SEDS using integration time $T = 120$ s. Ridges in these FTLE plots show the location of the repelling LCSs and wraps around the four screws. In particular, two small-area closed LCSs appear in the left and right central regions, respectively. With screw rotation, the area of the left central region decreases, corresponding to the decrease of the closed LCS areas, as shown in Figure 9b; however, the area of the right central region increases corresponding to the increase of the closed LCS areas. In comparison with the base BS model, there is only a relatively small blue zone in the two central regions of forward-time FTLE maps, implying better mixing efficiency than the BS. Moreover, the above periodic changes of the two central regions result in fluid in the periodic conditions of stretching and compression. This is beneficial to enhancing local mixing efficiency in the two central regions. More importantly, SEDS has five intermeshing regions corresponding to five hyperbolic fixed points, whereas the other two kinds of four-screw extruders only have four intermeshing regions. Due to the increase of the number of intermeshing regions, the mixing efficiency of the SEDS is improved to some degree. In addition, several kinks appear in the closed LCSs in the SEDS, implying perfect folding action of the fluid, and enhancing fluid exchange for both inward and outward of the closed LCSs.

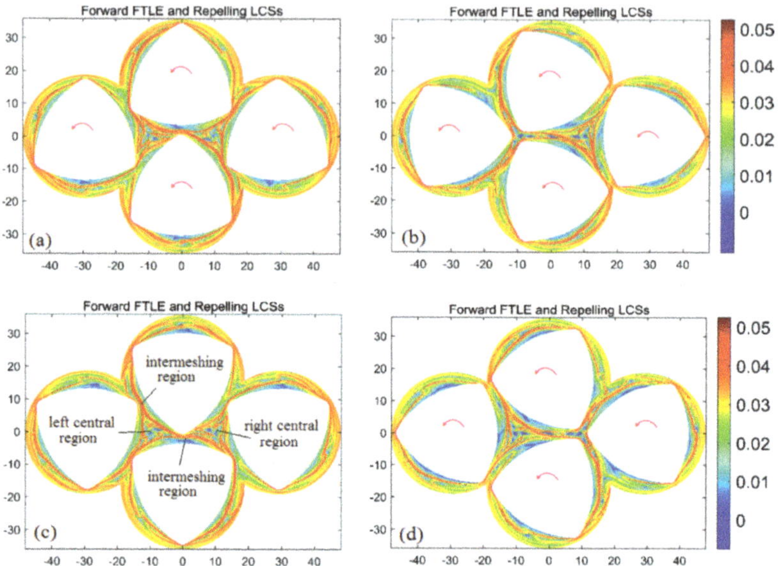

Figure 9. Forward-time FTLE maps in the SEDS with integration time T = a revolution and different initial time: (**a**) $t_0 = 0$ s; (**b**) $t_0 = 10$ s; (**c**) $t_0 = 20$ s; (**d**) $t_0 = 30$ s.

3.3. Comparisons of Poincaré Sections

The Poincaré section is a simple method for analyzing chaotic mixing flows. It is a powerful way to reveal regular zones and chaotic motions. The Poincaré section allows for a systematic reduction in the complexity of problems by reducing the number of dimensions. Previously, many researchers have investigated the relationship between a Poincaré section and chaotic flow [39]. Here, we used the same method to plot the Poincaré section. Initially, 441 points were distributed in a box region with a size of 0.5×0.5 mm, and tracked their positions over 500 periods for obtaining the Poincaré sections of three kinds of models, as shown in Figure 10.

As shown in Figure 10a, there was a large-size KAM island that can correspond to the closed diamond repelling LCS in the central region of the BS. This indicates the periodic property in the KAM island in which particles hardly mix with other regions in the BS. They served as the main barrier to better mix in the corotating four-screw extruders. In addition, several little-size KAM islands appeared near the screw walls. This is associated with the screw geometry. In comparison with the BS, there was no KAM island in the central region of the CRS, and only several little-size KAM islands appeared near the screw walls, as shown in Figure 10b. This phenomenon can be explained by the above analysis of LCS structures. This is different from the BS and SEDS, as the SEDS has two relatively small central regions. As shown in Figure 10c, there were several little-size KAM islands that appeared in the two central regions. Around these KAM islands, the fluid in the rest of the flow domain was in chaos. From the comparisons of the Poincaré sections in four kinds of four-screw extruders, we can conclude that KAM regions in the BS are the largest, followed by the SEDS, and those in the CRS were smallest. Therefore, it is deduced that the CRS has maximal mixing efficiency, followed by the SEDS, and the BS has minimal mixing efficiency.

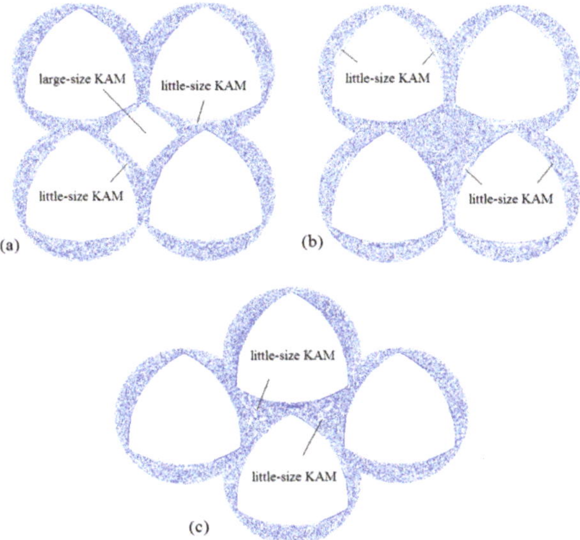

Figure 10. Comparisons of Poincaré sections of three types of FSEs: (**a**) BS; (**b**) CRS; (**c**) SEDS.

3.4. Quantitative Mixing Comparisons

In order to further understand the influences of the screw rotational manner and geometric configurations of four-screw extruders on chaotic mixing, the logarithm of stretching, segregation scale, and time-averaged efficiency were calculated and compared with the three types of four-screw extruders. Initially, 10,000 massless particles were injected in the flow domain, and particle statistics were employed to obtain the above parameters over five periods, as shown in Figure 11a–c.

The logarithmic of stretching is often used to evaluate fluid stretching [10]. Given motion $x = \chi(X,t)$, where initially $x = \chi(X,t)$ for an infinitesimal material line segment $x = \chi(X,t)$, located at position x at time t, the length of stretch of a material line is defined as:

$$\lambda = \frac{|dx|}{|dX|} \tag{9}$$

Figure 11a shows the plot of the logarithmic of stretching versus time for three kinds of four-screw extruders. The logarithmic of stretching in three models all increase exponentially over time due to the folding of the polymer melt between the four screws. The CRS showed better length of stretching, followed by SEDS, and then BS. This is due to the fact that the BS has a large-scale poor stretching mixing region in the central region, identified by the closed diamond LCSs, and the SEDS only has two little-area stretching mixing regions in the two central regions, but it has five intermeshing regions. The CRS, on the other hand, has no poor stretching mixing region in the central region.

The segregation scale is a measure of homogeneous concentration in flow regions [10] and can be expressed by:

$$S(t) = \int_0^{\zeta} R(r,t)dr \tag{10}$$

where $R(r,t)$ is the correlation coefficient for the concentration, and it gives the probability of finding a pair of random points with relative distance r with the same concentration [10]. The segregation scale is a measure of the size of the regions of homogeneous concentration, which decreases when mixing improves.

Figure 11b shows the segregation scale plotted over five revolutions for all three models. It can be seen that the three kinds of models showed a rapid drop during 1.5 periods (revolutions) due to the large segregated area. The CRS had a smaller segregation scale than the two other kinds of models, implying better distributive mixing efficiency due to no obvious mixing dead region. After 3.5 periods, the SEDS had a smaller segregation scale than the BS, because the SEDS has five intermeshing regions and a relatively small-sized central region. However, the relatively great magnitude of segregation scale in the BS had a little increase, from 3.5 to 5 revolutions, due to the main barrier of better mixing in the central region.

Time-average efficiency is often used to describe the stretching mixing efficiency during mixing. It is defined as:

$$\langle e_\lambda \rangle = \frac{1}{t} \int_0^t \frac{\lambda/\lambda}{(D:D)^{1/2}} dt \tag{11}$$

where $\langle e_\lambda \rangle$ is the time-average mixing efficiency, and D is the rate of strain tensor. It is clear from Figure 11c that the CRS showed better mean-time mixing efficiency, followed by the SEDS and BS. At five periods, the mean-time mixing efficiency of the CRS and SEDS was 2.2 times and 1.5 times of that of the BS, respectively. Therefore, to change the relative screw rotating direction and the positions of four screws can have great effects on the flow and mixing nature of four-screw extruders.

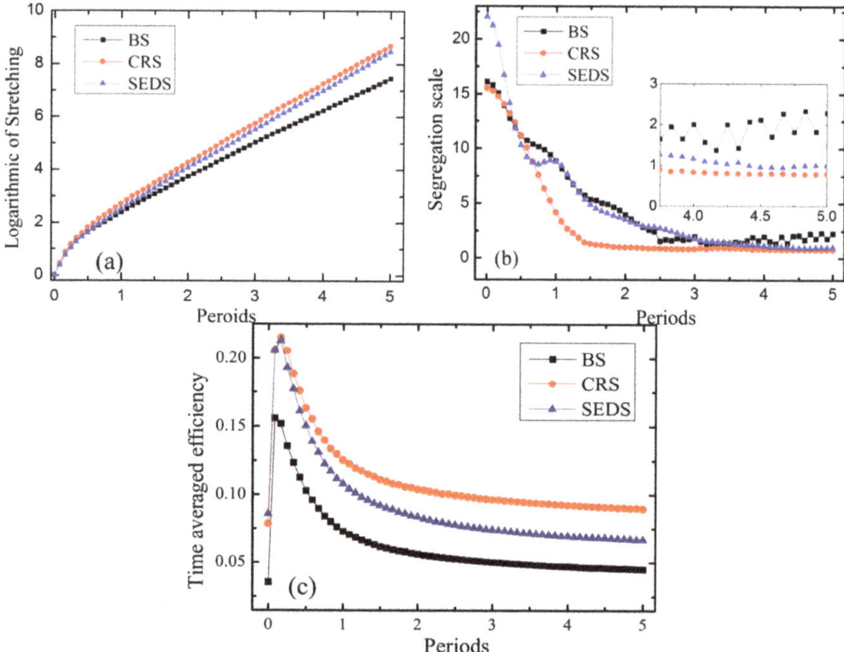

Figure 11. Comparisons of different mixing measurements between BS, SEDS, and SEDS: (a) logarithmic of stretching; (b) segregation scale; (c) time-average efficiency.

4. Conclusions

This paper explores the chaotic-mixing and fluid-transport characteristics within three kinds of novel industrial four-screw extruders by employing FTLE, LCSs, and Poincaré sections. We focused on two-dimensional chaotic advection and mixing in four-screw extruders using computational fluid dynamics. The manifold structures of the four-screw flow systems were visualized by extracting the hyperbolic LCSs from the FTLE maps. Hyperbolic LCSs as fluid-transport boundaries were used to analyze the material transport in two-dimension flow field of FSEs. Moreover, Poincaré sections were employed to indicate the chaotic and regular regions, which are key in improving local mixing.

Furthermore, quantitative mixing measures by means of the segregation scale, logarithmic of stretching, and mean-time mixing efficiency were employed to provide a rigorous method for comparison of the three FSE systems.

In the BS, namely, the corotating square-arranged four-screw extruder, the hyperbolic LCSs have four hyperbolic fixed points in the four intermesh regions, which are important to enhance mixing efficiency due to the fluid stretching and folding actions, which can help us better understand the better mixing performances in the intermeshing regions. Hyperbolic LCSs, as an underlying transportation boundary, separate the fluid flow of the mixer into three isolated regions, namely, the inner mixing region, outer mixing region, and central region. In the outer mixing region, the fluid was stretched strongly, indicating better mixing efficiency. In the inner mixing region, however, mixing was relatively poor. However, the central region, surrounded by the closed diamond LCSs, was a dead region, in which there was a large-size KAM island in the Poincaré sections of the BS. This obviously decreases the global mixing efficiency in the BS. Correspondingly, the mixing-efficiency measurement of the segregation scale, logarithmic of stretching, and average-time mixing efficiency was lower than the other two FSE systems. So, the central region should be considered to optimize the manifold geometry controlled by the BS profiles to enhance mixing efficiency.

In the CRS, namely, the counter-rotating square-arrayed four-screw extruder, the geometry of the flow field was the same as that of the BS. However, the corotating and the counter-rotating manners of the four screws was relatively different from that of the BS. This caused the dead region in the central region of the flow domain to disappear from the analysis of the repelling LCS and Poincaré sections. Local mixing efficiency in the central region was obviously enhanced. Furthermore, the CRS had a more complex manifold structure and great numbers of kinks in the hyperbolic LCSs, resulting in stronger fluid stretching and transport. Therefore, the CRS had a smaller segregation scale, and greater logarithmic of stretching and average-time mixing efficiency than the other four-screw systems. So, the screw rotating manner has a great influence on the manifold structure and mixing efficiency in a four-screw flow system.

The SEDS, namely, the equidistance-arrayed four-screw extruder, has five intermeshing regions and two central regions, which are different from the BS and CRS. The manifold structure of the SEDS has five hyperbolic fixed points in five intermeshing regions, implying better local mixing performance. In addition, several kinks appeared in the closed LCSs in the SEDS, implying perfect folding action of the fluid, and enhancing fluid exchange for both inward and outward of the closed LCSs. Therefore, the SEDS had a smaller segregation scale, and greater logarithmic of stretching and average time mixing efficiency than the BS. However, two small-area closed LCSs appeared in the left and right central regions, which corresponded to several little-size KAM islands. This resulted in relatively poor mixing in the SEDS compared to the CRS, because the CRS had no dead region in the central region. So, varying the relative positions of screws could change the manifold structure and enhance the mixing efficiency of the four-screw flow system.

This paper has shown that a fresh Lagrangian perspective is more feasible than the traditional Eulerian method in numerically investigating the evolution of two-dimensional mixing performance within a novel screw extruder. FTLE and LCS are useful tools for analyzing chaotic mixing flow. This method is robust and flexible, and can be applied to other types of polymer-processing equipment. It provides a better understanding of the mixing mechanism in screw extruders.

Author Contributions: Conceptualization, X.Z.Z.; methodology, X.Z.Z. and Y.T.; formal analysis, X.Z.Z. and Y.T.; investigation, X.Z.Z., Y.T., and Y.X.H.; data curation, Y.X.H.; writing—original draft preparation, Y.T.; writing—review and editing, X.Z.Z.; supervision, X.Z.Z.; project administration, X.Z.Z.; funding acquisition, X.Z.Z.

Funding: This research was funded by National Natural Science Foundation of China (Grant Nos. 51473073 and 50903042), the Program for Liaoning Excellent Talents in University (Grant No. LR2016022), and the Liaoning Province Natural Science Foundation (Grant No. 2015020142).

Conflicts of Interest: The authors declare no conflict of interest.

References

1. Lawal, A.; Kalyon, D.M. Mechanisms of mixing in single and co-rotating twin screw extruders. *Polym. Eng. Sci.* **1995**, *35*, 1325–1338. [CrossRef]
2. Yao, W.G.; Takahashi, K.; Koyama, K.; Dai, G.C. Experimental study on distributive mixing characteristics of a new type of pin mixing section. *Polym. Eng. Sci.* **1997**, *37*, 615–622. [CrossRef]
3. Nakayama, Y.; Takeda, E.; Shigeishi, T.; Kajiwara, T. Melt-mixing by novel pitched-tip kneading disks in a co-rotating twin-screw extruder. *Chem. Eng. Sci.* **2011**, *66*, 103–110. [CrossRef]
4. Ishikawa, T.; Amano, T.; Kihara, S.I.; Funatsu, K. Flow patterns and mixing mechanisms in the screw mixing element of a co-rotating twin-screw extruder. *Chem. Eng. Sci.* **2002**, *42*, 925–939. [CrossRef]
5. Zhu, X.Z.; Xie, Y.J.; Yuan, H.Q. Numerical simulation of flow characteristics of co-rotating intermeshing four-screw extruder. *J. Reinf. Plast. Compos.* **2008**, *27*, 321–334. [CrossRef]
6. Hutchinson, B.C.; Rios, A.C.; Osswald, T.A. Modeling the distributive mixing in an internal batch mixer. *Int. Polym. Process.* **1999**, *14*, 315–321. [CrossRef]
7. Bravo, V.L.; Hrymak, A.N.; Wright, J.D. Numerical simulation of pressure and velocity profiles in kneading elements of a co-rotating twin screw extruder. *Polym. Eng. Sci.* **2000**, *40*, 525–541. [CrossRef]
8. Erol, M.; Kalyon, D.M. Assessment of the degree of mixedness of filled polymers. *Int. Polym. Process.* **2005**, *20*, 228–237. [CrossRef]
9. Luo, P.; Jia, H.; Xin, C. An Experimental study of liquid mixing in a multi-orifice-impinging transverse jet mixer using plif. *Chem. Eng. J.* **2013**, *228*, 554–564. [CrossRef]
10. Connelly, R.K.; Kokini, J.L. Examination of the mixing ability of single and twin screw mixers using 2D finite element method simulation with particle tracking. *J. Food Eng.* **2007**, *79*, 956–969. [CrossRef]
11. Domingues, N.; Gaspar-Cunha, A.; Covas, J.A. A quantitative approach to assess the mixing ability of single-screw extruders for polymer extrusion. *J. Polym. Eng.* **2012**, *32*, 81–94. [CrossRef]
12. Zhang, X.M.; Feng, L.F.; Chen, W.X. Numerical simulation and experimental validation of mixing performance of kneading discs in a twin screw extruder. *Polym. Eng. Sci.* **2009**, *49*, 1772–1783. [CrossRef]
13. Nakayama, Y.; Takeda, E.; Shigeishi, T.; Kajiwara, T. Comparative study of internal batch mixer such as cam, boundary and roller. *Chem. Eng. Sci.* **2011**, *66*, 2502–2511.
14. Dhakal, P.; Das, S.R.; Poudyal, H.; Chandy, A.J. Numerical simulations of partially-filled rubber mixing in a 2-wing rotor-equipped chamber. *J. Appl. Polym. Sci.* **2016**, *134*, 44250. [CrossRef]
15. Das, S.R.; Poudyal, H.; Chandy, A.J. Numerical Investigation of Effect of Rotor Phase Angle in Partially-Filled Rubber Mixing. *Int. Polym. Process.* **2017**, *32*, 343–354. [CrossRef]
16. Wiggins, S.; Ottino, J.M. Foundations of chaotic mixing. *Math. Phys. Eng. Sci.* **2004**, *362*, 937–970. [CrossRef] [PubMed]
17. Xia, H.M.C.; Shu, S.; Wan, Y.; Chew, T. Influence of the reynolds number on chaotic mixing in a spatially periodic micromixer and its characterization using dynamical system techniques. *J. Micromech. Microeng.* **2006**, *16*, 53–61. [CrossRef]
18. Lee, T.H.; Kwon, T.H. A new representative measure of chaotic mixing in a chaos single-screw extruder. *Adv. Polym. Technol.* **1999**, *18*, 53–68. [CrossRef]
19. Hwang, W.R.; Kwon, T.H. Chaotic volumetric transports in a single-screw extrusion process. *Polym. Eng. Sci.* **2003**, *43*, 783–797. [CrossRef]
20. Hwang, W.R.; Kwon, T.H. Dynamical modeling of chaos single-screw extruder and its three-dimensional numerical analysis. *Polym. Eng. Sci.* **2000**, *40*, 702–714. [CrossRef]
21. Niu, X.; Lee, Y.K. Efficient spatial-temporal chaotic mixing in microchannels. *J. Micromech. Microeng.* **2003**, *13*, 454–462. [CrossRef]
22. Wang, W.; Manas-Zloczower, I.; Kaufman, M. Entropy time evolution in a twin-flight single-screw extruder and its relationship to chaos. *Chem. Eng. Sci.* **2005**, *192*, 405–423. [CrossRef]
23. Haller, G.; Yuan, G. Lagrangian coherent structures and mixing in two-dimensional turbulence. *Phys. D* **2000**, *147*, 352–370. [CrossRef]
24. Haller, G. Finding finite-time invariant manifolds in two-dimensional velocity fields. *Chaos* **2000**, *10*, 99–108. [CrossRef] [PubMed]
25. Haller, G. Lagrangian coherent structures. *Annu. Rev. Fluid Mech.* **2015**, *47*, 137–162. [CrossRef]

26. Santitissadeekorn, N.; Bohl, D.; Bollt, E.M. Analysis and modeling of an experimental device by finite-time lyapunov exponent method. *Int. J. Bifurcat. Chaos* **2011**, *19*, 993–1006. [CrossRef]

27. Robinson, M.; Cleary, P.W. The influence of cam geometry and operating conditions on chaotic mixing of viscous fluids in a twin cam mixer. *AIChE J.* **2011**, *57*, 581–598. [CrossRef]

28. Robinson, M.; Cleary, P.W. Flow and mixing performance in helical ribbon mixers. *Chem. Eng. Sci.* **2012**, *84*, 382–398. [CrossRef]

29. Conti, G.; Badin, G. Hyperbolic Covariant Coherent Structures in Two-Dimensional Flows. *Fluids* **2017**, *2*, 50. [CrossRef]

30. O'Farrell, C.; Dabiri, J.O. A lagrangian approach to identifying vortex pinch-off. *Chaos* **2010**, *20*, 261–300. [CrossRef] [PubMed]

31. Bozorgmagham, A.E.; Ross, S.D. Atmospheric lagrangian coherent structures considering unresolved turbulence and forecast uncertainty. *Commun. Nonlinear Sci.* **2015**, *22*, 964–979. [CrossRef]

32. Prant, S.V. Chaotic lagrangian transport and mixing in the ocean. *Eur. Phys. J. Spec. Top.* **2015**, *223*, 2723–2743. [CrossRef]

33. Mukiibi, D.; Badin, G.; Serra, N. Three dimensional chaotic advection by mixed layer baroclinic instabilities. *J. Phys. Oceanogr.* **2016**, *46*, 1509–1529. [CrossRef]

34. Green, M.A.; Rowley, C.W.; Smits, A.J. Using hyperbolic lagrangian coherent structures to investigate vortices in bioinspired fluid flows. *Chaos* **2010**, *20*, 017510. [CrossRef] [PubMed]

35. Borgogno, D.; Rubino, G.; Veranda, M.; Cappello, S.; Grasso, D. Detection of magnetic barriers in a chaotic domain: first application of finite time lyapunov exponent method to a magnetic confinement configuration. *Plasma Phys. Control. Fusion* **2015**, *57*, 401–406.

36. Connelly, R.K.; Kokini, J.L. The effect of shear thinning and differential viscoelasticity on mixing in a model 2D mixer as determined using FEM with particle tracking. *J. Non-Newton. Fluid Mech.* **2004**, *123*, 1–17. [CrossRef]

37. Shadden, S.C.; Lekien, F.; Marsden, J.E. Definition and properties of lagrangian coherent structures from finite-time lyapunov exponents in a two-dimensional periodic flows. *Phys. D* **2005**, *212*, 271–304. [CrossRef]

38. Ortega-Casanova, J. CFD study on mixing enhancement in a channel at a low Reynolds number by pitching a square cylinder. *Comput. Fluids* **2017**, *145*, 141–152. [CrossRef]

39. Ottino, J.M. Mixing, chaotic advection, and turbulence. *Annu. Rev. Fluid Mech.* **1990**, *22*, 207–254. [CrossRef]

 *materials*

Article

3D Numerical Simulation of Reactive Extrusion Processes for Preparing PP/TiO$_2$ Nanocomposites in a Corotating Twin Screw Extruder

Dapeng Sun [1,2], Xiangzhe Zhu [1,*] and Mingguang Gao [1]

[1]   School of Mechanical Engineering, Liaoning Shihua University, Fushun 113001, China; xzzhu@126.com; dpsun2019@163.com (D.S.); mggao1987@163.com (M.G.)
[2]   College of Pipeline and Civil Engineering, China University of Petroleum (East China), Qingdao 266580, China
*   Correspondence: xzzhu@126.com; Tel.: +86-2456-8650-42

Received: 30 January 2019; Accepted: 19 February 2019; Published: 23 February 2019

Abstract: To better understand the relationship between flow, mixing and reactions in the process of preparing PP/TiO$_2$, a 3D numerical simulation in a co-rotating twin screw extruder (TSE) was firstly employed using commercial CFD code, ANSYS Polyflow. The effects of rotating speed of screws, stagger angle of knead blocks, inlet flow rate and initial temperature of barrel on the mixing and reaction process in the TSE were investigated. The results reveal that the studied operational and geometric parameters, which determine mixing efficiency, residence time distribution, and temperature of the flows in the TSE, affect the local species concentration, reaction time and reaction rate, and hence have great influences on the conversion rate. The results show that increasing the rotating speed and inlet flow rate can decrease the time for sufficient mixing, which is not conducive to intensive reaction, and increasing the stagger angle has the opposite effect. Moreover, the conversion rate greatly affected by the initial temperature of barrel.

Keywords: twin screw extruder; simulation; residence time distribution; PP/TiO$_2$ nanocomposites; conversion

---

1. Introduction

The application of Polypropylene (PP) covers many industrial fields, including film, thermoforming and automotive fields, due to its easy processability, relatively high mechanical properties and great recyclability [1]. Nevertheless, there are disadvantages of net PP, such as low polarity, poor ultraviolet, low thermal stability and non-bacterial resistance [2]. One way to improve the PP properties mentioned above is to disperse titanium dioxide (TiO$_2$) into PP matrix [3]. In situ sol–gel reactions in melt are the most suitable method for producing dispersed TiO$_2$ nanoparticles in PP matrix without solvent, and it can be applied to industrial processes in a twin screw extruder. Polymer matrix-based nanocomposites have attracted a lot of attention in the recent literature [4,5]. Meanwhile, twin screw extruders (TSEs) have been broadly applied in the processing of polymer nanocomposites due to their excellent mixing performance and modular design characteristic. With the great improvement of computing power, the simulation methods have been intensively used to better understand the factors influencing the preparation of nanocomposites by reactive extrusion.

The twin screw extruder (TSE) is a key equipment in polymer processing, which is mainly used as a conveyor, mixer or reactor for particles, granules and viscous fluids [6]. As polymer mixing equipment, the mechanism of melt flow and mixing is very complex. With the development of the computational method, more and more researchers are studying the isothermal or non-isothermal flow mechanisms in TSEs by means of numerical simulation [7–10]. Recently, Zhang et al. [11] analyzed the mixing of generalized Newtonian fluid in kneading zones of a TSE using finite element method (FEM) and experiment. Distributive mixing and overall efficiency of different types of kneading blocks were studied by calculating the mixing parameters, such as area stretch ratio, instantaneous efficiency, and time-average efficiency. Robinson et al. [12] carried out numerical simulations of three-dimensional flow in a co-rotating TSE to reveal the influences of geometry and fill level on the transport and mixing behavior. Reitz et al. [13] studied the residence time distribution of hot melt extrusion processes experimentally and mathematically. Kruijt et al. [14–16] proposed a mapping method of polymer melt in a co-rotating TSE in terms of the concentration distribution, residence time distribution and separation strength to characterize the distribution mixing.

Generally, TSE is taken as a continuous reactor in which the reactions and extrusion process occur simultaneously [17,18]. Consequently, reactive extrusion is a complex process that involves many aspects, because of the high number of operating variables and their interactions during the process [19]. For example, with the proceeding of mixing and reaction process, the properties of the molten polymers will change along the axial direction in TSEs. The residence time distribution, pressure, segregation scale and mixing efficiency are important indexes to measure the mixing performances, which also have great influence on the reaction, and that will in turn modify the material properties. Moreover, the reactive extrusion processes are often accompanied by heat transfer, viscous dissipation and reaction heat.

To investigate the effects of operating parameters and screw geometry on the mixing performance during reactive extrusion process in TSEs, lots of experimental and numerical simulation studies have been employed in recent years [20–22]. Typically, Rozen [23] experimentally and numerically studied the product distribution of two parallel chemical reactions to characterize mixing in a corotating TES. The correlation between the rate of viscous deformation of liquid elements and screw rotational speed was investigated. Zhu et al. [24] analyzed the polymerization of e-caprolactone in a co-rotating TSE with finite element method. The effects of screw rotational speed, geometry of screw element, and initial conversion at the channel inlet on polymerization was discussed. Zhang et al. [25] investigated the effects of screw configuration, rotating speed and silica on the mechanical and foaming properties of PP/wood-fiber composites. Kim et al. [26] studied the process of synthesizing polyamide polyester-based diblock, triblock, and random block copolymers in a modular, corotating TSE via reactive extrusion. Rigoussen et al. [27] investigated the effect and role of cardanol in the compatibilization of PLA/ABS immiscible blends by reactive extrusion. Berzin et al. [28,29] investigated the effects of different oxidant concentrations and basic conditions on the molecular weight distribution and rheological properties of PP, then compared the theoretical model with the experimental data.

However, the above studies were focused on the reactive extrusion processes, such as monomer polymerization, polymer chemical modification, rheological modification, etc.; few studies have been conducted on the sol-gel reaction occurring in the polymer matrix from dynamic simulation viewpoint. As for research on PP/TiO$_2$ nanocomposites, most of it has focused on the properties of experimental products using several techniques, such as XRD, TEM micrographs, Raman spectroscopy, X-ray diffraction and mechanical analysis [5,30,31]. However, studies on the mixing and reaction process of PP/TiO$_2$ nanocomposites by means of numerical simulation based on Computational Fluid Dynamics (CFD) are few and are limited to 1D or 2D models [32,33]. In the present paper, the reaction process of preparing PP/TiO$_2$ nanocomposites in a corotating TSE with two combined screws was investigated via 3D numerical simulation based on FEM. The reactive patterns of nanocomposites in the corotating TSE were investigated firstly. Moreover, the effects of screw rotational speed, stagger angle of kneading blocks, inlet flow rate and initial temperature of barrel on the mixing and reaction process were discussed. The aim of this study is to better understand the effects of geometric and operating parameters on the reactive extrusion process and investigate the relationship between the mixing characteristics and the reactive process in a corotating TSE.

2. Materials and Methods

2.1. Geometrical and FE Models

The corotating TSE investigated in this study consisted of a flow channel and two combined screw elements, as shown in Figure 1a,b, which consisted of thread sections and kneading block sections. Meanwhile, the stagger angle of the kneading block section is 90°, as shown in Figure 1b. The lengths of kneading blocks and thread sections were 32 mm and 25 mm, respectively. Gambit software was used to establish the geometric models of channel and screws, respectively.

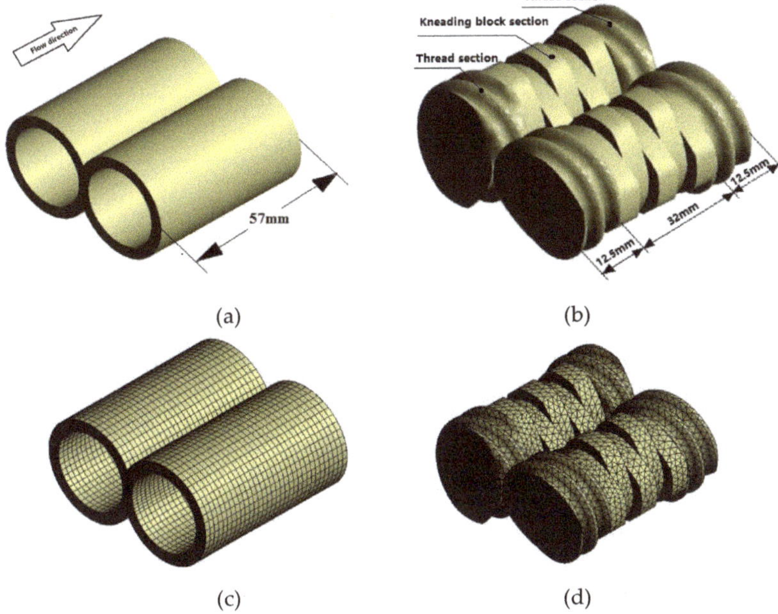

Figure 1. Geometrical and FE models: (**a**) Geometry of flow channel; (**b**) geometry of combined screws; (**c**) FE model of flow channel; (**d**) FE model of two screws.

The mesh superposition technique is used to put the meshes of flow channel and screws together without remeshing for the periodical geometric changes. Figure 1c,d illustrates the three-dimensional finite element (FE) models of the flow channel and combined screws of the TSE. The quadrilateral cells are used to mesh the flow channel. To achieve accurate simulation results in the small clearances and near the walls, four boundary layer grids are employed in the FE model of the flow channel. The FE model of the flow channel consists of 34,300 cells and 38,232 nodes. For the FE model of the screws, the tetrahedral cells are used for partition, different screw structures with different stagger angles (see Figure 12) adopt the same meshing method and mesh density. The geometric dimensions of the TSE are listed in Table 1.

Table 1. Parameters of geometric model of the extruder.

Parameter	Value
Inner diameter of barrel	33 mm
Screw tip diameter	32 mm
Screw root diameter	26 mm
Centerline distance of screws	30 mm
Screw clearance	0.5 mm

Table 1. *Cont.*

Parameter	Value
Clearance of screw and barrel	0.5 mm
Leads of screws	2
Length of kneading blocks	32 mm
Length of thread sections	12.5 mm
Total length of combined screws	57 mm

2.2. Mathematical Models

Due to the specific conditions of the extrusion process and the characteristics of polymer, the following assumptions are made in order to ensure the accuracy of calculation of the three-dimensional flow field: (1) The channel is fully filled with polymer melt at all times. (2) The fluid is incompressible. (3) The flow is laminar flow. (4) There is no slip between the fluid and the wall. (5) The force of inertia and the force of gravity are ignored. Based on above assumptions, the flow chart of the numerical simulation is shown in Figure 2.

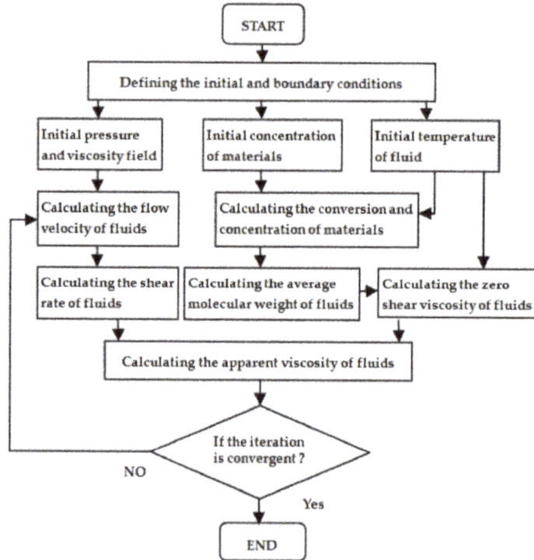

Figure 2. Flow chart of numerical simulation.

The governing equations used in this paper are as follows:

The continuity and momentum equations are respectively expressed as follows [34]:

$$\nabla \cdot \mathbf{V} = 0 \tag{1}$$

$$\rho\left(\frac{\partial \mathbf{V}}{\partial t} + \mathbf{V} \cdot \nabla \mathbf{V}\right) = -\nabla p + \eta \nabla^2 \mathbf{V} + \rho g \tag{2}$$

where **v** denotes the velocity vector; p denotes the pressure; ρ denotes the density, t and η denotes time, and shear viscosity, respectively.

The energy conservation obeys the following equation [34]:

$$\rho C_P \frac{\partial T}{\partial t} + \rho C_P \mathbf{V} \cdot \nabla T = \boldsymbol{\tau} : \nabla \cdot \mathbf{V} + \mathbf{r} - \nabla \cdot (k\nabla T) \tag{3}$$

where T denotes the absolute temperature, C_P denotes heat capacity, r denotes the heat source, k denotes the kinetic constant, $k\nabla T$ denotes the heat flux, $\boldsymbol{\tau}$ is the stress tensor, the form of $\tau = 2$ denotes the viscous heating.

The stress tensor in Equation (3) is written as follows:

$$\boldsymbol{\tau} = 2\eta\left(\dot{\gamma}, T\right)\mathbf{D} \tag{4}$$

In which $\mathbf{D}$ denotes the deformation tensor rate, and $\mathbf{D} = \frac{1}{2}\left[\Delta\mathbf{V} + (\Delta\mathbf{V})^{T}\right]$; η denotes the shear viscosity; $\dot{\gamma}$ denotes the effective shear rate and can be expressed as:

$$\dot{\gamma} = \sqrt{2(\mathbf{D}:\mathbf{D})} \tag{5}$$

The convection-diffusion equation can be defined as:

$$\frac{\partial}{\partial t}(\rho\omega_i) + \nabla\cdot(\rho\omega_i\mathbf{v}) + \nabla\cdot\mathbf{j}_i = R_i + S_i \tag{6}$$

where ω_i represents the mass fraction of the component i, R_i is the source term of the reaction, S_i represents a user-defined source item, $\mathbf{j}_i$ represents mass diffusion flux.

In this work, it is assumed that the nanocomposites have the viscosity of the PP matrix. This is predicted by the Carreau-Yasuda law:

$$\eta = H(T)\cdot\eta_0\left[1 + (\lambda\dot{\gamma})^a\right]^{\frac{n-1}{a}} \tag{7}$$

where λ is the characteristic time, $\dot{\gamma}$ is the shear rate, a represents the Yasuda parameter, n is the power-law index. According to the Arrhenius law, $H(T)$ can be defined as:

$$H(T) = \exp\left(\frac{E}{R}\left(\frac{1}{T} - \frac{1}{T_r}\right)\right) \tag{8}$$

In which E is an activation energy for flow, R is the gas constant, and T_r is the reference temperature.

2.3. Reaction Kinetics

The formation of TiO_2 in this paper was based on the sol-gel method, and tetrabutyl titanate $(Ti(OC_4H_9)_4)$ was used as the alkyd precursor with high chemical activity. It is assumed that after the precursor material with a mass equivalent to a certain percentage of PP melt is uniformly mixed with PP melt, the precursor material tetrabutyl titanate will undergo hydrolysis and condensation reaction in the PP matrix. The principle is as follows [32]:

$$Ti(OR)_4 + H_2O \rightleftharpoons Ti(OR)_3OH + ROH \tag{9}$$

The condensation reactions lead to the formation of oxo bridges, Ti–O–Ti represented by the following reactions:

Alcoxolation or alcohol elimination:

$$Ti(OR)_3OH + Ti(OR)_4 \rightleftharpoons (OR)_3 - Ti - O - Ti - (OR)_3 + ROH \tag{10}$$

Oxalation or water elimination:

$$Ti(OR)_3OH + Ti(OR)_3OH \rightleftharpoons (OR)_3Ti - O - Ti - (OR)_3 + H_2O \tag{11}$$

The overall generalized reaction can be written as:

$$Ti(OR)_4 + 2H_2O \rightleftharpoons TiO_2 + 4ROH \tag{12}$$

When a quantity of inorganic precursor, Ti(OR)$_4$, is introduced into the molten PP matrix, the dilution effect on the PP matrix must be considered in the modeling. Therefore, based on a mixing law, the zero-shear viscosity η_0 of mixing fluid can be defined as:

$$\eta_0 = \eta_{0\_Ti(OR)_4}[Ti(OR)_4] + \eta_{0\_PP}(1 - [Ti(OR)_4])\tag{13}$$

where $[Ti(OR)_4]$ is the inorganic precursor concentration, $\eta_{0\_Ti(OR)_4}$ is the zero-shear viscosity of the inorganic precursor, η_{0_PP} is the zero-shear viscosity of PP melt.

According to the Arrhenius law, the kinetic constant of reaction can be written as:

$$k = k_0 \exp\left(-\frac{E_a}{RT}\right)\tag{14}$$

where k is the kinetic constant, k_0 is the pre-exponential factor, E_a is the activation energy for reaction, R is the gas constant, and T is the absolute temperature.

According to Equation (13), the calculated zero-shear viscosity of the mixing fluid with 10wt% precursors and other data are shown in Table 2.

Table 2. The material properties at different temperature.

$[Ti(OR)_4]$ wt%	$k(s^{-1})$ 453.15 K	$k(s^{-1})$ 493.15 K	$k(s^{-1})$ 523.15 K	Ea (KJ·mol^{-1})	η_0 (Pa·S)
10 wt%	0.0017	0.0102	0.0152	65	3500

The conversion rate can be expressed as:

$$K_a = \left(1 - [Ti(OR)_4]/[Ti(OR)_4]_{initial}\right) \times 100\%\tag{15}$$

where K_a is conversion rate, $[Ti(OR)_4]$ is the inorganic precursor concentration at some time, $[Ti(OR)_4]_{initial}$ is the initial concentration of the inorganic precursor.

2.4. Quantitative Mixing

The area stretch ζ is the ratio of the deformed surface da at time Δt to initial infinitesimal surface dA:

$$\varsigma = \frac{|da|}{|dA|}\tag{16}$$

Average mixing efficiency is often used to describe the stretching mixing efficiency during mixing. It is defined as:

$$\langle e_\varsigma \rangle = \frac{1}{t}\int_0^t \frac{\dot{\varsigma}/\varsigma}{(D:D)^{1/2}}dt\tag{17}$$

where $\langle e_\varsigma \rangle$ is the time-average mixing efficiency, and $\mathbf{D}$ is the rate of strain tensor.

2.5. Grid Independence Validation

The effects of cell numbers in the FE model on the numerical results are validated by the grid-independence test. Therefore, two finite element mesh models with different densities are established, as shown in Figure 3, where the model of Mesh A and Mesh B have 34,300 and 57,600 cells, respectively. Figure 4 indicates that the magnitudes of velocity in two kinds of FE models are almost same, which explains that the FE model with 34300 cells meets the requirement of solutions and is selected to study the mixing and reactive mechanisms in the corotating TSE.

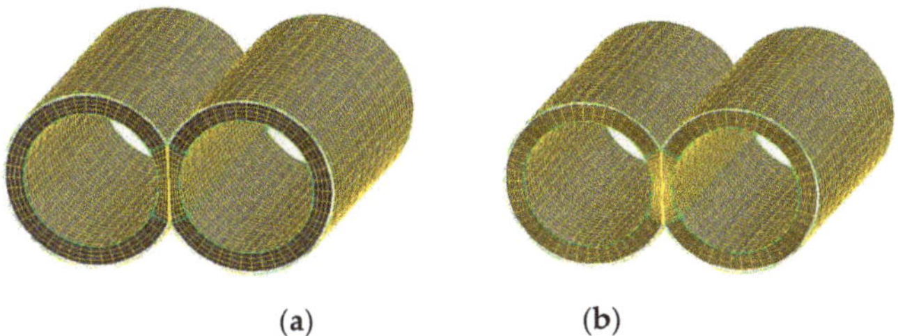

(a)   (b)

Figure 3. Finite element models with different mesh densities: (**a**) mesh A; (**b**) mesh B.

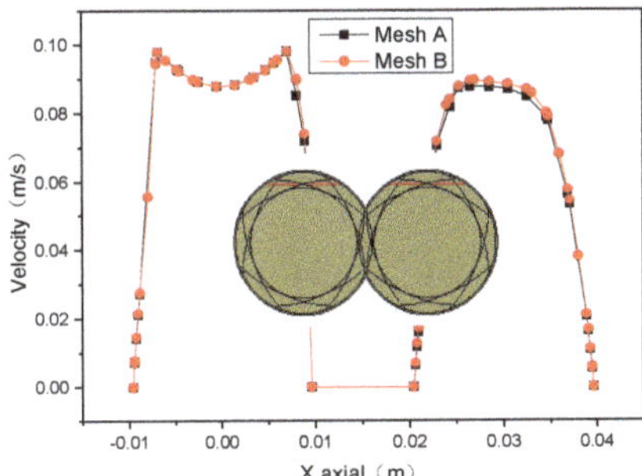

Figure 4. Velocity magnitude profiles and position of line in the TSE.

3. Results and Discussion

3.1. Concentration Patterns of Different Species

To learn about the details of mixing and reactive performances in the corotating TSE, some monitor planes and lines are firstly selected. Figure 5 illustrates a sketch of the sample planes and monitor lines of the TSE. The origin of coordinates is at the center of the left screw and three planes are chosen as sample planes, i.e., plane A (y = 13.6 mm), plane B (z = 28.5 mm) and plane C (x = 0 mm), where D = 34 mm is the inner diameter of barrel and L = 57 mm is the axial length of screw. Additionally, Line 1 (x = 16 mm, y = 0 mm), Line 2 (x = 15 mm, y = 0 mm) and Line 3 (x = 46 mm, y = 0 mm) are selected as monitor lines to investigate the local mixing and reactive properties along the axial direction.

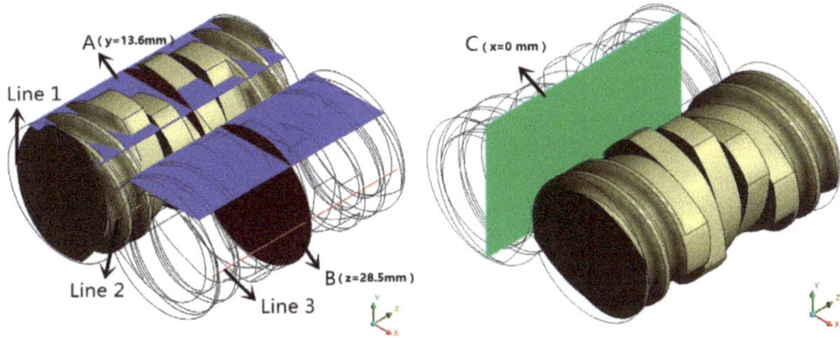

Figure 5. Sketch of sample planes and monitor lines in the TSE.

Figure 6 shows the concentration distribution of reactant and product at plane A in the corotating TSE, in which the rotational speed is 60 rpm, the stagger angle of the knead blocks is 45°, the initial temperature is 493.15 K and the inlet flow rate is 2×10^{-6} m^3/s. It can be seen from Figure 5, that the concentration of reactant at plane A gradually decreases along the combined screws from inlet to outlet, and the concentration of product tend to go in the opposite trend. In the thread sections, the degree of reaction in the root of thread groove is higher than that in other regions, because the mixing degree of materials in the root of thread groove is relatively high. In the section of kneading blocks, due to the great amplitude of shear rate, the materials blend well, and the distribution of the reaction becomes more uniform.

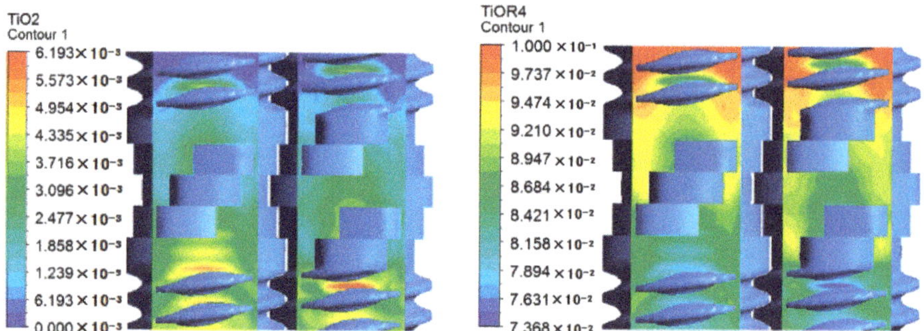

Figure 6. Concentrations of reactant and product at plane A.

Figure 7 shows the contours of axial velocity, reactant and product at cross plane C of the TSE, in which the left end is the inlet and the right end is the outlet. As can be seen from Figure 7a, the maximum axial velocity occurs at both ends of the kneading block section, resulting in great materials conveyance. The negative axial velocity in the middle area of the kneading block section indicates the presence of back flow, which can enhance the axial mixing capacity and promote the reaction. In Figure 7b,c, the concentration of $Ti(OR)_4$ decreased gradually from the inlet to outlet, and the concentration of TiO_2 has the opposite trend to $Ti(OR)_4$. In the kneading block section, due to the presence of back flow, the concentration distributions of reactant and product are more uniform than the thread sections.

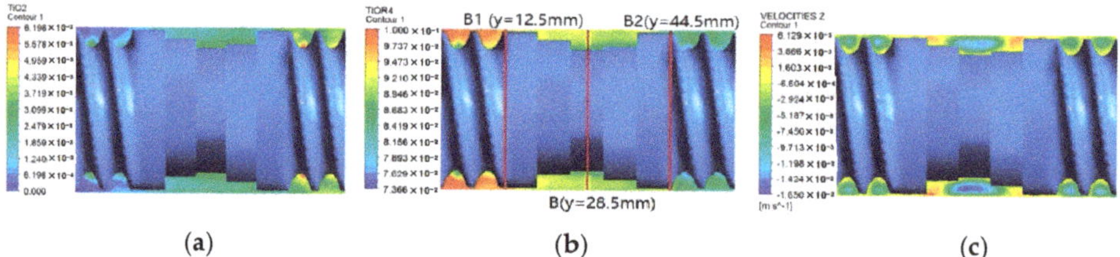

(a)   (b)   (c)

Figure 7. Contours of different fields at cross plane C: (**a**) product; (**b**) reactant; (**c**) axial velocity.

To accurately study the reaction degree of materials in the axial direction, three planes B1, B, B2 are selected, as shown in Figure 7b. As can be seen from Figure 8, from plane B1 to plane B2, the maximum concentration of reactants decreased from about 0.0976 to 0.0917, and the maximum concentration of TiO_2 increased from about 0.0021 to 0.0054. Since both planes B1 and B2 are located in the connection between kneading blocks and thread sections, the reaction distribution is roughly the same. Plane B is located in the middle of the kneading block, where the back-flow phenomenon is relatively prominent, so the concentration contours of reactant and product are more uniform than the other two sections.

Figure 9 shows the profiles of temperature and conversion at different monitor lines of the TSE. Figure 9a illustrates that the temperature increases gradually from the inlet to the outlet along the axial direction. Since Line 1 and Line 3 are at similar positions, the trends of temperature distributions are roughly the same. Line 2 is located at the intermeshing region of the TSE, where the exchange of material is more frequent and change of temperature is more gradual. It can be seen from Figure 9b that Line 1 and Line 3 have basically the same change trend because their positions are symmetrical in the TSE. Materials in the intermeshing region exchange frequently, leading to good mixing ability and higher temperature amplitude. As a result, the conversion rate at Line 2 is higher than that at Line 1 and Line 3.

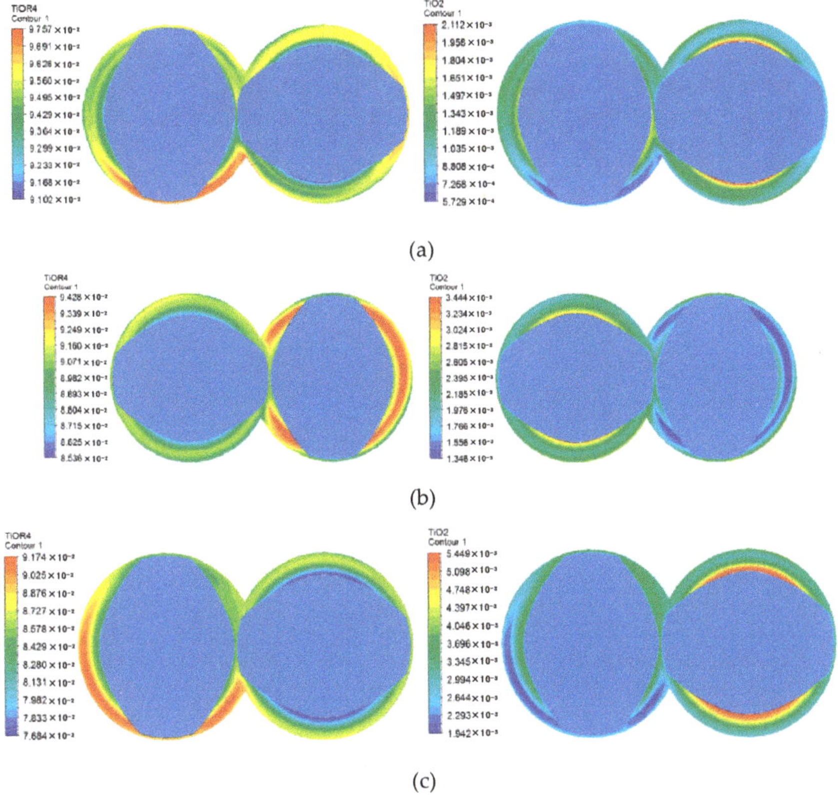

Figure 8. Concentration contours of reactant and product at different axial sections: (**a**) cross plane B1; (**b**) cross plane B; (**c**) cross plane B2.

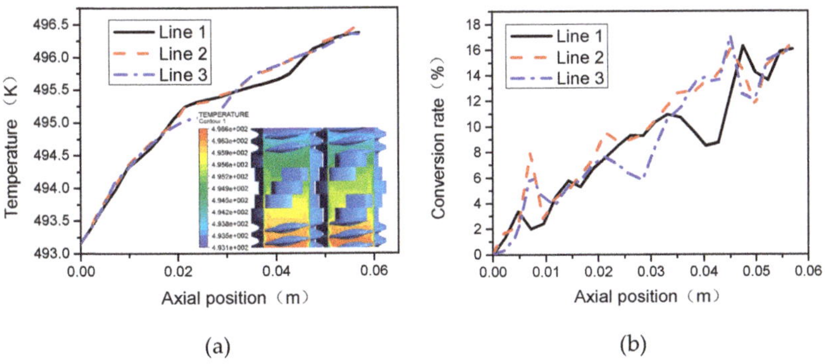

Figure 9. Profiles of temperature and conversion at different monitor lines: (**a**) temperature; (**b**) local conversion rate.

3.2. Influence of Rotating Speed

The rotational speed of the screw has a direct influence on the conveying ability, mixing effect and conversion rate of the materials in corotating TSEs. Figure 10 illustrates the local conversions at Line 1, Line 2 and Line 3 with different screw speeds, namely 30 rpm, 40 rpm and 60 rpm, in the corotating TSE, in which the stagger angle of the knead blocks is 45°, the initial temperature is 493.15 K and the inlet flow rate is 2×10^{-6} m^3/s. It can be seen that the conversion rates have roughly the same change trend, increasing along the axial distance. Interestingly, the increase of the rotational speed does not promote the reaction. At the same axial position, conversion rate decreases with the increase of speed, this is due to the fact that the increase of screw speed induces the material pass through the TES faster, not leaving enough time for mixing, at the same time reducing the probability of reaction.

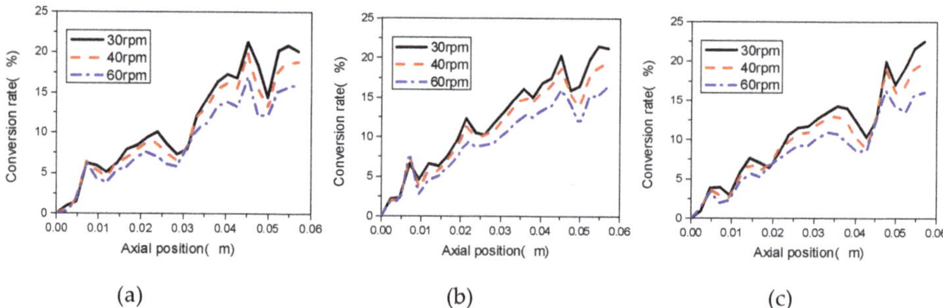

Figure 10. Effect of screw speeds on conversion rate at (**a**) Line 1; (**b**) Line 2; (**c**) Line 3 in the TSE.

To statistically analyze the movement of materials in the composite section of the TSE, the particle trace method (PTM) is adopted based on flow field calculation. 5000 tracer particles without volume, mass and influence on each other were released at the inlet of the corotating TSE and statistical processing is performed using POLYSTAT code embedded into POLYFLOW software.

Residence time distribution (RTD) is an important parameter for the reactive extrusion process, which directly depends on the rotational speed of screws. Figure 11a illustrates the cumulative RTD distribution of materials in the TSE with different rotational speeds. From Figure 11a, we can see that with the increase of rotational speed, more particles flow out in the same period of time. For example, with 20 s, the cumulative RTD is 0.55 at the rotational speed of 30 rpm, and 0.7, 0.87 at rotational speeds of 40 rpm and 60 rpm, respectively. Figure 11b shows the RTD distribution in the TSEs with different rotating screw speeds. As can be seen from Figure 11b, with the increase of rotational speed, the RTD curves shift from right to left and the transverse width between the inflection points of the RTD curves become narrower gradually, and the average residence time reduced to 21.6 s, 18.16 s and 13.4 s with the increase of rotational speed. This is due to the fact that the increase of screw rotational speed will cause a greater pressure difference between the inlet and outlet of the TSE, as shown in Figure 11c. As is well known, greater pressure difference will decrease the residence of materials, which is not good for the mixing and reaction.

Figure 11d shows the effect of screw speeds on the distributions of average shear rates along the extrusion direction. It can be seen from Figure 11d that the average shear rates at kneading block section is lower than that in the thread section. With the increase of rotating speed, the shear rate increases gradually at the same axial position, taking the axial position of 0.05 m as example, the average shear rate is about 62 s^{-1} at rotating speed of 60 rpm, 43 s^{-1} at rotating speed of 40 rpm, and 32 s^{-1} at rotating speed of 30 rpm, respectively. In Figure 11e, it can be found that increasing rotating speed improves the average temperature and the variance of average temperature increases dramatically along the axial direction. When the speed is 60 rpm, the maximum temperature magnitude reaches 496 K, about 1 K and 2 K higher than that at speeds of 40 rpm and 30 rpm, respectively. This is due to the fact that the viscous dissipation increases with the increase of rotational speed, which is caused by the increased magnitude of average shear rate, as shown in Figure 11d. The magnitude of average shear rate is important parameters to evaluate the dispersive mixing in TSE.

According to above analysis, the temperature amplitude of the nanocomposites in the TSE increases with the increase of rotational speed of screws, and reaction rate is positively affected by temperature, so higher rotational speed can promote the progress of the reaction process. However, the average conversion rates decrease with the increase of screw rotational speed, as shown in Figure 11f. This is due to the fact that the improvement of rotation speeds reduces the residence time of materials, which is not conducive to the average conversion rate.

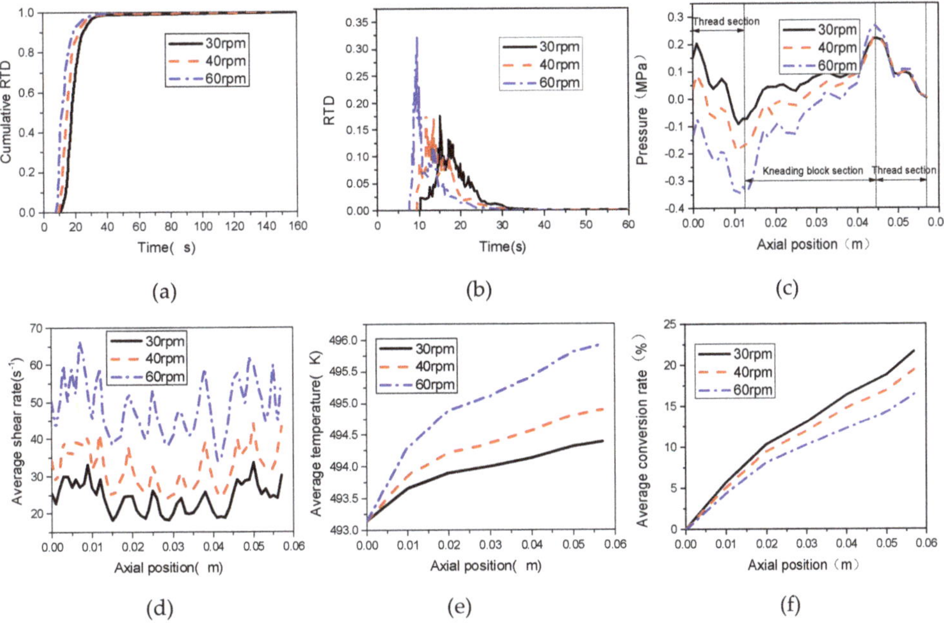

Figure 11. Effect of screw speeds on (**a**) cumulative RTD; (**b**) RTD; (**c**) pressure; (**d**) average shear rate; (**e**) average temperature; (**f**) conversion rate.

3.3. Effect of Stagger Angle

The kneading block sections of the combined screws are the dominant ones for mixing efficiency of the whole TSE, greatly influencing the reaction. The stagger angles of kneading blocks have a great impact on the mixing efficiency and conversion rate. In this section, the typical stagger angles of 45°, 60° and 90° of the kneading blocks, namely S45, S60 and S90, as shown in Figure 12, are selected to study the effect of stagger angle on the mixing and reactive characteristics in the TSE. In addition, the rotational speed is 60 rpm, the initial temperature is 493.15 K, and the inlet flow rate is 2×10^{-6} m^3/s.

Figure 13a,b illustrates that increasing the stagger angle of kneading blocks can decelerate the pass through of materials in the TSE. The mean residence times are 13.4 s, 18.97 s and 25.07 s with the increasing of stagger angle, which is conducive to the intensive mixing of the material and the promotion of reaction.

The mixing efficiency increases with the increase of stretching strength. It can be seen from Figure 13c that the magnitudes of mean logarithmic of area stretching in the three kinds of TESs with stagger angles of 45°, 60° and 90° gradually increase along the axial direction, indicating that they all have good mixing capacity. In the inlet section, the mean logarithmic of area stretching are almost same in three kinds of TSEs. However, in the kneading block section, the magnitude of logarithmic of area stretching in the TSE with stagger angle of 90° is superior to that in the other two TSEs with stagger angles of 45° and 60°, respectively, indicating that screws with a stagger angle of 90° have better tensile capacity and better mixing efficiency, which is advantageous to the reaction.

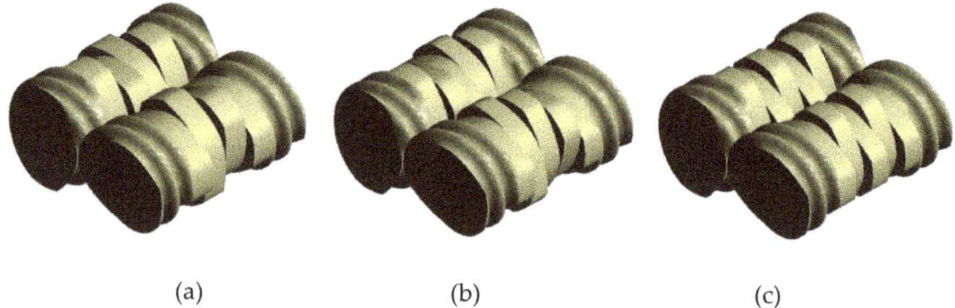

Figure 12. Geometries of screws with different stagger angles of kneading blocks: (**a**) S45; (**b**) S60; (**c**) S90.

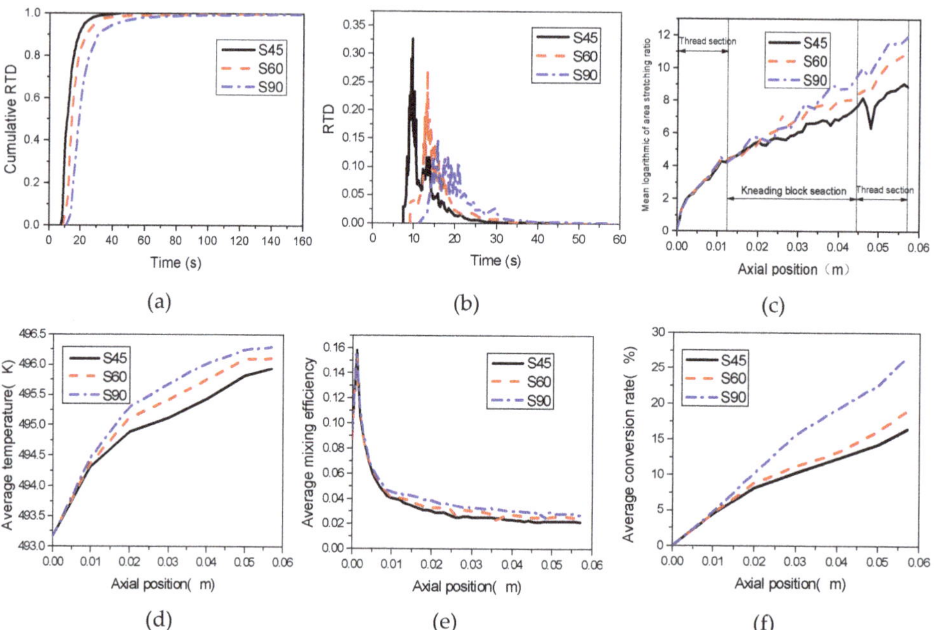

Figure 13. Effect of stagger angle on (**a**) cumulative RTD; (**b**) RTD; (**c**) logarithmic of area stretching; (**d**) mixing efficiency; (**e**) temperature; (**f**) conversion rate.

Under the action of shear rate, viscous dissipation of polymer melt will occur, which will gradually increase the material temperature. As can be seen from Figure 13d, at the same axial position, the temperature increases with the increase of stagger angle. At axial position of 0.04 m, the temperature is 496.1 K when the stagger angle is 90°; as for the kneading blocks with stagger of 60° and 45°, the corresponding temperatures are 495.4 K and 495.8 K, respectively. According to the reaction rate constant of Equation (14), the reaction rate increases with the increase of temperature in the TSE, which is good for the reaction. In addition, Figure 13f provides evidence for this view, at the axial position of 0.04 m, the average conversion rate is about 20% with the stagger angle of 90°, which is much higher than the average conversions of kneading blocks with the stagger angle of 60° and 45°. In addition, the average mixing efficiency increase with the increase of stagger angle, as shown in Figure 13e. Moreover, the variance of the conversion rate is not consistent with the variance of temperature, suggesting that temperature can affect the reaction process, but it is not the only factor which determines the reaction process in the TSE.

3.4. Effect of Inlet Flow Rate

The inlet flow rate has an impact on the flow and mixing performances of materials in TSE, and then affects the reaction processes. In this section, we select the TSE with stagger angle of 45°, assuming the rotational speed of screw is 60 rpm, to analyze the effect of inlet flow rate on the preparation process of PP/TiO$_2$ nanocomposite.

Figure 14a,b illustrates the effect of inlet flow rate on the cumulative RTD and RTD, respectively. We can see that increasing the inlet flow rate will decrease the residence time of materials, which is not good for mixing and reaction. The mean residence time is 17.91 s, 13.4 s and 11.55 s with increasing flow rate, as shown in Figure 14b. Figure 14c illustrates the effect of inlet flow rate on the mean logarithmic of area stretching. It indicates that the mean logarithmic of area stretching reduces with the increase of inlet flow rate, which is averse to mixing of materials. Figure 14d depicts that the average temperature increases along the axial direction, and increasing the inlet flow rate will reduce the average temperature. According to Equation (14), the reaction rate constant increases with the increase of temperature magnitude, indicating that higher inlet flow rate is not conducive to the proceeding of reaction. As can be seen from Figure 14e, the average conversion rate decreases with the increase of inlet flow rate, which further prove the conclusion above. Figure 14f shows the effect of inlet flow rate on local conversion at the monitor Line 2 of the TSE.

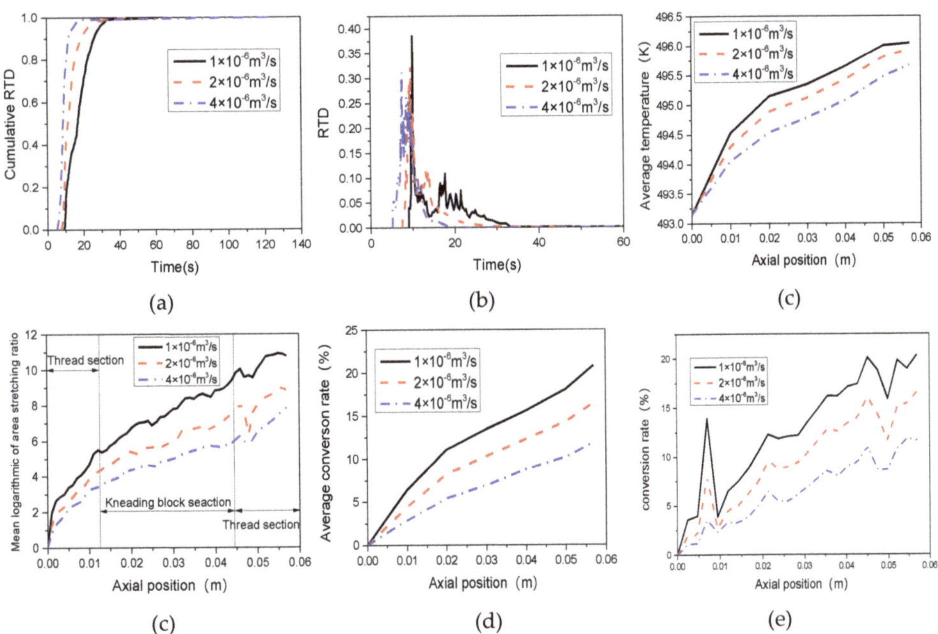

Figure 14. Effect of inlet flow rate on (**a**) cumulative RTD; (**b**) RTD; (**c**) logarithmic of area stretching; (**d**) temperature; (**e**) conversion rate; (**f**) local conversion at Line 2.

3.5. Effect of Initial Temperature

Temperature has a significant influence on material diffusion, convection and reaction rate. We select the TSE with a stagger angle of 45°, and a speed of screw of 60 rpm, to analyze the effect of initial temperature of barrel on the preparation process of PP/TiO$_2$ nanocomposite. Line 2 is selected as the local analysis object, and the initial temperature of barrel in the corotating TSE is set as 453.15 K, 493.15 K and 523.15 K, respectively.

Figure 15a shows the change curve of local conversion rate at Line 2 of the TSE at different initial temperatures of barrel. As can be seen from the figure, the change trends of the conversion rate at the monitoring Line 2 are roughly same. Along the axial direction of the screws, the local conversion rate of Line 2 gradually increases from the inlet to outlet of the TES, which fluctuates at both ends of the combined screws, and steadily rises in the middle meshing block area. At the same axial position, the higher the initial temperature is, the higher the conversion rate obtains. When the initial temperature in the TSE is 453.15 K, the reaction is relatively slow, and the highest conversion rate is only about 3%. However, under the other two initial temperatures, the conversion rate at Line 2 reaches 17% and 23% respectively, with a large difference.

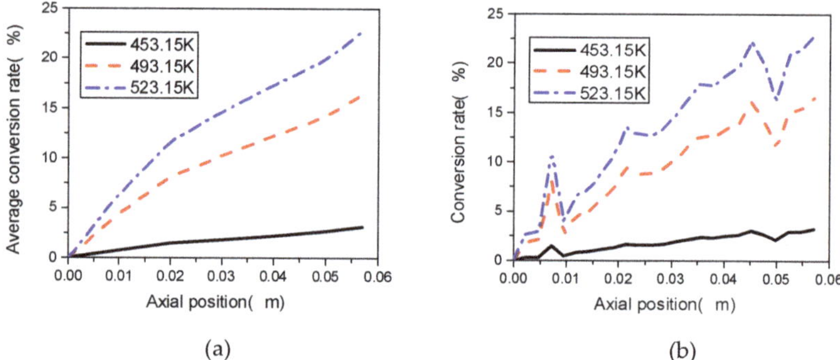

Figure 15. Effect of initial temperature of barrel on conversion rate at (**a**) Line 2; (**b**) the whole TSE.

Figure 15b shows the distribution of average conversion rate change in the TSE at different initial temperatures of barrel. It can be seen from this figure that the reaction is slow at low initial temperature magnitude. For example, when the initial temperature magnitude is 523.15 K, the average conversion rate reaches 23%, which is about 20% higher than that when the initial temperature is 453.15 K. This phenomenon is mainly due to the fact that the reaction rate constant is greatly affected by temperature distributions in the TSE.

4. Conclusions

In this work, the sol-gel reaction process of preparing PP/TiO$_2$ nanocomposites in a corotating TSE with two combined screws of conveying and mixing screw elements is first investigated via 3D numerical modeling based on FEM. The reactive patterns of nanocomposites in the corotating TSE are firstly investigated. Moreover, to control reactions in extrusion and optimize the geometric parameters of TSE, the effects of screw rotational speed, stagger angle, inlet flow rate and initial barrel temperature on the mixing and reaction process are investigated, and the relationship between the mixing characteristics and the reactive process in a corotating TSE is discussed. The following conclusions are drawn.

In the reaction of PP/TiO$_2$ nanocomposites prepared by a corotating TSE, the conversion increases gradually from the inlet to the outlet along the axial position. Although increasing the rotating speed of the screw can increase the average temperature of materials, as the axial velocity of materials increases, the residence time of the materials in the TSE gradually becomes shorter, which is not conducive to the full mixing and reaction process of materials. Increasing the stagger angle of the kneading blocks gradually increase the temperature of the material which is good to improve the conversion rate.

At the same time, increasing stagger angle will improve the stretching efficiency of the material in TES, and prolong the residence time of the material in the extruder, which is conducive to the intensive mixing and reaction. Increasing the flow rate at the entrance of the screw extruder in the combination sections will shorten the residence time of the material in the extruder, which is not conducive to mixing.

Moreover, the temperature change is inversely proportional to the inlet flow rate, and the conversion rate decreases with the increase of the inlet flow rate. Increasing the initial temperature in the TSE will reduce the viscosity of the material while increasing the reaction rate constant, speed up the reaction process, and gradually increase the conversion of PP/TiO_2 nanocomposites. Generally, besides the parameters mentioned above in this paper, concentration of the precursor and the length of the screw will greatly influence the mixing and reaction processes in the TSEs. Consequently, study of these two parameters will provide reference for the control and optimization of reactive extrusion processes in TSEs.

Author Contributions: Conceptualization and methodology, X.Z.Z.; validation, M.G.G.; formal analysis, M.G.G. and S.D.P.; investigation, X.Z.Z., S.D.P. and M.G.G.; resources and data curation, M.G.G.; writing—original draft preparation, S.D.P.; writing—review and editing, X.Z.Z.; visualization and supervision, X.Z.Z; project administration and funding acquisition, X.Z.Z.

Funding: This work was supported by National Natural Science Foundation of China (grant No. 51473073, 50903042 and 51303075); Program for Liaoning Excellent Talents in University (grant No. LR2016022) and Liaoning Province Natural Science Foundation (grant No. 2015020142).

Conflicts of Interest: The authors declare no conflict of interest.

References

1. Maddah, H.A. Polypropylene as a promising plastic: A review. *Am. J. Polym. Sci.* **2016**, *6*, 1–11.
2. El-Dessouky, H.M.; Lawrence, C.A. Nanoparticles dispersion in processing functionalised PP/TiO_2 nanocomposites: Distribution and properties. *J. Nanopart. Res.* **2010**, *13*, 1115–1124. [CrossRef]
3. Bahloul, W.; Mélis, F.; Bounor-Legaré, V.; Cassagnau, P. Structural characterisation and antibacterial activity of PP/TiO_2 nanocomposites prepared by an in situ sol–gel method. *Mater. Chem. Coll. Phys.* **2012**, *134*, 399–406. [CrossRef]
4. Dou, Q.; Zhu, X.; Peter, K.; Demco, D.E.; Möller, M.; Melian, C. Preparation of polypropylene/silica composites by in-situ sol–gel processing using hyperbranched polyethoxysiloxane. *J. Sol-Gel Sci. Technol.* **2008**, *48*, 51–60. [CrossRef]
5. Soares, I.L.; Chimanowsky, J.P.; Luetkmeyer, L.; Silva, E.O.d.; Souza, D.d.H.S.; Tavares, M.I.B. Evaluation of the influence of modified tio2 particles on polypropylene composites. *J. Nanosci. Nanotechnol.* **2015**, *15*, 5723–5732. [CrossRef] [PubMed]
6. Sakai, T. Screw extrusion technology - past, present and future. *Polimery* **2013**, *58*, 847–857. [CrossRef]
7. Sardo, L.; Vergnes, B.; Valette, R. Numerical modelling of the non-isothermal flow of a non-newtonian polymer in a co-kneader. *Int. Polym. Proc.* **2017**, *32*, 425–433. [CrossRef]
8. Malik, M.; Kalyon, D.M.; Golba, J.C. Simulation of co-rotating twin screw extrusion process subject to pressure-dependent wall slip at barrel and screw surfaces at barrel and screw surfaces 3d fem analysis. *Int. Polym. Proc.* **2014**, *29*, 52–63. [CrossRef]
9. Salahudeen, S.A.; AlOthman, O.; Elleithy, R.H.; Al-Zahrani, S.M.; Rahmat, A.R.B. Optimization of rotor speed based on stretching, efficiency, and viscous heating in nonintermeshing internal batch mixer: Simulation and experimental verification. *J. Appl. Polym. Sci.* **2013**, *127*, 2739–2748. [CrossRef]
10. Salahudeena, S.A.; Rabeh, H.E.; AlOthmanbc, O.; Alzahrani, S.M. Comparative study of internal batch mixer such as cam, banbury and roller: Numerical simulation and experimental verification. *Chem. Eng. Sci.* **2011**, *66*, 2502–2511. [CrossRef]
11. Zhang, X.M.; Feng, L.F.; Chen, W.X.; Hu, G.H. Numerical simulation and experimental validation of mixing performance of kneading discs in a twin screw extruder. *Polym. Eng. Sci.* **2009**, *49*, 1772–1783. [CrossRef]
12. Robinson, M.; Cleary, P.W. Effect of geometry and fill level on the transport and mixing behaviour of a co-rotating twin screw extruder. *Comput. Part. Mech.* **2018**. [CrossRef]

13. Reitz, E.; Podhaisky, H.; Ely, D.; Thommes, M. Residence time modeling of hot melt extrusion processes. *Eur. J. Pharm. Biopharm.* **2013**, *85*, 1200–1205. [CrossRef] [PubMed]

14. Galaktionov, O.S.; Anderson, P.D.; Kruijt, P.G.M.; Peters, G.W.M.; Meijer, H.E.H. A mapping approach for three-dimentional distributive mixing analysis. *Comput. Fluids* **2001**, *30*, 271–289. [CrossRef]

15. Galaktionov, O.S.; Anderson, P.D.; Peters, G.W.M.; Meijer, H.E.H. Mapping approach for 3d laminar mixing simulations application to industrial flows. *Int. J. Numer. Meth. Fluids* **2002**, *40*, 345–351. [CrossRef]

16. Kruijt, P.G.M.; Galaktionov, O.S.; Peters, G.W.M.; Meijer, H.E.H. The mapping method for mixing optimization. *Int. Polym. Proc.* **2001**, *16*, 161–171. [CrossRef]

17. Berzin, F.; Vergnes, B. Modeling of twin screw reactive extrusion. **2017**, 37–70. [CrossRef]

18. Gug, J.I.; Tan, B.; Soule, J.; Downie, M.; Barrington, J.; Sobkowicz, M.J. Analysis of Models Predicting Morphology Transitions in Reactive Twin-Screw Extrusion of Bio-Based Polyester/Polyamide Blends. *Int. Polym. Proc.* **2017**, *32*, 363–377. [CrossRef]

19. Cailloux, J.; Hakim, R.N.; Santana, O.O. Reactive extrusion: A useful process to manufacture structurally modified PLA/o-MMT composites. *Composites* **2016**, *88*, 106–115. [CrossRef]

20. Fel, E.; Massardier, V.; Mélis, F.; Vergnes, B.; Cassagnau, P. Residence time distribution in a high shear twin screw extruder. *Int. Polym. Proc.* **2014**, *29*, 71–80. [CrossRef]

21. Connelly, R.K.; Kokini, J.L. The effect of shear thinning and differential viscoelasticity on mixing in a model 2D mixer as determined using fem with particle tracking. *J. Non-Newtonian. Fluid Mech.* **2004**, *123*, 1–17. [CrossRef]

22. Fard, A.S.; Hulsen, M.A.; Meijer, H.E.H.; Famili, N.M.H.; Anderson, P.D. Tools to simulate distributive mixing in twin-screw extruders. *Macromol. Theory Simul.* **2012**, *21*, 217–240. [CrossRef]

23. Rożeń, A.; Bakker, R.A.; Bałdyga, J. Effect of operating parameters and screw geometry on micromixing in a co-rotating twin-screw extruder. *Chem. Eng. Res. Des.* **2001**, *79*, 938–942. [CrossRef]

24. Zhu, L.J.; Narh, K.A.; Hyun, K.S. Investigation of mixing mechanisms and energy balance in reactive extrusion using three-dimensional numerical simulation method. *Int. J. Heat Mass Transf.* **2005**, *48*, 3411–3422. [CrossRef]

25. Zhang, Z.X.; Gao, C.; Xin, Z.X.; Kim, J.K. Effects of extruder parameters and silica on physico-mechanical and foaming properties of pp/wood-fiber composites. *Composites* **2012**, *43*, 2047–2057. [CrossRef]

26. Kim, I.; White, J.L. Reactive copolymerization of various monomers based on lactams and lactones in a twin-screw extruder. *J. Appl. Polym. Sci.* **2005**, *96*, 1875–1887. [CrossRef]

27. Rigoussen, A.; Verge, P.; Raquez, J.M. In-depth investigation on the effect and role of cardanol in the compatibilization of PLA/ABS immiscible blends by reactive extrusion. *Eur. Polym. J.* **2017**, *93*, 272–283. [CrossRef]

28. Berzin, F.; Vergnes, B.; Dufossé, P.; Delamare, L. Modeling of peroxide initiated controlled degradation of polypropylene in a twin screw extruder. *Polym. Eng. Sci.* **2000**, *40*, 344–356. [CrossRef]

29. Berzin, F.; Vergnes, B.; Canevarolo, S.V.; Machado, A.V.; Covas, J.A. Evolution of the peroxide-induced degradation of polypropylene along a twin-screw extruder: Experimental data and theoretical predictions. *J. Appl. Polym. Sci.* **2006**, *99*, 2082–2090. [CrossRef]

30. Silva, R.P.; Oliveira, R.V.B. Non-isothermal degradation kinetics and morphology of pp/tio2 nanocomposites using titanium n-butoxide precursor. *Int. J. Plast. Technol.* **2016**, *20*, 364–377. [CrossRef]

31. Esthappan, S.K.; Kuttappan, S.K.; Joseph, R. Thermal and mechanical properties of polypropylene/titanium dioxide nanocomposite fibers. *Mater. Des.* **2012**, *37*, 537–542. [CrossRef]

32. Bahloul, W.; Oddes, O.; Bounor-Legaré, V.; Mélis, F.; Cassagnau, P.; Vergnes, B. Reactive extrusion processing of polypropylene/tio$_2$ nanocomposites by in situ synthesis of the nanofillers: Experiments and modeling. *AIChE J.* **2011**, *57*, 2174–2184. [CrossRef]

33. Zhu, X.Z.; Sun, D.P. Numerical study of reaction process for preparing PP/TiO$_2$ nanocomposites in an internal batch mixer. *J. Reinf. Plast. Comp.* **2014**, *33*, 977–990. [CrossRef]

34. Estanislao, O.R. Numerical simulations of reactive extrusion in twin screw extruders. Ph.D. Thesis, University of Waterloo, Waterloo, ON, Canada, January 2009.

Article

Hierarchical Structure of iPP During Injection Molding Process with Fast Mold Temperature Evolution

Vito Speranza, Sara Liparoti *, Roberto Pantani and Giuseppe Titomanlio

Department of Industrial Engineering, University of Salerno–via Giovanni Paolo II, 132, 84084 Fisciano (SA), Italy; vsperanza@unisa.it (V.S.); rpantani@unisa.it (R.P.); gtitomanlio@unisa.it (G.T.)
* Correspondence: sliparoti@unisa.it; Tel.: +39-089-964-007

Received: 3 January 2019; Accepted: 25 January 2019; Published: 30 January 2019

Abstract: Mold surface temperature strongly influences the molecular orientation and morphology developed in injection molded samples. In this work, an isotactic polypropylene was injected into a rectangular mold, in which the cavity surface temperature was properly modulated during the process by an electrical heating device. The induced thermo-mechanical histories strongly influenced the morphology developed in the injection molded parts. Polarized optical microscope and atomic force microscope were adopted for morphological investigations. The combination of flow field and cooling rate experienced by the polymer determined the hierarchical structure. Under strong flow fields and high temperatures, a tightly packed structure, called shish-kebab, aligned along the flow direction, was observed. Under weak flow fields, the formation of β-phase, as cylindrites form, was observed. The formation of each morphological structure was analyzed and discussed on the bases of the flow and temperature fields, experienced by the polymer during each stage of the injection molding process.

Keywords: morphology; injection molding; cylindrites; mold temperature

---

1. Introduction

Injection molding is a widespread process for the mass transformation of polymeric materials into final usable objects. During such a process, the polymer is melted and injected into a cavity with the shape of the desired objects, where additional material is fed into the cavity to compensate for the shrinkage due to the solidification. After that, cooling up to the extraction temperature takes place [1]. During the injection molding stages, the polymer chains are subjected to intensive shear and elongational flow fields that, on their turn, influence the solidification process and the morphology developed both along the flow and transverse directions [2]. The shear experienced by the polymer is not homogeneous along the cavity thickness; in particular, it is known that the shear is larger at the cavity wall than in the inner parts of the cavity. The shear and the molecular stretch undergone by the polymer chains determine, together with the cooling rate, the morphology developed in a certain area of the molded objects. Under high shear stresses, anisotropic shish-kebab morphology is observed, whereas in quiescent conditions, isotropic spherulitic morphology is found [3–5]. Different morphological structures give rise to a distribution of mechanical properties. Thus, understanding the mechanism that regulates the formation of any morphology with the aim of modulating mechanical properties of the polymeric objects is of great interest [6–10].

Fujiyama et al [11] proposed a model for the formation of the shish-kebab morphology. When a flow field is applied, coiled polymer chains undergo an extension, attaining a high degree of alignment that induces the crystallization into shish. The chains that do not reach a high degree of alignment, the kebabs, crystallize epitaxially on the shish, giving rise to chain–folded structures.

Pogodina et al. observed that structures aligned with the flow, the shish, appear in isotactic polypropylene (iPP) already at 148 °C, applying a step shear of 10 s^{-1} [12]. Both the molecular orientation in the melt, and the crystallization degree in the solid increase with shear rate and shear duration. Short duration high shear rate is more effective in orienting molecules than long duration small shear rate.

Jeu et al suggested that under flow the formation of shish precursors occurs, namely a mesophase, promoting the nucleation process, which induces the successive formation of shish [13].

All these works suggested that when the polymer chains undergo a huge flow field, they crystallize in the form of shish; namely, structures aligned along the flow direction, at temperatures significantly higher than those observed in quiescent conditions. Even if many works were carried out to analyze the mechanism that induces the formation of different morphological structures [4,14,15], the conditions that allow their formation during industrial polymer processes are still under debate [16].

In this paper, a deep characterization of morphology has been carried out on injection molded samples. The samples have been obtained coupling the injection molding with a system that allows modulating the cavity surface temperature during the process. Such a system causes, due to joule effect, the increase of cavity surface temperature up to the selected values within a few seconds. Different cavity surface temperatures and heating times have been adopted during the production of the molded samples, with the aim of analyzing the effects of the temperature, and the flow fields, on the morphology developed along the sample thickness. Also, mechanisms that allow the formation of shish-kebab morphology have been proposed, and related with the flow and temperature fields undergone by polymer chains during the process.

The formation of different morphological structures is also related to the formation of different crystalline phases when the polymer chains experience certain temperature and flow fields. There are three types of crystal phases found in iPP; namely α, β, and γ-phase [17]. γ-phase can be found only when pressures higher than 2000 bar are achieved [18]. α-phase is characteristic of most of the parts of the molded samples, since it is the most stable one, whereas β-phase is thermodynamically metastable, and more difficult to be found [19]. The high shear stresses experienced by the polymer chains during the injection molding process mainly induces the transformation of β-phase into α-phase [20]. However, the β-phase formation is highly desired, since it enhances the toughness of the molded samples. Generally, this increase of toughness has been achieved by adding β-nucleating agents during the process [21].

In this paper, the possibility to drive the formation of different crystalline phases, with higher attention to the formation of β-phase, has been discussed on the bases of the adopted operating conditions.

2. Materials and Methods

The T30G polypropylene (iPP) commercial grade (Basell, Ferrara, Italy) was adopted for the injection molding tests. This material was characterized concerning rheology, and crystallization in previous works [22–25].

Figure 1 schematizes the polymer flow path during the process. The path includes the sprue, the runner, the gate and the cavity. The first 70 mm of the cavity, downstream the gate, undergo temperature cycles due to the presence of a heating device located just below the cavity surface. The heating device, characterized elsewhere [26,27], is mainly composed of a conductive layer (80 μm thickness) that induces the increase of temperature up to values (T_{cs}) intermediate between the mold temperature and the injection temperature; the electrically conductive layer is sandwiched by two insulating layers (140 μm thickness on the mold side, and 20 μm thickness on the cavity side). A steel layer (100 μm thickness) protects the heating device from the incoming melt. The cavity dimensions are 110 mm length, 12.7 mm width, and 1.50 mm thickness, as shown in Figure 1.

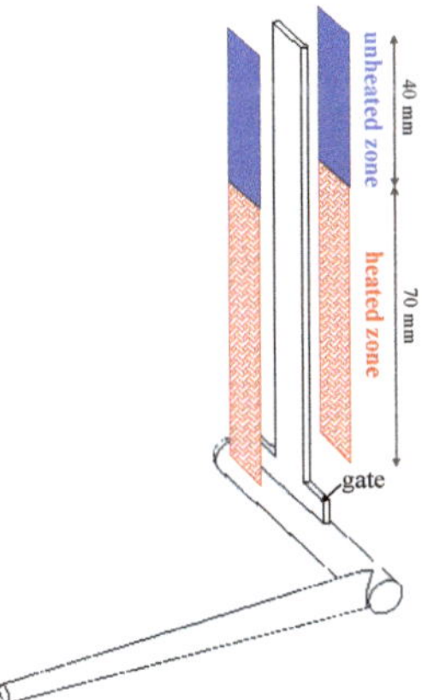

Figure 1. Sketch of the cavity adopted for the injection molding tests, with the heating device.

The injection molding tests were performed with 220 °C melt injection temperature, 2.9 cm^3 s^{-1} average volumetric flow rate (the cavity filling time was about 0.7 s), and 25 °C mold temperature. 720 bars were adopted during the holding stage. Table 1 summarizes all the operating conditions. The tests named Passive were performed without activating the heater, the Steel tests were performed replacing the heating device with a steel layer of the same thickness. In Table 1, the name of each test is composed by the temperature T_{cs} in °C measured on the cavity surface after 6 s heating time (T_{cs} = 80 °C and 150 °C for the tests considered in this work), and the heating time t_h in seconds. Examples of the temperature evolutions acquired during tests with 4 W/cm^2 are shown in Figure S1 of the Supplementary Materials.

Table 1. Operating conditions of the injection molding tests. (P = electrical power density; T_{cs} = temperature measured on the cavity surface 6 s after the melt entrance into the cavity; the activation time, t_a, and the heating time, t_h, represent the time during which the heating device was kept active before the first contact of the melt with the cavity surface, and the time during which the heating device was active after the contact of melt with the cavity surface, respectively).

Test Name	P (W/cm^2)	T_{cs} (°C)	t_h (s)	t_a (s)
Steel	0	25	0	0
Passive	0	25	0	0
150-07	9.5	150	0.7	2
150-6	9.5	150	6	2
80-07	4	80	0.7	2
80-1	4	80	1.3	2
80-6	4	80	6	2

The activation time, t_a, and the heating time, t_h, represent the time during which the heating device was kept active before the first contact of the melt with the cavity surface, and the time during which the heating device was active after the contact of melt with the cavity surface, respectively.

The comparison reported in this work concerning the morphology developed along the sample thickness refers to the position at 15 mm downstream the cavity entrance.

The samples were cut along the flow direction, parallel to the flow-thickness plane, and chemically etched, following a procedure reported elsewhere [28]. Atomic Force Microscope (AFM) investigations were conducted in air and at room temperature with a Bruker Dimension instrument (Multimode Dimension V, Veeco, Santa Barbara, CA, USA) coupled with a Nanoscope V controller operating in tapping mode. Commercial probe tips with nominal spring constants of 42 N m^{-1}, resonance frequencies of 300 kHz, and tip radius of 7 nm were used.

Sample slices were also analyzed by optical microscopy in polarized light by an Olympus BX51 microscope (OM, Olympus Italia S.R.L., Segrate, Italy) with crossed polarizer-analyzer. Optical micrographs were taken with sample slices oriented at 45° with respect to the analyzer.

The simulation software for the injection molding process was developed at University of Salerno (Fisciano, SA, Italy) [24], adapted to evaluate the flow rate during the injection molding process.

3. Results

Figure 2 shows the morphology developed during the injection molding process, with the heating device adapted as an insulating layer, namely for the Passive test. The polarized optical micrograph (OM) and the AFM topographic maps for each position along the sample thickness are shown in the same figure. For each AFM map, the distance from the sample surface is reported. The OM displays a distinct skin-core morphology, characteristic of the injection molding process. The AFM maps allow obtaining more detailed information about the structure developed along the sample thickness. Within a small distance from the sample surface, 0.03 mm, a poorly organized structure was detected. From 0.03 mm to 0.24 mm from the sample surface, AFM shows a highly-oriented layer composed of structures aligned along the flow direction, having an average thickness in the range 100–300 nm. The layer that appears brown in the OM, between 0.40 mm and 0.47 mm, is composed of thick (more than 300 nm thickness) aligned structures characterized by significant lateral growth. The core is composed of randomly dispersed spherulites, as clearly visible in the AFM map at 0.67 mm, with a mean diameter of 20 ± 5 μm.

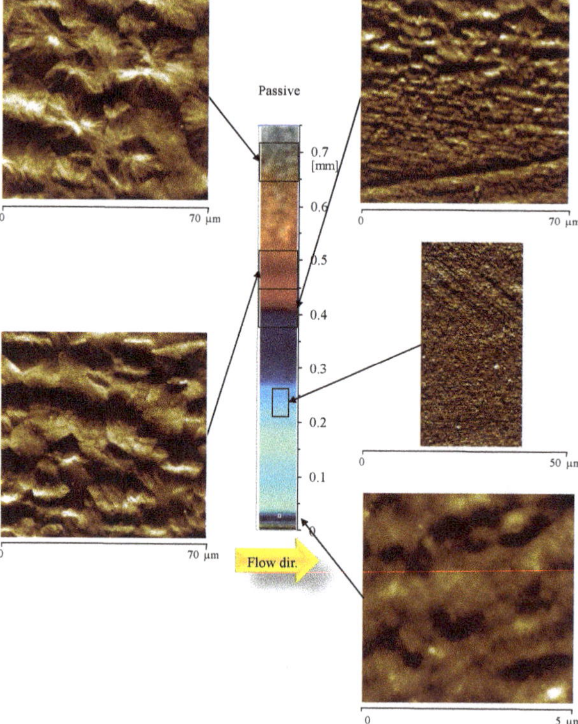

Figure 2. Optical micrograph and AFM height maps, at different distances from the sample surface, of the sample Passive, at 15 mm downstream the gate.

Figure 3 shows a magnification of the oriented zone at 0.24 mm and at 0.40 mm from the sample surface.

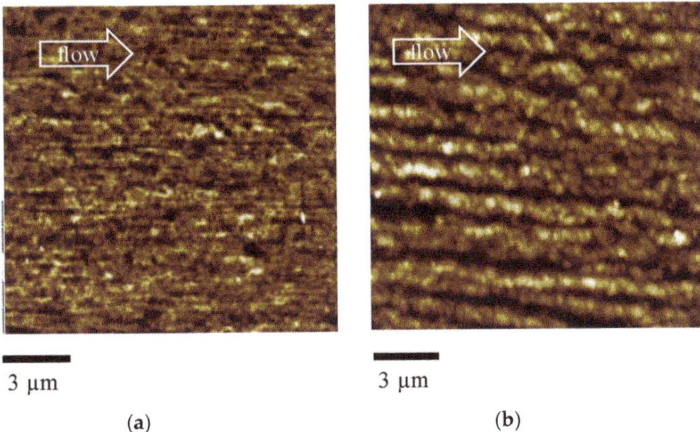

Figure 3. Enlargement of the AFM height map at (**a**) 0.24 mm and (**b**) 0.4 mm distance from the sample surface for the Passive sample.

Figure 3a mainly shows tightly packed structures aligned along the flow direction. Figure 3b also shows thick structures aligned along the flow direction, with average thickness ranging between 300 and 1000 nm. Comparison between the two figures clearly confirms that the thickness of the structures aligned along the flow direction increases with the distance from the sample surface.

Figure 4 shows the OM and the AFM height maps of the sample 80-07, obtained adopting 80 °C as cavity surface temperature during the cavity filling (0.7 s).

The OM displays, again, a skin-core morphology; however, the width of colored bands is different with respect to the Passive sample, shown in Figure 2. This is mainly due to the different distribution of the orientation levels. Adopting high temperatures on the cavity surface during the filling, structures oriented along the flow direction were detected already at the sample surface. The AFM map at 0.25 mm shows that thin (smaller than 300 nm thickness) structures aligned along the flow direction can be observed up to 0.30 mm from the sample surface. Again, the thickness of these structures increases as the distance from the sample surface increases. Comparison of OM and AFM images shows that the structures aligned along the flow direction are also detectable in the layers where OM seems to show spherulitic structures. The AFM map at 0.50 mm from the sample surface shows thick (up to 6 µm thickness) structures aligned along the flow direction, characterized by significant lateral growth in the transverse direction (orthogonal to the flow front). The aligned structures disappear at 0.57 mm from the surface, where only randomly distributed spherulites can be observed. A magnification of the oriented layers at 0.25 mm, and at 0.50 mm is shown in Figure 5a,b, respectively.

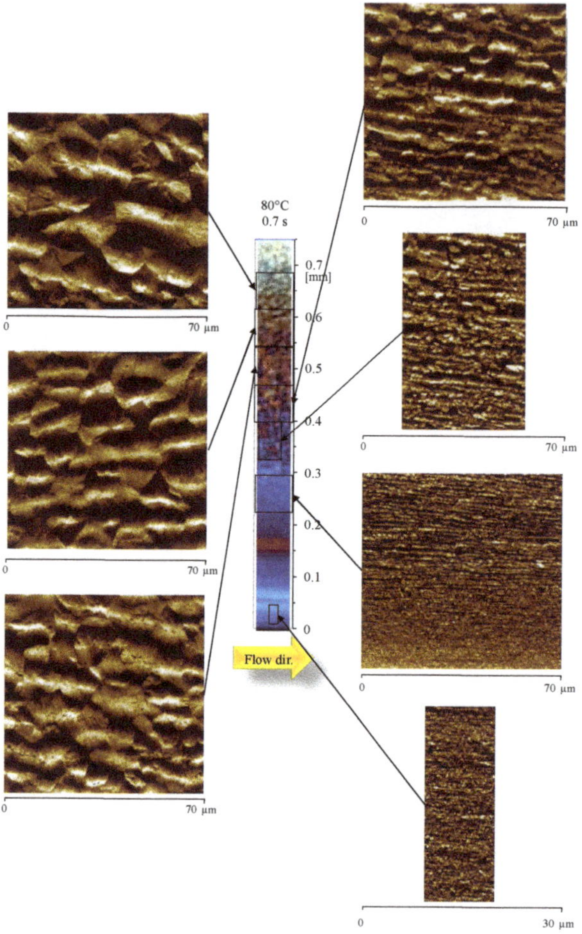

Figure 4. Optical micrograph and AFM height maps, at different distances from the sample surface, of the sample 80-07, at 15 mm downstream the gate.

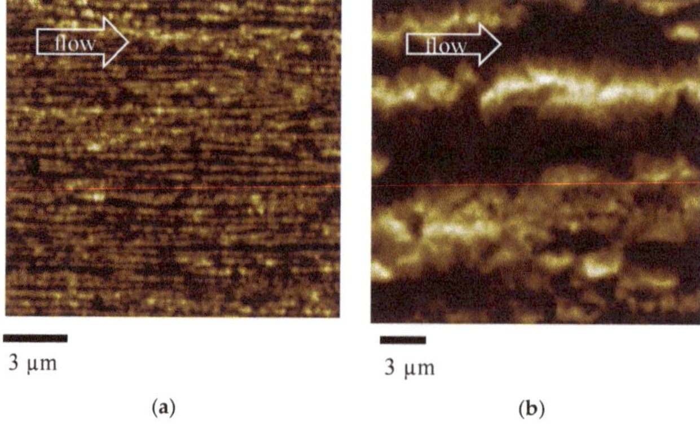

Figure 5. Magnification of the AFM height map at (**a**) 0.25 mm and (**b**) 0.50 distance from the sample surface for 80-07 sample.

As already observed for the Passive case, the larger the distance from the sample surface, the larger the lateral growth (namely, the thickness of the aligned structures increases).

Figure 6 shows the morphology developed with higher cavity surface temperatures, during the filling (Test 150-07).

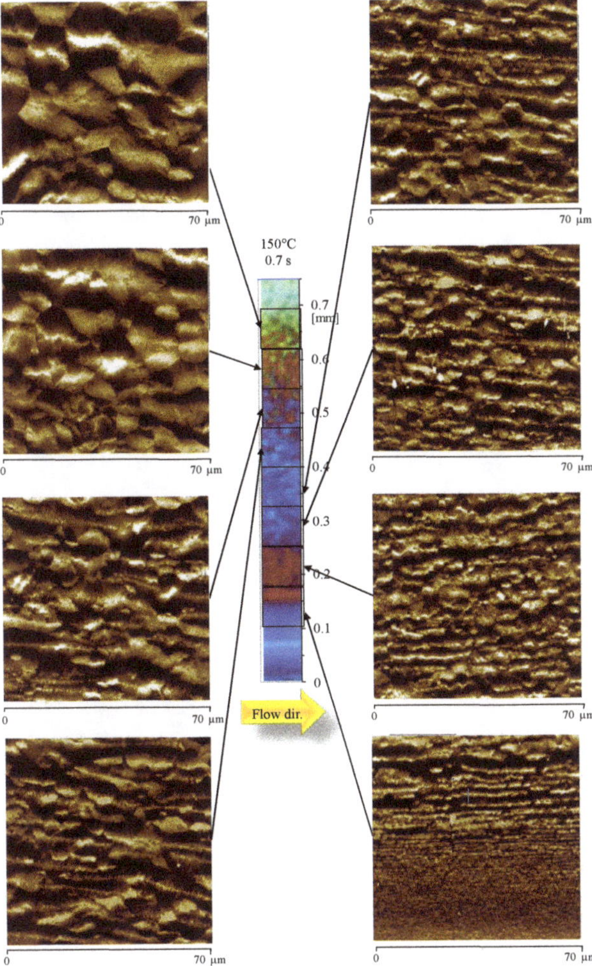

Figure 6. Optical micrograph and AFM height maps, at different distances from the sample surface, of the sample 150-07, at 15 mm downstream the gate.

The distribution of colored band in the OM image is similar to the one observed for the sample 80-07. The analysis of the AFM maps confirms that, also in this case, structures aligned along the flow direction are already present close to the sample surface, although the sample was obtained with a cooling rate significantly smaller than the Passive case. The layer characterized by thin aligned structures (with thickness smaller than 300 nm) appears to be smaller with respect to the case 80-07. In particular, the AFM map shows that the layer with tightly packed aligned structures ends at 0.15 mm. After this distance, the lateral growth becomes considerable, and the thickness suddenly becomes larger than 300 nm. The layer characterized by a considerable lateral growth is wide, and it ends almost at the same distance (0.50 mm from the sample surface, as shown by AFM map) observed in the case 80-07.

Figure 7 shows the morphology developed in the sample obtained with 80 °C cavity surface temperature, and 1.3 s heating time. At 0.04 mm from the sample surface, the AFM map shows tightly packed structures aligned along the flow direction. Tightly packed structures, with thicknesses up to 300 nm, are present up to 0.18 mm (i.e. on the border of the OM blue layer). After that, the lateral growth becomes significant (higher than 300 nm thickness), and the structures aligned along the flow direction have a width ranging between 2–4 μm (up to 0.22 mm from the sample surface). Structures aligned along the flow direction can be recognized up to 0.3 mm from the sample surface; after that position, only randomly dispersed spherulites can be recognized.

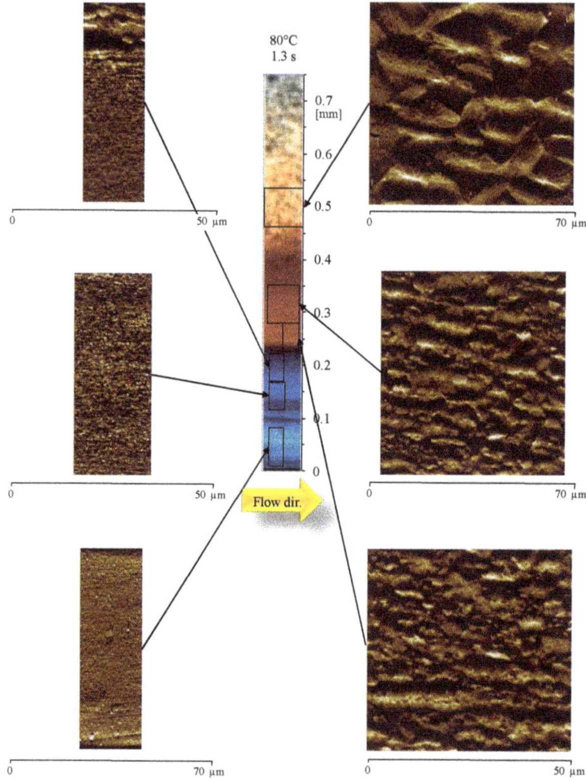

Figure 7. Optical micrograph and AFM height maps, at different distances from the sample surface, of the sample 80-1, at 15 mm downstream the gate.

Figure 8a,b shows OM of the samples obtained with 6 s heating time and different cavity surface temperatures, 150 °C and 80 °C respectively. During these tests, the cavity surface temperature was kept high for times comparable with the holding time. In Figure 8a, the AFM map shows that the layer characterized by tightly packed structures (up to 300 nm thickness), aligned along the flow direction, extends up to 0.15 mm, which is the same distance observed for the sample 150-07. The morphology evolves from aligned structures to randomly dispersed spherulites within a short distance. At 0.25 mm from the sample surface, the AFM map only shows randomly dispersed spherulites. The sample 80-6 (Figure 8b) shows a similar distribution: the width of the layer with thin and packed structures extends up to 0.2 mm, the same distance observed for the samples 80-07, and 80-1. The layer in which the lateral growth is significant has a width close to the one observed for the sample 80-1 (see the AFM map at 0.32 mm). After 0.30 mm from the sample surface, only spherulites can be recognized.

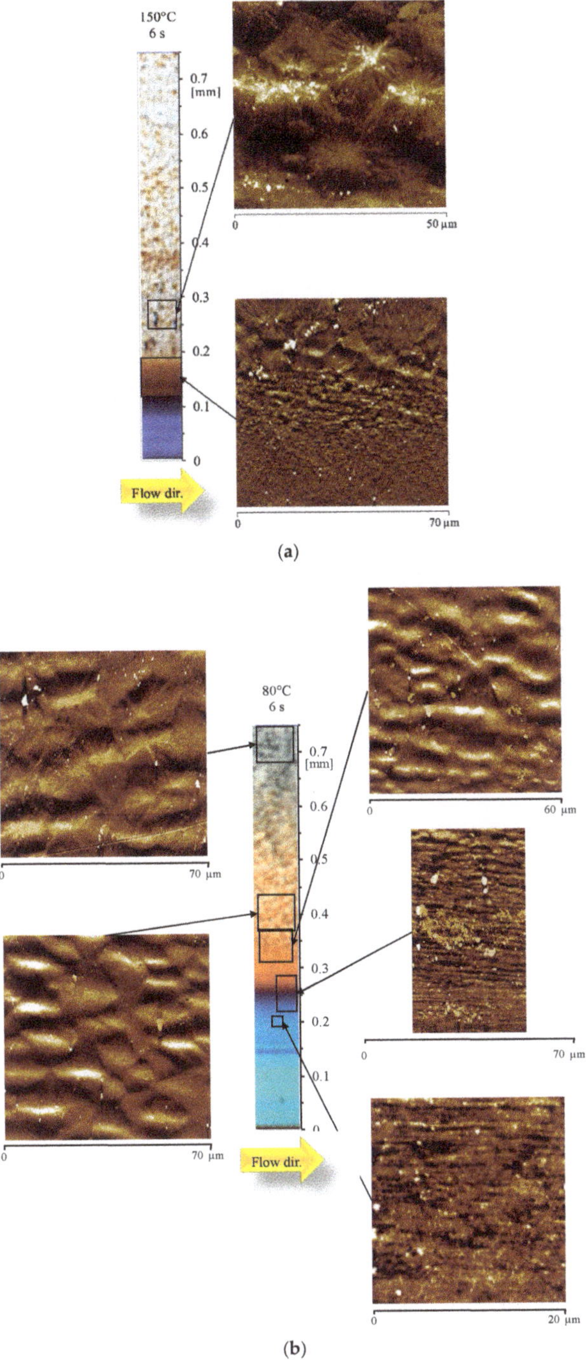

Figure 8. Optical micrograph and AFM height maps, at different distances from the sample surface, of the sample (**a**) 150-6 and (**b**) 80-6 (at 15 mm downstream the gate).

4. Discussion

Several studies, in the literature, report morphological results obtained when the polymer chains undergo homogeneous flow field [4,14,29–31]. However, few papers discuss the effect of inhomogeneous flow and temperature fields characteristic of the injection molding process [32–34]. As mentioned above, the hierarchical structures in polymer products are determined by the flow and temperature fields, and by the macromolecular characteristics. The amount of the stretched/oriented chains induced by the flow in the mold significantly decreases from the sample surface to the sample core. In the sample core, slow cooling and weak flows allow a more complete molecular relaxation of polymer chain, with respect to the layers close to the sample surface. Close to the sample surface, the stronger flow fields favor crystallization in the form of structures aligned along the flow direction, namely shishes, at temperatures significantly higher than in the almost quiescent conditions of the core [4,35,36]. It has been hypothesized that the presence of micellar nuclei, considered precursors for shish formation, is a prerequisite for the prolific formation of shishes [14]; however, the mechanisms that regulate shish formation are still under debate [14,37–39]. Keller and coworkers [36] suggest that two critical strain rates determine the final morphology: one below which the polymer chains rest in the coiled conformation, another one above which the polymer chains extend and shish formation occurs. Between these two critical values a significant lateral growth is allowed. Mackley et al [40] ascribed the formation of shishes to the extension of polymer chains, due to high flow intensity; with less intense flow, a significant lateral growth was found.

Recent studies relate the formation of shishes to the stretched chain network, instead than the extension of a single polymer chain [15,41]. Hsiao and co-workers [4,29] hypothesized that the polymer chains stretched more than a certain critical value could aggregate to form extended chains, the shishes, and the remaining coil polymer chains could then crystallize on the shishes in a folded periodic manner, forming the shish–kebab morphology. Kornfield et al [14] hypothesized that the formation of shishes takes place in two steps: the initiation, and the propagation. During the first step, the high shear stress induces the formation of precursors/nuclei, whose concentration increases with the flow intensity. Thanks to the flow, the long chains are transported on the precursors and attached to them. After that, the flow induces the extension of long chains, that achieve high orientation levels. During the propagation, the attachment of long chains, one upon the other, allows for the growth and formation of the so-called shish [14]. Mykhaylyk et al [42] supposed that the flow promotes the alignment and the aggregation of precursor/nuclei along the flow direction in the form of shishes.

All of these theories associate flow intensity with the density of precursors/nuclei, and the arrangement of precursors/nuclei in the space. Under strong flow field, the density of the nuclei is so high and the growth rate of these nuclei is so fast that the crystals get compact very quickly [43] and lateral growth is significantly limited due to the impingement. In these cases, the nucleation of α-phase was found to be predominant with respect to the other crystalline phases [32,34], and shishes were found. This kind of crystallization is based on the unidirectional propagation of a growth front [44], and it characterizes the layers close to the surface of the samples analyzed in this work. Obviously, the larger the distance from the sample surface, the more significant the lateral growth, and subsequently the higher the thickness of the shishes. The smaller the flow intensity, the smaller also the density of precursors/nuclei. As the distance from the sample surface increases, the flow intensity and the cooling rate decreases. As a consequence, the density of α-row nuclei decreases, and lateral growth is allowed. In this case, the growing of other crystalline phases, i.e. β-phase [39,45], in the form of cylindrites, was also observed. In particular, the formation of β-phase, with temperatures close to the ones adopted in this work, and with a shear rate of $1\ \mathrm{s}^{-1}$, was already observed for the grade of iPP adopted in this work [46]. Figure 9a,b shows structures that appear aligned along the flow direction, and formed by shish-like core structures with secondary growth of lamellae from the shish-like core in the transverse direction, as also reported in the sketch of Figure 9c. Thus, one can conclude that the structures shown in Figure 9a,b, found in the samples 80-07 and 150-07, can be attributed to the growth of β-phase in the cylindritic form [34,47].

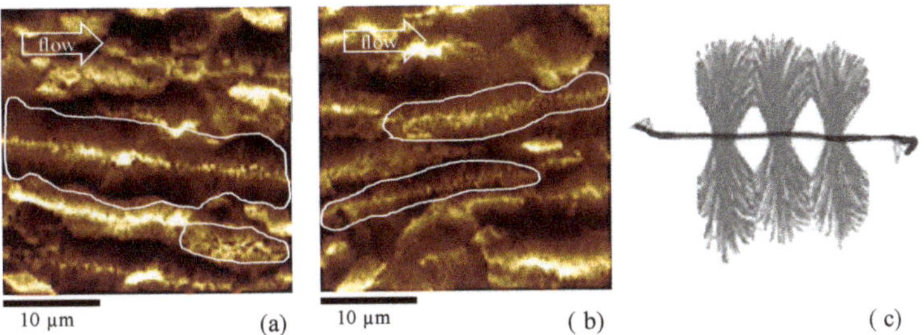

Figure 9. AFM height maps of the samples (**a**) 80-07; and (**b**) 150-07, at 0.50 mm from the sample surface; (**c**) Sketch of cylindritic structure adapted from An et al [34].

The cylindrites are characterized by smaller length than the fibers, as also shown in the figure, since the impingement takes place in both directions with respect to the flow front, probably due to weak flow. The impingement along the flow direction is pointed out in the figure by white boundaries. Also, the cylindrites appear to be less oriented along the flow direction with respect to the structures observed in the layers close to the sample surface. It is very interesting that properly changing the injection molding operating conditions makes it possible to detect β-cylindrites, since, generally, their presence is induced by β-nucleating agent in order to improve the toughness of the sample [48,49].

The AFM maps obtained for the samples produced in this work were analyzed taking into account all of the mentioned theories on the formation of different kinds of morphologies. In particular, on the basis of the height profiles, six areas were identified:

- the area with poorly formed structures close to the sample surface;
- the area with tightly packed structures, shishes with thickness up to 300 ± 50 nm;
- the area with structures aligned along the flow direction, which includes structures presenting a thickness larger than 300 nm and up to about 2 µm, namely shish-kebab;
- the area characterized by cylindritic structures (up to 6 µm thickness);
- the area characterized by randomly dispersed spherulites;
- the overlapping zone, in both the shishes and cylindrites can be also observed.

Figure 10 shows the results.

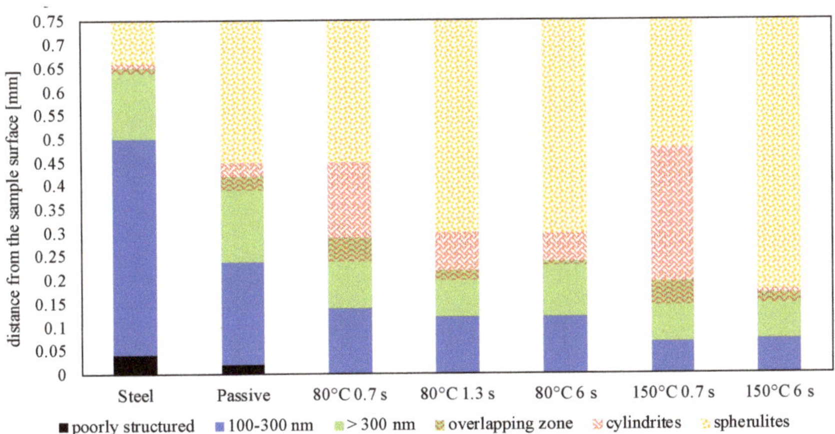

Figure 10. Representation of the layers characterized by the presence of: poorly structured (black), tightly packed shishes (thickness up to 300 ± 50 nm) (blue), less packed shishes (green), cylindrites (red) and spherulites (yellow) as evaluated from AFM acquisitions. There is also an overlapping zone in which different kinds of oriented structures can be detected.

In Figure 10, the sample (named Steel) obtained by replacing the heating device with a steel layer of the same thickness—in other words, the sample obtained in conventional injection molding conditions—is also reported. The Steel sample shows the widest layer characterized by tightly packed shishes, up to 0.5 mm from the sample surface. The layer characterized by thicker aligned structures extends up to 0.66 mm. The width of layers characterized by shishes mainly depends on the fast cooling (the fastest one in this work) that the polymer experienced already during the filling. The Passive sample shows a wide layer with tightly packed shishes (it ends at about 0.25 mm from the sample surface). Such a layer becomes 50% and 70% thinner as the cavity surface temperature T_{cs} increases up to 80°C and 150°C, respectively. The effect of the heating time on the width of this layer is negligible. Again, Passive sample shows the widest layer with less packed shishes (green), such a layer becomes thinner as the T_{cs} increases, although the thickness of such a layer is less sensitive to the cavity surface temperature than the layer with tightly packed shishes (300 nm thickness).

At a certain distance from the sample surface, which in the case of the Passive sample corresponds with 0.42 mm, cylindritic structures were also found. These structures were found in all the analyzed samples; however, a larger amount was found in the samples obtained with 0.7 s heating time. Generally, the layer characterized by cylindrites is the most sensitive one to the heating time: the longer is the heating time, the smaller is the width of the cylindritic layer. For instance, with $T_{cs} = 150$ °C and 6 s heating time, this layer becomes very thin (almost disappearing). The β-phase, characteristic of cylindritic structures is a metastable, and thus small changes in the flow and temperature fields would induce transition into α-phase. In particular, an increase of the heating time promotes the transition into spherulites.

Figure 11a shows the flow rate at the cavity entrance, evaluated by the software code for the injection molding process developed at the University of Salerno, during both the filling and the packing stages. Figure 11b shows the flow rate, namely the integral over the cross section of the velocity profile, plotted versus the average cross section temperature, for the tests Steel, Passive, 80-07, 150-07, and 150-6. Obviously, the polymer experienced the highest temperature at the beginning of the process. For each test, the flow rate was high and essentially constant during the filling stage (namely when temperature is close to the injection temperature). Soon after the cavity filling, the packing stage took place. During such a stage, additional feeding of melt compensated for the shrinkage. In the first tenths of a second of the packing stage, the flow rate decreased from the high values of the filling to much smaller values.

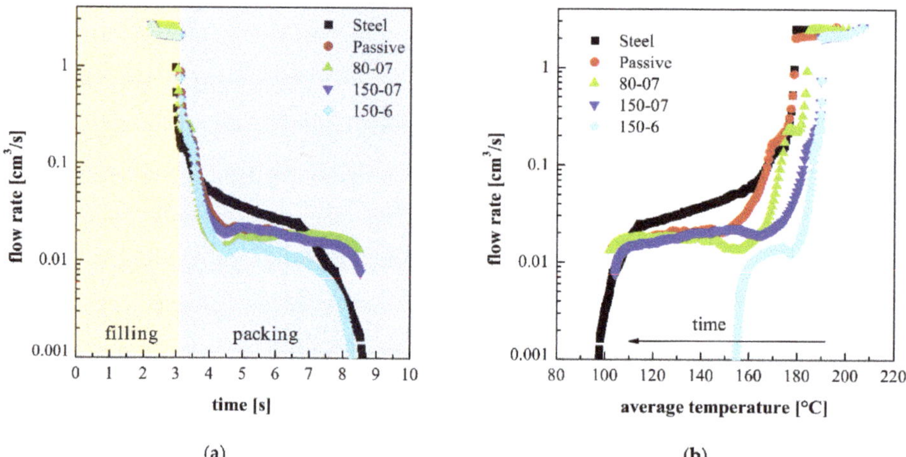

Figure 11. Flow rate vs time (**a**), and average cross section temperature (**b**), for the test Steel, Passive, 80-07, 150-07 and 150-6.

The boundary condition modification on the cavity surface induces a shift of the flow rates, during each stage of the injection molding process, toward high temperatures. The tests Steel and Passive showed the same flow rate, in the same temperature range, during the filling, and the early stage of the packing. During the packing, the Steel test shows a significantly higher flow rate than the Passive test, since this last test shows a slow heat exchange due to the presence of the heating device acting as an insulator. The tests 80-07 and 150-07 show flow rates similar to the one shown by the Passive test; however, they occur within higher temperatures, depending on the adopted temperature T_{cs}. The test 150-6 shows the same flow rate of the test 150-07 during the filling, whereas the packing flow rate decreased monotonically to lower values with respect the other one, as the polymer average temperature in the cross section becomes uniform, and close to the cavity surface temperature (150 °C). For a given initial temperature, the higher the polymer average temperature, the smaller the amount of material needed to compensate for the volume reduction. Thus, the flow rate during the packing, for the sample 150-6, resulted in being smaller with respect to the flow rate of the other tests.

On the bases of AFM investigations and the calculation of flow rates, it can be hypothesized that the high flow rates during the filling stage, and also during the early stage of packing, are able to induce high molecular orientation. In this case, the formation of the tightly packed shishes is promoted. During the packing stage, the flow rate, certainly smaller than the flow rate during the filling, is able to orient the molecules because the temperatures are smaller. In the inner part of the sample, the intensity of the flow, being smaller with respect to the layers close to the sample surface, allowed a certain lateral growth, and the morphology evolved toward shish-kebab. The thickness of the shish-kebab depends on the impingement of the growing structures.

For the Steel test, the high flow rate during the packing stage was responsible for the widest extension of the layer characterized by tightly packed shishes. When 0.7 s was adopted as the heating time, the cooling rate was smaller with respect to the Passive case: during the packing step, the same flow rates are attained at a temperature about 20 °C larger than the Passive case (see Figure 11). This fact allows a significant lateral growth, and also the formation of cylindritic structures whose thickness resulted in becoming intermediate between the thickness of the shish-kebab and the mean diameter of the spherulites. With longer heating times, the cooling rate was additionally decreased, and thus, the molecular orientation was expected to be smaller due to the smaller relaxation times at high temperatures. As a result, the layer characterized by cylindrites was thinner, and spherulitical growth was allowed.

5. Conclusions

In this paper, a deep analysis of the morphology developed along the thickness of the samples obtained by injection molding was performed. The samples were obtained adopting different temperatures and times for the cavity surface heating, thus the polymer underwent different thermo-mechanical histories that influence the morphology developed along the sample thickness. Several morphological structures were detected. A poorly structured layer, a layer characterized by tightly packed structures, with a mean thickness smaller than 300 nm, aligned along the flow direction; a layer characterized by less packed aligned structures; a layer characterized by cylindritic structures; and a spherulitical layer, from the sample surface to the core. There is also an area in which both shish and cylindrites could be observed. The formation of these layers was determined by both the cooling rate and the flow field experienced by the polymer during the three main stages of the process, namely the filling, the packing, and the cooling. If the flow field was strong, but the cooling rate is very high, the sudden solidification of the melt occurred, and the polymer chains hadn't enough time for hierarchical structuring. As a consequence, the poorly structured layer, characteristic of the zones close to the sample surface, formed. If the cooling rate was not fast enough to induce sudden solidification of the polymer, the polymer chains had enough time for structuring. In this case, under a strong flow field, characteristic of the filling stage and the early stage of the packing, tightly packed structures form, namely shishes, with a mean thickness ranging between 100-300 nm. The formation of such packed structures is mainly due to the high density of precursors/nuclei, due to the strong shearing flow. The growth rate of these nuclei is so fast that the crystals become compact very quickly, and the lateral growth is essentially prevented. The larger is the distance from the sample surface, the lower is the flow intensity, and consequently the density of precursor/nuclei. Thus, as the distance from the sample surface increases, the lateral growth becomes significant, and shish-kebab morphology can be observed. The thickness of the layer with packed shishes also depends on the modulation of the cavity surface temperature. The higher the cavity surface temperature, the thinner the layer with packed shishes. In the presence of slow cooling and weak flow, the growth of β-phase is allowed. In these conditions, cylindritic structures appear to be formed by shish-like core structures aligned along the flow direction, with secondary growth of lamellae from the shish-like core in the transverse direction. Generally, the formation of such structures is induced by nucleating agents in order to achieve high toughness. It is very interesting that, in the cases proposed in this work, the formation of such structures was achieved by a proper change in cavity surface temperature. The width of the layer characterized by cylindritic structures was affected by the adopted heating time. The longer the heating time, the thinner the layer with cylindritic structures, and spherulites can be observed for the most of sample thickness. This is because polymer chains had enough time to relax and lose orientation.

Supplementary Materials: The following are available online at http://www.mdpi.com/1996-1944/12/3/424/s1, Figure S1: Temperature evolutions measured by means of a thermocouple type T located on the cavity surface at 20 mm from the cavity entrance during the tests 80-07, 80-1 and 80-6.

Author Contributions: Conceptualization, S.L. and G.T.; methodology, R.P.; software, V.S.; validation, V.S. and R.P.; investigation, S.L.; writing—original draft preparation, S.L.; writing—review and editing, R.P.; supervision Giuseppe Titomanlio.

Funding: This research received no external funding.

Conflicts of Interest: The authors declare no conflict of interest.

References

1. Kamal, M.R.; Isayev, A.I.; Liu, S.-J. Injection molding: Technology and fundamentals. In *Progress in Polymer Processing*; Hanser, Ed.; Carl Hanser Verlag GmbH & Co. KG: Munich, Germany, 2009; p. 737.

2. Janeschitz-Kriegl, H.; Ratajski, E. Kinetics of polymer crystallization under processing conditions: Transformation of dormant nuclei by the action of flow. *Polymer* **2005**, *46*, 3856–3870. [CrossRef]

3. Hashimoto, T.; Murase, H.; Ohta, Y. A New Scenario of Flow-Induced Shish-Kebab Formation in Entangled Polymer Solutions. *Macromolecules* **2010**, *43*, 6542–6548. [CrossRef]

4. Somani, R.H.; Yang, L.; Zhu, L.; Hsiao, B.S. Flow-induced shish-kebab precursor structures in entangled polymer melts. *Polymer* **2005**, *46*, 8587–8623. [CrossRef]

5. D'Haese, M.; Mykhaylyk, O.O.; Van Puyvelde, P. On the Onset of Oriented Structures in Flow-Induced Crystallization of Polymers: A Comparison of Experimental Techniques. *Macromolecules* **2011**, *44*, 1783–1787. [CrossRef]

6. Pan, Y.; Shi, S.; Xu, W.; Zheng, G.; Dai, K.; Liu, C.; Chen, J.; Shen, C. Wide distribution of shish-kebab structure and tensile property of micro-injection-molded isotactic polypropylene microparts: A comparative study with injection-molded macroparts. *J. Mater. Sci.* **2014**, *49*, 1041–1048. [CrossRef]

7. Liparoti, S.; Sorrentino, A.; Speranza, V.; Titomanlio, G. Multiscale mechanical characterization of iPP injection molded samples. *Eur. Polym. J.* **2017**, *90*. [CrossRef]

8. Mark, J.E. Mechanical Properties of Polymers Based on Nanostructure and Morphology. Edited by Georg H. Michler and Francisco J. Balta'-Calleja. *Angew. Chem. Int. Ed.* **2006**, *45*, 6080. [CrossRef]

9. Bai, H.; Deng, H.; Zhang, Q.; Wang, K.; Fu, Q.; Zhang, Z.; Men, Y. Effect of annealing on the microstructure and mechanical properties of polypropylene with oriented shish-kebab structure. *Polym. Int.* **2012**, *61*, 252–258. [CrossRef]

10. Fujiyama, M. Polypropylene Morphology-mechanical property relationships in injection molding. In *Polypropylene*; Karger-Kocsis, J., Ed.; Springer: Dordrecht, Switzerland, 1999; pp. 519–526.

11. Fujiyama, M.; Wakino, T.; Kawasaki, Y. Structure of skin layer in injection-molded polypropylene. *J. Appl. Polym. Sci.* **1988**, *35*, 29–49. [CrossRef]

12. Pogodina, N.V.; Lavrenko, V.P.; Srinivas, S.; Winter, H.H. Rheology and structure of isotactic polypropylene near the gel point: Quiescent and shear-induced crystallization. *Polymer* **2001**, *42*, 9031–9043. [CrossRef]

13. Li, L.; de Jeu, W.H. Shear-Induced Smectic Ordering in the Melt of Isotactic Polypropylene. *Phys. Rev. Lett.* **2004**, *92*, 075506. [CrossRef] [PubMed]

14. Fernandez-Ballester, L.; Thurman, D.W.; Zhou, W.; Kornfield, J.A. Effect of Long Chains on the Threshold Stresses for Flow-Induced Crystallization in iPP: Shish Kebabs vs Sausages. *Macromolecules* **2012**, *45*, 6557–6570. [CrossRef]

15. Roozemond, P.C.; Peters, G.W.M. Flow-enhanced nucleation of poly(1-butene): Model application to short-term and continuous shear and extensional flow. *J. Rheol.* **2013**, *57*, 1633–1653. [CrossRef]

16. Chen, Y.; Yang, S.; Yang, H.; Zhong, G.; Fang, D.; Hsiao, B.S.; Li, Z. Deformation behavior of oriented β-crystals in injection-molded isotactic polypropylene by in situ X-ray scattering. *Polymer* **2016**, *84*, 254–266. [CrossRef]

17. van der Meer, D.W. *Structure-Property Relationships in Isotactic Polypropylene*; Twente University Press: Enschede, The Netherlands, 2003.

18. Van Erp, T.B.; Balzano, L.; Peters, G.W.M. Oriented gamma phase in isotactic polypropylene homopolymer. *ACS Macro Lett.* **2012**, *1*, 618–622. [CrossRef]

19. Chen, Y.H.; Mao, Y.M.; Li, Z.M.; Hsiao, B.S. Competitive growth of α-and β-crystals in β-nucleated isotactic polypropylene under shear flow. *Macromolecules* **2010**, *43*, 6760–6771. [CrossRef]

20. Rizvi, S.J.A. Effect of injection molding parameters on crystallinity and mechanical properties of isotactic polypropylene. *Int. J. Plast. Technol.* **2017**, *21*, 404–426. [CrossRef]

21. Labour, T.; Vigier, G.; Séguéla, R.; Gauthier, C.; Orange, G.; Bomal, Y. Influence of the β-crystalline phase on the mechanical properties of unfilled and calcium carbonate-filled polypropylene: Ductile cracking and impact behavior. *J. Polym. Sci. Part B Polym. Phys.* **2002**, *40*, 31–42. [CrossRef]

22. Pantani, R.; Coccorullo, I.; Volpe, V.; Titomanlio, G. Shear-Induced Nucleation and Growth in Isotactic Polypropylene. *Macromolecules* **2010**, *43*, 9030–9038. [CrossRef]

23. Pantani, R.; Speranza, V.; Titomanlio, G. Evolution of iPP Relaxation Spectrum during Crystallization. *Macromol. Theory Simul.* **2014**, *23*, 300–306. [CrossRef]

24. Pantani, R.; Speranza, V.; Titomanlio, G. Effect of flow-induced crystallization on the distribution of spherulite dimensions along cross section of injection molded parts. *Eur. Polym. J.* **2017**, *97*, 220–229. [CrossRef]

25. De Santis, F.; Adamovsky, S.; Titomanlio, G.; Schick, C. Scanning Nanocalorimetry at High Cooling Rate of Isotactic Polypropylene. *Macromolecules* **2006**, *39*, 2562–2567. [CrossRef]

26. Liparoti, S.; Landi, G.; Sorrentino, A.; Speranza, V.; Cakmak, M.; Neitzert, H.C. Flexible Poly(Amide-Imide)-Carbon Black Based Microheater with High-Temperature Capability and an Extremely Low Temperature Coefficient. *Adv. Electron. Mater.* **2016**, *2*, 1600126. [CrossRef]

27. Liparoti, S.; Sorrentino, A.; Titomanlio, G. Fast cavity surface temperature evolution in injection molding: control of cooling stage and final morphology analysis. *RSC Adv.* **2016**, *6*, 99274–99281. [CrossRef]

28. Kalay, G.; Bevis, M.J. Processing and physical property relationships in injection-molded isotactic polypropylene. 2. Morphology and crystallinity. *J. Polym. Sci. Part B Polym. Phys.* **1997**, *35*, 265–291. [CrossRef]

29. Hsiao, B.S.; Yang, L.; Somani, R.H.; Avila-Orta, C.A.; Zhu, L. Unexpected Shish-Kebab Structure in a Sheared Polyethylene Melt. *Phys. Rev. Lett.* **2005**, *94*, 117802. [CrossRef] [PubMed]

30. Mykhaylyk, O.O.; Chambon, P.; Impradice, C.; Fairclough, J.P.A.; Terrill, N.J.; Ryan, A.J. Control of Structural Morphology in Shear-Induced Crystallization of Polymers. *Macromolecules* **2010**, *43*, 2389–2405. [CrossRef]

31. Mykhaylyk, O.O.; Chambon, P.; Graham, R.S.; Fairclough, J.P.A.; Olmsted, P.D.; Ryan, A.J. The Specific Work of Flow as a Criterion for Orientation in Polymer Crystallization. *Macromolecules* **2008**, *41*, 1901–1904. [CrossRef]

32. An, Y.; Gu, L.; Wang, Y.; Li, Y.-M.; Yang, W.; Xie, B.-H.; Yang, M.B. Morphologies of injection molded isotactic polypropylene/ultra high molecular weight polyethylene blends. *Mater. Des.* **2012**, *35*, 633–639. [CrossRef]

33. Yang, H.-R.; Lei, J.; Li, L.; Fu, Q.; Li, Z.-M. Formation of Interlinked Shish-Kebabs in Injection-Molded Polyethylene under the Coexistence of Lightly Cross-Linked Chain Network and Oscillation Shear Flow. *Macromolecules* **2012**, *45*, 6600–6610. [CrossRef]

34. An, Y.; Bao, R.-Y.; Liu, Z.-Y.; Wu, X.-J.; Yang, W.; Xie, B.-H.; Yang, M.B. Unusual hierarchical structures of mini-injection molded isotactic polypropylene/ultrahigh molecular weight polyethylene blends. *Eur. Polym. J.* **2013**, *49*, 538–548. [CrossRef]

35. Roozemond, P.C.; van Drongelen, M.; Ma, Z.; Hulsen, M.A.; Peters, G.W.M. Modeling flow-induced crystallization in isotactic polypropylene at high shear rates. *J. Rheol.* **2015**, *59*, 613–642. [CrossRef]

36. Keller, A.; Kolnaar, H. Flow-Induced Orientation and Structure Formation. *Mater. Sci. Technol.* **1997**. [CrossRef]

37. Ogino, Y.; Fukushima, H.; Matsuba, G.; Takahashi, N.; Nishida, K.; Kanaya, T. Effects of high molecular weight component on crystallization of polyethylene under shear flow. *Polymer* **2006**, *47*, 5669–5677. [CrossRef]

38. Zhang, C.; Hu, H.; Wang, D.; Yan, S.; Han, C.C. In situ optical microscope study of the shear-induced crystallization of isotactic polypropylene. *Polymer* **2005**, *46*, 8157–8161. [CrossRef]

39. Zhou, Q.; Liu, F.; Guo, C.; Fu, Q.; Shen, K.; Zhang, J. Shish–kebab-like cylindrulite structures resulted from periodical shear-induced crystallization of isotactic polypropylene. *Polymer* **2011**, *52*, 2970–2978. [CrossRef]

40. Mackley, M.R.; Keller, A. Flow Induced Polymer Chain Extension and Its Relation to Fibrous Crystallization. *Philos. Trans. R. Soc. A Math. Phys. Eng. Sci.* **1975**, *278*, 29–66. [CrossRef]

41. Wang, Z.; Ma, Z.; Li, L. Flow-Induced Crystallization of Polymers: Molecular and Thermodynamic Considerations. *Macromolecules* **2016**, *49*, 1505–1517. [CrossRef]

42. Mykhaylyk, O.O.; Fernyhough, C.M.; Okura, M.; Fairclough, J.P.A.; Ryan, A.J.; Graham, R. Monodisperse macromolecules—A stepping stone to understanding industrial polymers. *Eur. Polym. J.* **2011**, *47*, 447–464. [CrossRef]

43. Huo, H.; Jiang, S.; An, L.; Feng, J. Influence of Shear on Crystallization Behavior of the β Phase in Isotactic Polypropylene with β-Nucleating Agent. *Macromolecules* **2004**, *37*, 2478–2483. [CrossRef]

44. Balzano, L.; Ma, Z.; Cavallo, D.; van Erp, T.B.; Fernandez-Ballester, L.; Peters, G.W.M. Molecular Aspects of the Formation of Shish-Kebab in Isotactic Polypropylene. *Macromolecules* **2016**, *49*, 3799–3809. [CrossRef]

45. Wang, Y.; Meng, K.; Hong, S.; Xie, X.; Zhang, C.; Han, C.C. Shear-induced crystallization in a blend of isotactic polypropylene and high density polyethylene. *Polymer* **2009**, *50*, 636–644. [CrossRef]

46. Coccorullo, I.; Pantani, R.; Titomanlio, G. Spherulitic Nucleation and Growth Rates in an iPP under Continuous Shear Flow. *Macromolecules* **2008**, *41*, 9214–9223. [CrossRef]

47. Pan, Y.; Guo, X.; Zheng, G.; Liu, C.; Chen, Q.; Shen, C.; Liu, X. Shear-Induced Skin-Core Structure of Molten Isotactic Polypropylene and the Formation of β-Crystal. *Macromol. Mater. Eng.* **2018**, *303*, 1800083. [CrossRef]

48. Li, X.; Guo, C.; Zhang, Y.; Liu, K.; Zhang, J. The Morphology and Mechanical Properties of Isotactic Polypropylene Injection-Molded Samples with the Presence of β-Nucleation Agent and Periodical Shear Field. *J. Macromol. Sci. Part B* **2015**, *54*, 215–229. [CrossRef]

49. Zhang, B.; Chen, J.; Ji, F.; Zhang, X.; Zheng, G.; Shen, C. Effects of melt structure on shear-induced β-cylindrites of isotactic polypropylene. *Polymer* **2012**, *53*, 1791–1800. [CrossRef]

 *materials*

Article

Process Induced Morphology Development of Isotactic Polypropylene on the Basis of Molecular Stretch and Mechanical Work Evolutions

Sara Liparoti, Vito Speranza *, Roberto Pantani and Giuseppe Titomanlio

Department of Industrial Engineering, University of Salerno—via Giovanni Paolo II, 132, 84084 Fisciano (SA), Italy; sliparoti@unisa.it (S.L.); rpantani@unisa.it (R.P.); gtitomanlio@unisa.it (G.T.)
* Correspondence: vsperanza@unisa.it; Tel.: +39-089-964-145

Received: 9 January 2019; Accepted: 5 February 2019; Published: 7 February 2019

Abstract: It is well known that under high shear rates polymers tend to solidify with formation of morphological elements oriented and aligned along the flow direction. On the other hand, stretched polymer chains may not have sufficient time to undergo the structuring steps, which give rise to fibrillar morphology. In the last decades, several authors have proposed a combined criterion based on both a critical shear rate and a critical mechanical work, which guaranties adequate time for molecular structuring. In this paper, the criterion, reformulated on the basis of critical values of both molecular stretch and mechanical work and adjusted to account for the unsteady character of the polymer processing operations, is applied to the analysis of a set of isotactic polypropylene injection molded samples obtained under very different thermal boundary conditions. The evolutions of molecular stretch and mechanical work are evaluated using process simulation. The results of the model reproduce the main characteristics of the morphology distribution detected on the cross sections of moldings, obtained under very different thermal boundary conditions, assuming that the critical work is a function of temperature.

Keywords: morphology; injection molding; numerical simulation; morphology prediction; shear layer

---

1. Introduction

The crystallization morphology of polymers solidifying from flowing melts has been the object of several studies [1–8]. Many of them indicate that there is a critical shear rate, of the order of the inverse of Rouse time of the longest chains in the polymer, above which the morphology of the solidified polymer shows an ensemble of shish kebabs oriented along the flow direction, if the mechanical work accumulated afterwards is larger than a critical value [6,9,10]. The critical work guaranties the time required for kinetic processes finalized to create stable nuclei and to align the stable nuclei into rows, extending them along the flow direction until the microstructure is saturated. The higher the stress the faster those processes are [7].

It is well known that, during the injection molding process, oriented shish-kebab layers (also called shear layers) form within the final object cross section, and that the thickness of the shear layer changes remarkably depending on the processing conditions [11,12].

The results of the criterion for the formation of shish-kebab morphology were already favorably considered in comparison with the cross-section morphology distribution reported in the literature for isotactic polypropylene (iPP) injection molding samples, obtained under conventional injection molding conditions (CIM) [13]. In those cases, a critical molecular stretch parameter was recognized as more appropriate than the critical shear rate in order to obtain reliable predictions of the shear layer width. The evolutions of both the molecular stretch parameter and the mechanical work, accumulated after the critical stretch had been reached, were predicted adopting appropriate field equations for the description of temperature and flow fields. The effect of a stretch relaxation, while the mechanical work is accumulated, was disregarded in the previous work [13]. The constitutive equations for the rheological and crystallization behaviors (and their interactions) specifically identified for the adopted resin were also included in the model [14–16].

The aim of this paper is to identify a criterion based on the critical values of both a molecular stretch and the mechanical work and to apply it to the complete processing of the material to a final product. The aim of this paper is to identify a criterion based on the critical values of both a molecular stretch, and the mechanical work and to apply it to the complete processing of the material up to a final product accounting of unsteady conditions which in the injection molding can not be neglected since at the end of the filling step even a relaxation takes place. In order to verify the criterion against a set of injection molding conditions as wide as possible, the criterion was applied to moldings obtained with a procedure that allows a fast evolution of the cavity surface temperature during the process [17–19]. Such a procedure (by adopting appropriate cavity surface heating powers and times) allows the calibration of the morphology of the moldings all the way from a shear layer over most of the cross section to nearly completely spherulitic cross sections [19,20]. In order to evaluate the predictions of the model for crystallization into an oriented shish-kebab morphology, the evolutions of both the molecular stretch parameter and the mechanical work were evaluated using simulation of the injection molding process, accounting for cavity surface heating during the process. The overall objective was the optimization of the morphology inside the molded parts through modulation of the process parameters.

2. Materials and Methods

The simulation of the injection molding process was performed describing the evolution of the temperature field along both flow and thickness directions; to that purpose, convection, transverse conduction, and both crystallization heat and viscous heat generations [14,15,21] were accounted for. As mentioned in the introduction, the cavity surface was kept, during the process and until cooling, at temperatures intermediate between injection and mold temperatures. To this purpose, thin heating devices, made of several thin layers (one of which being the heating element operated by joule effect), were adopted. The heat transfer inside the heating devices was also simulated.

Momentum balance was solved adopting lubrication approximation with a viscosity depending upon shear rate, $\dot{\gamma}$, through a Cross equation with zero-shear rate viscosity η_0 (function of temperature, T, pressure, P, and crystallinity, χ) [14].

$$\eta(T,P,\dot{\gamma},\chi) = \frac{\eta_0(T,P,\chi)}{1 + \left(\frac{\eta_0(T,P,\chi)\dot{\gamma}}{\tau_R}\right)^{1-n}} \tag{1}$$

In Equation (1), n and τ_R are material parameters. The viscoelastic nature of the polymer was considered using a relaxation time λ [14].

$$\lambda(T,P,\chi,\Delta) = \frac{\lambda_0(T,P,\chi)}{1 + (a\Delta)^b} \tag{2}$$

The function of molecular stretch parameter Δ, whereas a, b, and λ_0 are material parameters. The molecular stretch parameter was evaluated as the difference between the largest and the smallest eigenvalues of the molecular conformation tensor [22,23]

$$\underline{\underline{A}} = 3\frac{(\langle RR \rangle - \langle RR \rangle_0)}{\langle R_0^2 \rangle}, \tag{3}$$

where $\underline{R}$ is the end-to-end vector of a molecular sub chain. The evolution of the conformation tensor describing the sub chain population was described using Maxwell-type equation

$$\frac{D}{Dt}\underline{\underline{A}} - (\nabla \underline{v})^T \times \underline{\underline{A}} - \underline{\underline{A}} \times (\nabla \underline{v}) = -\frac{1}{\lambda}\underline{\underline{A}} + (\nabla \underline{v})^T + (\nabla \underline{v}), \qquad (4)$$

where $\nabla\,\underline{v}$ is the velocity gradient.

Mesomorphic and α crystallization processes, both competing for the same amorphous, were considered. The Nakamura equation [24] was adopted for the kinetics of the mesophase, the and Hoffman-Lauritzen equation [25] was adopted for the crystallization toward the α phase.

$$G(T(t), P, \Delta) = G_0 \exp\left(-\frac{U}{R(T - T_\infty)}\right) \exp\left(-\frac{K_g(T + T_m(P, \Delta))}{2T^2(T_m(P, \Delta) - T)}\right) \qquad (5)$$

The effect of both pressure and flow on crystallization kinetics were accounted for by considering the crystallization temperature T_m function of both pressure, and molecular stretch parameter, Δ [14,26].

The temperature and pressure evolutions obtained by the simulations were satisfactorily compared with experimental results. In addition, final crystallinity distributions inside moldings obtained with the iPP T30G were satisfactorily described [22]. In this work, the simulations were performed with the aim of analyzing the morphology distribution along the cross sections of moldings previously obtained and reported in [20]. The experimental information relevant to the simulation of those tests is summarized below.

Injection molding experiments were carried out adopting the iPP grade T30G (supplied by Lyondell Basell, Ferrara, Italy) previously characterized for rheology and crystallization kinetic [13,14,16]. The polymer was injected into a cavity having the thickness, width, and length of 1.50 mm, 12.7 mm, 110 mm, respectively. Five pressure transducers were placed along the flow path: one (P0) in the injection chamber, one (P1) in the runner close to the gate, and three (P2, P3 and P4) inside the cavity, 15 mm, 60 mm, and 105 mm downstream from the gate position, respectively. Thin heating devices were placed inside the mold, layered very close to the cavity surface (0.1 mm from it) in order to achieve a fast evolution of the cavity surface temperature during the injection cycle; the heating devices were made of several polymeric layers, one of them (thickness 0.05 mm) was electrically conductive being made of carbon black loaded poly(amide-imide). A fast thermocouple was allocated on the cavity surface in position P2. The injection molding tests were performed with 2.3 cm^3/s average flow rate, 220 °C melt temperature, 25 °C overall mold temperature, 720 bar holding pressure, and 10–13 s holding time.

The electrical power, P_e, supplied to the heating devices determined the evolution of the cavity surface temperature toward a temperature level intermediate between mold and injection temperatures. In particular, with 4 W/cm^2 or 9.5 W/cm^2, the cavity surface reached about 80 °C or 150 °C, after about 6 s; in the following, these temperatures will be denoted as cavity surface temperature, T_{cs} (for each injection molding test). In order to avoid a cold contact of the polymer with the cavity surface, the electrical power in the heating devices was supplied in advance, in particular of a time $t_a = 2$ s, with respect to the time at which the flow front reaches the position P2 inside the cavity. The cooling steps were determined by de-activation of the heating devices.

Tests were performed adopting, for each heating power, the following heating times, t_h, (after the cavity surface pre-heating): 0.7 s (namely the cavity filling time), 1.3 s, and 6 s. In addition, tests without activating the heating devices, namely Passive test, and tests where the heating devices were replaced by steel layers of the same geometry (namely Steel test) were carried out. All injection molding tests considered in this work are reported in Table 1.

Polarized optical micrographs (OM) of sample slices cut along the flow—thickness planes were obtained using a Olympus BX51 microscope (Olympus Italia S.r.l., Segrate, Italy) with crossed analyzer-polarizer. The flow direction of sample slices was rotated 45° with respect to the analyzer [20].

The slices were chemically etched following procedure proposed by Bassett [27], and analyzed with Atomic Force Microscopy (Bruker Dimension coupled with Nanoscope V controller) operating in tapping mode with a probe tip having 42 N/m spring constant, 300 kHz resonance frequency, and 7 nm radius (Bruker, Billerica, MA, USA) [20].

Table 1. Injection molding operating conditions.

Test Run	P_e (W/cm^2)	T_{cs} (°C)	t_h (s)	t_a (s)
Steel	0	25	0	0
Passive	0	25	0	0
80-07	4	80	0.7	2
80-1	4	80	1.3	2
80-6	4	80	6	2
150-07	9.5	150	0.7	2
150-1	9.5	150	1.3	2
150-6	9.5	150	6	2

Figures 1 and 2 show comparisons between some simulation and experimental results for evolution of surface temperature in position P2 and for pressure distributions during some of the tests listed in Table 1.

As mentioned above, the heating of the mold surface started in advance with respect to the first contact of the polymer in position P2, and, in Figure 1, the value of time is set to zero at the polymer contact in position P2. The electrical power was selected on the basis of the desired final temperature. When the polymer reaches position P2, it undergoes a jump up by effect of the temperature of the hot polymer, which was injected at 220 °C. After about 6 s, a nearly steady temperature on the cavity surface was recorded. Sample cooling takes place by effect of the temperature difference between the sample and the cavity surface; cavity surface temperature decreases to the value of the mold at the heating device de-activation.

The main characteristics of the experimental temperature evolutions on the cavity surface were reproduced by the simulation. The differences are mainly related to the description of the sample cooling and, in particular to the inflection zones along the experimental curves, at temperatures below 80 °C. These inflections have to be related to concentrated development of the latent heat of crystallization, occurring at the detachments of the sample from the cavity surface, which were not accounted for in the simulation.

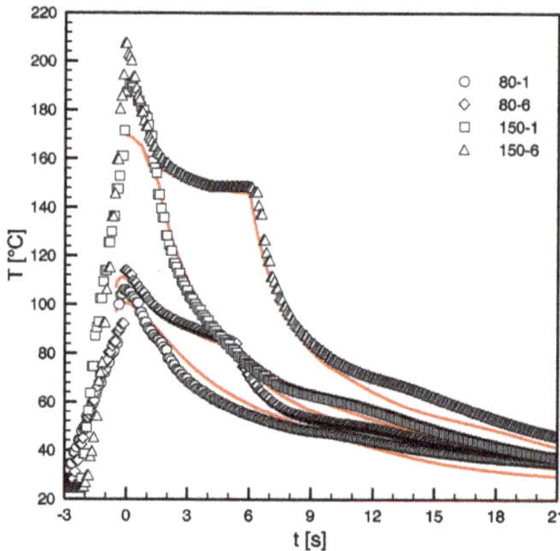

Figure 1. Comparison between experimental (symbols) and simulated (red full lines) cavity surface temperatures evolutions obtained for the tests 80-1, 80-6, 150-1 and 150-6.

Figure 2 shows the comparison between experimental and simulated pressure evolutions along the flow path, for some of the tests considered in this paper. Similarly to the temperature evolutions reported in Figure 1 and in the plots reported in Figure 2, the time t = 0 s corresponds to the first contact of the melt in position P2. At the end of the filling step, the injection chamber pressure, P0, was found to range between 450 bars and 340 bars upon a change of the cavity heating power from

4 W/cm^2 or 9.5 W/cm^2. At the end of the filling step, the pressure undergoes a jump toward the holding pressure (which was kept at 720 bars for all tests).

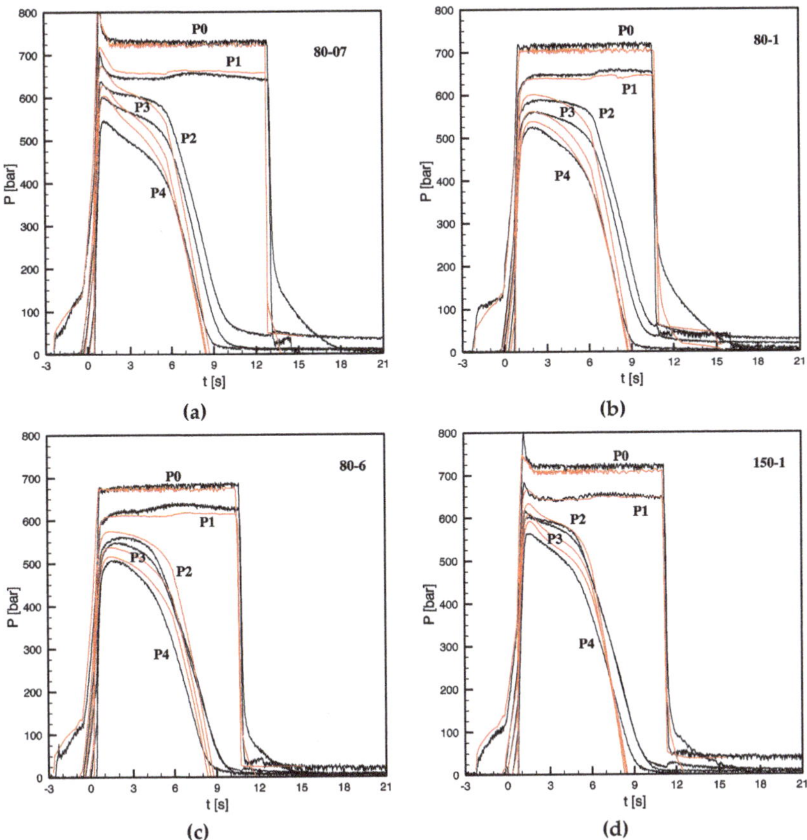

Figure 2. Comparison between experimental (black lines) and simulated (red lines) pressure evolutions for the tests carried out with heating powers 4 W/cm^2 and 9.5 W/cm^2 with different heating time as specified in single plot. (**a**) 80 °C and 0.7 s, (**b**) 80 °C and 1.3 s, (**c**) 80 °C and 6 s, (**d**) 150 °C and 1.3 s.

After filling, the density increase, determined by both cooling and crystallization processes taking place inside the cavity, is compensated by the packing flow. The packing flow essentially keeps the pressure constant inside the cavity, as long as it is not hindered by the gate sealing process. When the gate sealing process starts to have an impact on the packing flow, the pressure drops through the gate (essentially between P1-P2). When the gate sealing process starts, to have an impact on the packing flow, the pressure drops through the gate (essentially between P1–P2) undergoes a fast increase of the enhancement rate process, which for all tests considered in this work takes place between 5 s and 6.5 s.

The pressure in position P1 drops to zero when the holding pressure is released in the injection chamber, unless, somewhere between positions P0 and P1, the process of cross-section solidification is already sufficiently advanced to significantly slow down the back flow from P1 to P0. Indeed, for all the plots reported in Figure 2 the pressure in P1 drops instantaneously to a value small but still considerable (about 100 bars), soon after it slowly reduces to zero.

The main experimental characteristics of pressure evolutions, in particular, the pressure growth during the filling step, both pressure levels at beginning of the packing step, and gate sealing times, are also essentially nicely reproduced by the simulations. The slopes of the pressure curves during cooling are only slightly overestimated. The existence of an advanced (not complete) process of cross-section solidification between the injection chamber and position P1, at the end of the holding time, is also reproduced, as demonstrated by the fact that the pressure does not completely drop to zero at the holding pressure release.

In conclusion, the comparisons between experimental acquisitions and simulation results show that the main characteristics of the thermo-mechanical histories, and thus of the phenomena involved during the process, are reproduced by the simulation. This is a strong indication that relevant phenomena taking place during the process were satisfactorily described, thus accrediting the use of the simulation results for the application of the criterion for the formation of fibrillar morphology.

3. The Criterion for Crystallization into Fibrillar Morphology

The idea that the criterion for the attainment of fibrillar morphology, as found in the shear layer of injection molded objects, has to be based on the achievement of an adequate (critical) amount of mechanical work spent with shear rates above a critical value was proposed several years ago [9,10,13,28–30]. The critical value of shear rate was identified to be of the order of the inverse of the Rouse number, because it has to be able to sufficiently stretch the macromolecules. The critical mechanical work stands there to allow for kinetic processes related to the stretching of long molecules, creating stable shish nuclei aligned into rows, upon which the bulk of the material can crystallize as kebabs, transforming it into a fibrillar structure; the higher the stress the faster are those processes.

Reformulating the criterion in terms of the molecular stretching parameter Δ [13] is equivalent, or even a step ahead, with respect to the initial formulation in terms of shear rate. In other words, for each polymer, there is a critical amount of mechanical work, W_c, which, once accumulated in the presence of a molecular stretch, in other words, for each polymer, there is a critical amount of mechanical work, W_c, which, once accumulated in presence of a molecular stretch, Δ equal or larger than the critical stretch, Δ_c, determines the crystallization into the shish—kebab morphology. The authors believe that the larger the value of the stretch (above the critical value), the larger the density of molecules will be that will be able to act as nucleation seeds for shish formation. If the molecules have sufficient available time (assured by the accounting of mechanical work), they will arrange themselves into a mechanically efficient morphology, and the elements thickness would be determined by the density of the active nucleation seed, after lateral growth.

If while the mechanical work increases, there is some relaxation, which brings the stretch parameter below its critical value, Δ_c, the formation of the fibers remains, at least partially, compromised: even if the formation of a structure propaedeutic to the ordered and aligned shish-kebab morphology had begun, a morphology more or less different (depending on the extent of the stretch deviation), would form. If the stretch recovers its critical value, the accounting of the mechanical work will have to start again and, only if it reaches its critical value, W_c, a well-organized, aligned, and ordered structure will form. One would expect that the amount of deviation from a wel-organized and aligned structure would depend upon the amount of deviation from the criterion.

4. Stretch and Mechanical Work Distributions

The distributions of both the molecular stretch, Δ, and the integral of the mechanical work, W, both at the end of the filling step, and at the end of the process, are reported in Figure 3 for some of the injection conditions considered in this paper; in particular, the following conditions are considered: Steel, Passive, 80-07, 80-6, 150-07 and 150-1. In addition, the corresponding optical micrographs of the final molding cross-section morphologies in position P2 are reported on the top of each plot. The optical micrographs give some information about the cross-section morphology distribution, but several aspects need deeper analysis to be identified. Certainly, the surface layer, appearing as a dark layer close the sample surface, when present, can be identified in the optical micrographs. In the literature, it was pointed out that this layer, unlike shear layer, is characterized by the presence of unaligned elements [31]. Furthermore, an area, often rather wide, with spherulitic character can be identified on the symmetry plane of the sample. More difficult to detect using optical microscopy is the shish-kebab morphology, especially if (and when) the oriented elements are very thin. Under some processing conditions, between the shish kebab and the spherulitic morphology, there is a transition region, where elongated elements, having thickness ranging from about 4 μm up to 6 μm, are aligned (although sometimes not perfectly) along the flow direction. Depending upon operating conditions, the thickness of the transition region can be very wide or it can even disappear; sometimes the morphology of this region can (especially where the elements are thick) be detected using optical microscopy. What is difficult to identify using optical microscopy are both the borders of this region, which, therefore, remain without a quantitative definition.

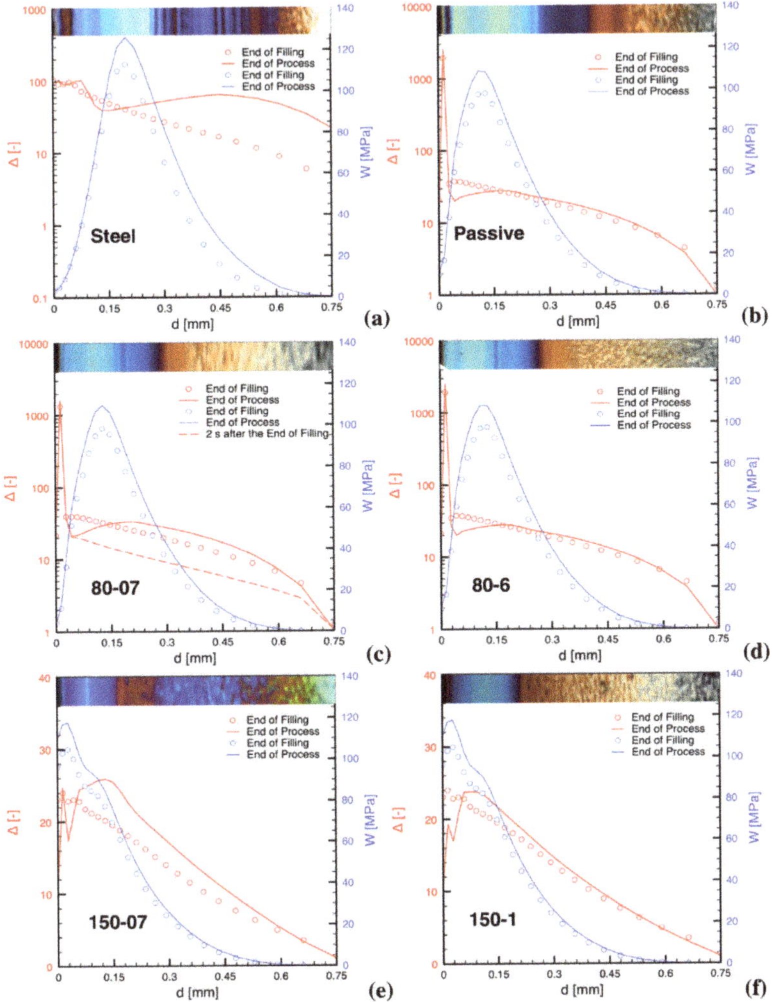

Figure 3. Molecular stretch distribution (namely Δ in red) and integral of mechanical work (namely W in blue) as calculated by the simulation for tests with different T_{cs} and heating times: Steel (**a**); Passive (**b**); 80 °C and 0.7 s (**c**); 80 °C and 6 s (**d**); 150 °C and 0.7 s (**e**); 150 °C and 1.3 s (**f**). Cross section distributions in position P2 at the end of filling (symbols) and at the end of the process (full lines). The Δ distribution 2 s after the end of filling (red dashed line) is also reported for the test 80-07 (**c**).

A comparison of Δ and W distributions for the Steel and Passive tests shows that main differences concerning the plot of Δ can be ascribed to the higher cooling rate of the Steel test with respect to the Passive test. Close to the sample surface, the material, due to the very high cooling rate that occurs during its first contact with the mold, cannot be deformed before solidifying, thus the peak showed in the Passive test vanishes. Moreover, the higher cooling rate in the Steel test enhances the effect on the stretch of the packing flow, thus the final stretch curve shows a more pronounced maximum and higher stretch values in the region close to the sample midplane than in the Passive test.

The boundary conditions difference between the Passive and the 80-07 tests is small, being limited to the cavity surface temperature only during the filling step; consequently, the corresponding two plots of W and Δ, reported in Figure 3, are very similar. By comparing the distribution curves of the same variable in the two tests, small differences can be identified; they are consistent with a slightly faster cooling for the Passive sample. The plot corresponding to the condition 80-6 is similar to 80-07, except for the final Δ distribution, which undergoes an additional relaxation with respect to the 80-07 test.

In order to point out that soon after the end of the filling step there is a stretch/stress relaxation, a fifth curve (the dashed red one) reporting the Δ distribution, calculated about two seconds after the end of the filling step, has been included in the plot related to the condition 80-07. Considerable relaxation, with respect to the Δ distribution, at the end of the filling step is clearly detectable. During such a relaxation, an additional stretch build up starts to set in as a consequence of the packing flow. The final stretch distributions also include eventual further relaxation determined by long cavity surface heating, if present. The packing steps and the eventual final relaxation determine the shape of the final stretch curve.

The tests 150-07 and 150-1, obtained with the highest heating electrical power, show values of the final stretch distribution much smaller than those evaluated in the other conditions. This is due to a packing flow poorly effective in stretching molecules, whereas the relaxation is faster. The stretch peaks at the surface are still present, although their values are orders of magnitude smaller than in the other tests due to the high temperature (150 °C) during both the filling and the packing steps.

The comparison of the stretch distributions obtained for the tests that adopt different heating powers confirms that the higher the heating power is the faster the stretch relaxation rate is. Indeed, the stretch relaxation rate with the power of 4 W/cm^2 (which allows reaching 80 °C) certainly appears slower than with the heating power of 9.5 W/cm^2 (which allows reaching 150 °C); in this case in fact, a large decrease of Δ occurs in a very short time (only about 0.6 s).

As far as the integral of the mechanical work is concerned, close to the cavity surface the contribution of the filling step appears dominant; the contribution of the packing step (although still minor) is larger in the region of the maximum; beyond the maximum the packing step gives a contribution which continuously decreases toward the sample midplane, and the decrease is faster when either the cavity surface temperature or the heating time increase.

It is worth pointing out that the plots reported in Figure 3 are not sufficient to clarify if, according to the criterion illustrated above, at the end of the process, in a given position on the cross section, the morphology will be fibrillar or not. Indeed, the criterion for fibrillar morphology specified above clarifies that the fraction of mechanical work accumulated before the stretch reaches its critical value Δ_c has to be disregarded in the account and, furthermore, if by effect of a relaxation the stretch decreases below its critical value, the accounted mechanical work disappears (has to be cancelled).

5. Final Cross Section Morphology as Determined by Molecular Stretch and Mechanical Work Evolutions

By essentially applying the criterion aforementioned with Δ_c and W_c equal to 7 and 10 MPa respectively, the morphologies of some moldings obtained using CIM with the same iPP adopted in this paper was correctly predicted [13]; therefore, these critical values will be the starting point of the following analysis.

Let us now combine the molecular stretch and the mechanical work evolutions for a case far from the surface, in particular, in position P2 and at 0.44 mm from the cavity surface for the test 80-07. For this case, the molecular stretch and the mechanical work evolutions are reported in Figure 4. The red line is the molecular stretch, whereas the full blue line is the total work accumulated from the beginning of the process. The horizontal red and blue dotted lines represent the critical value of the molecular stretch and the mechanical work respectively, namely 7 and 10 MPa. However, according to the criterion adopted in this paper, the mechanical work accumulated before the molecular stretch has reached its critical value has to be disregarded. We will denote as "effective mechanical work", W_e, the work done when molecular stretch is beyond its critical value. W_e is also plotted in Figure 4 as a dashed blue line. Since the molecular stretch at about 1.4-1.5 s, as a consequence of the relaxation taking place after filling, drops below its critical value of 7, according to the criterion adopted in this paper, starting from that time, the effective work already accumulated has to be disregarded. Thus, its accounting drops to zero, and, from that time, it starts to increase again at about 3 s when the molecular stretch again overcomes its critical value. However, the work accumulated during the packing step is not sufficient to reach an effective critical work, in this case 10 MPa. Consequently, the condition for fibrillar morphology will not be achieved, although the stretch, and the integral of the overall mechanical work at the end of the process both reach their critical values.

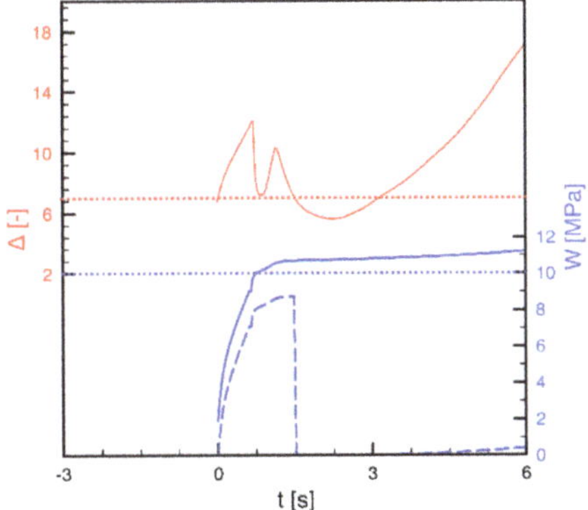

Figure 4. Predicted evolution of the molecular stretch (solid red line) and mechanical work (solid blue line) at 0.44 mm from the cavity surface for the test 80-07. The dashed blue line represents the effective mechanical work, W_e.

For some of the tests considered in this paper, the results, obtained applying the criterion clarified in Figure 4 (with Δ_c and W_c equal to 7 and 10 MPa respectively), for the thickness of the layer characterized by ordered and aligned elements are reported in Figure 5 as red triangles above corresponding optical micrographs.

The main effects of the operating conditions on the different characteristics of the cross section experimental morphology distribution of the same moldings considered in this paper have recently been investigated in detail using Atomic Force Microscopy (AFM) [20]. On the basis of that AFM analysis, a layer including both shish kebab and oriented and aligned morphological elements of thickness not larger than 4 μm was identified; its boundary is reported as a white mark in each of the optical micrographs of Figure 5. Moreover, on the optical micrograph of the Passive test an additional grey mark is reported to identify the surface layer.

Examples of different morphologies developed along the sample cross section are reported in Figure 6a–c. The AFM height images reported in Figure 6a–c were acquired for the 80-07 test at a distance of 0.25 mm, 0.38 mm (indicated by the white mark in the optical micrograph of Figure 5b), and 0.585 mm from the sample surface, respectively. Figure 6a shows that the blue layer in Figure 5b is a highly oriented layer characterized by tightly packed thin structures aligned along the flow direction; their thickness increases with the distance from the surface. Figure 6b shows that in correspondence of the white mark of Figure 5b there is a thickness transition (from about 4 μm to about 6 μm) of the oriented and aligned morphological structures. Figure 6c shows that on increasing the distance from the sample surface only un–oriented spherulitic structures can be detected.

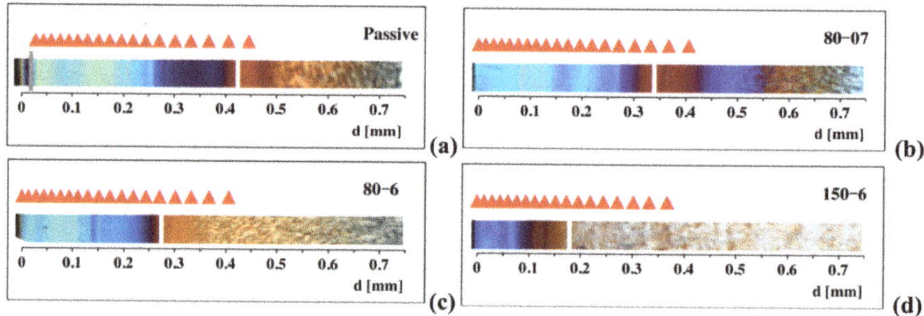

Figure 5. Comparison between experimental observation for the layers with oriented morphological elements of thickness not larger than 4 μm [20] (white marks) and predictions of the criterion obtained with Δ_c and W_c equal to 7 and 10 MPa respectively (red triangles).(**a**) Passive test; (**b**) 80-07 test; (**c**) 80-6 test; (**d**) 150-6 test.

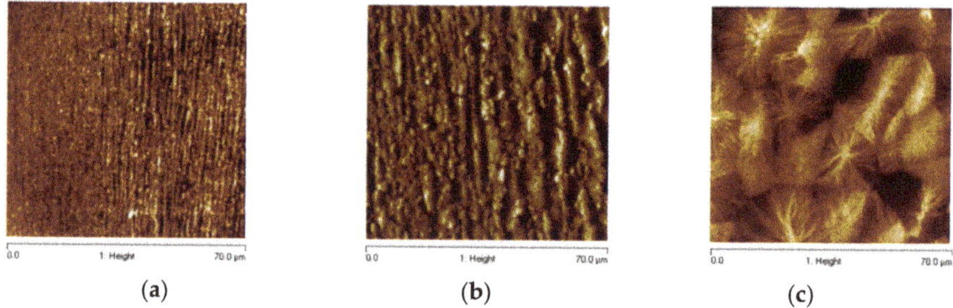

Figure 6. Atomic Force Microscopy (AFM) images over an area of 70 μm × 70 μm at 0.25 mm (**a**), at 0.38 mm (**b**), and 0.7 mm (**c**) from the sample surface for the test 80-07. At 0.25 mm, morphological elements aligned along the flow direction tightly packed can be observed. At 0.38 mm, the average diameter of the morphological elements aligned along the flow direction range from about 4 μm to about 6 μm. At 0.585 mm, spherulitic structures cover the whole investigated area.

The comparison between predictions of the criterion and experimental results for the positions and width of the layers presenting shish-kebab and elements ordered and oriented along the flow direction can be considered positive for the Passive test and the 80-07 test (Figure 5a,b respectively). Vice versa, when the heating time and the heating power increase the comparison cannot be considered satisfactory, as shown in Figure 5d for the test 150-6, for which the results of the criterion gave rise to a width more than double with respect to the experimental results.

On the other hand, experimental results reported by the Group of Sheffield University [10] indicated an increase of the critical work with the temperature for a low density polyethylene and a polypropylene-ethylene random copolymer. Furthermore, the results of an experimental analysis of the effect of shearing on the shish-kebab crystallization carried out at 140 °C using a Linkam cell on the same iPP adopted in this paper gave rise to a critical work of 20 MPa, rather than the 10 MPa adopted for the comparison reported in Figure 5. In addition, it has been shown [13] that the value of the critical work, W_c, determines not only the thickness of the surface layer (when present), but often it is also the controlling parameter for the thickness of the oriented layer (for instance, for the case 80-07 considered in Figure 4). One can thus expect that the increase of the mechanical work is consistent with the objective of obtaining predictions of thinner oriented layers at larger temperatures, as shown by the experimental results. In conclusion, the predictions for thickness and location of oriented layers of the injection molded samples considered in this paper were calculated again adopting critical stretch Δ_c equal to 7 (as for the comparisons reported in Figure 5), and a critical work W_c function of the temperature as shown in Figure 8, namely, equal to 10 MPa at low temperature, and linearly increasing for temperatures larger than 130 °C, considering that W_c was experimentally found to be 20 MPa at 140 °C (reported as a red dot in Figure 8).

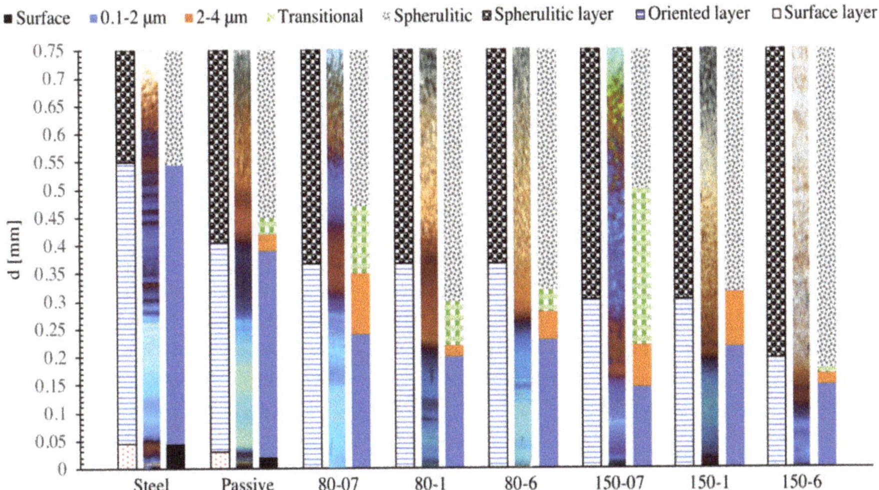

Figure 7. Comparison between experimental morphology distribution along the cross sections of the injection molded samples under the conditions listed in Table 1. Three bars are reported for each of the injection conditions, all of them refer to the morphology distribution in position P2: the central bar is the optical cross section micrograph; the bar on the left reports the predictions of the criterion adopted in this paper for the thicknesses of the surface layer, when present, of the oriented morphological layer and of the spherulitic layer; the bar on the right reports the width and characteristics of the morphological layers as detected using AFM analysis reported in [20]. Each color of the bar on the right corresponds to the thickness of the morphological elements as specified in the legend.

For each of the eight injection molded tests considered in this work, the predictions of the criterion (based on $\Delta_c = 7$ and W_c as shown in Figure 8) for the morphology along the cross section in position P2 are compared in Figure 7 with a synthesis of the corresponding experimental morphology distribution as detected by the AFM analysis reported in [20]. The corresponding polarized optical micrographs are reported in the figure. In particular, in the synthesis of the experimental morphology, layers of different colors, as indicated in the legend, correspond (after the surface layer which is black) to oriented morphological elements having thicknesses growing with the distance from the sample surface in the intervals 0.1–2 μm (blue) or 2–4 μm (orange) or, sometimes, a transitional layer toward a spherulitic morphology with elements of thickness ranging in the interval 4–6 μm (green); finally, a spherulitic layer was always present up to the sample midplane. Only the layer 0.1–2 μm and the spherulitic layer were found in all injection molding conditions, details of which are given below. Within the interval 2–4 μm, and especially within the interval 4–6 μm, when present, the morphology characteristics were found to gradually change their order and homogeneity. In particular, the order and the thickness of the elements were found to decrease, and the element thicknesses were found to increase with the distance from the sample surface including more and more cylindritic and sometimes small spherulitic structures.

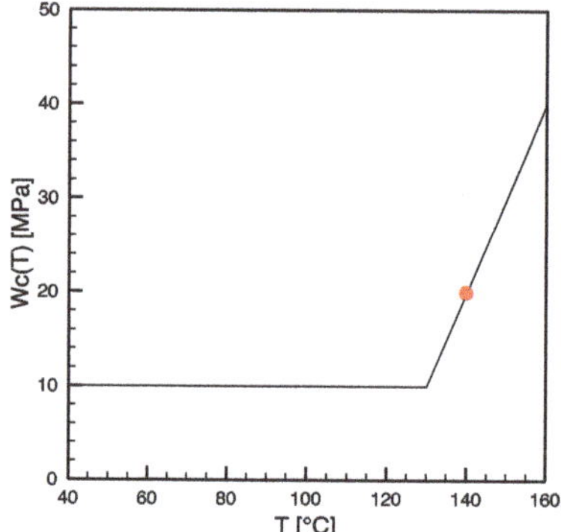

Figure 8. Critical work as function of temperature adopted for the results of simulations reported in Figure 7; the red dot represents the experimental results at 140 °C obtained using the iPP T30G adopted for the injection molding tests considered in this paper.

In most of the cases, the evolution from oriented elements to the spherulitic morphology was found to take place within a transitional layer of width of the order of only 70 µm. For the Steel test the evolution took place even and directly from the layer which includes aligned structures with thicknesses ranging between 0.1 µm–2 µm to the spherulitic layer. Only the case of 150-07 showed a much more extended (nearly one half of the sample thickness) transitional layer; this has to be related to the intensity of the packing flow, which, in turn, is determined by the amount and rate of cooling and crystallization after filling. Indeed, in the case 150-07, the packing flow is expected to have had the maximum intensity since during filling only a moderate cooling takes place, and crystallization has not started yet (except for small percentages at the surface). To a smaller extent, similar observations can be made in relation to the 80-07 sample, which is second as far as width of the transitional layer.

As far as the comparison, shown in Figure 7, between experimental results and the results of the criterion for oriented morphology distribution on the molding cross section, it appears that, with the choices described above for the critical stretch and for the critical work, the surface layer widths for the Steel and Passive tests are correctly predicted, and the absence of the surface layer in all the other tests is also properly described. Furthermore, simulations reproduce, within a reasonable approximation, the additional widths from the surface layer (when present) up to the positions within which oriented and aligned morphological elements of average diameter up to 4 µm were identified [20]. The discrepancy is smaller than 15% (i.e., smaller than 40 µm) for all cases, except for the cases 80-1, 80-6, and 150-07, for which the discrepancies are 30% for the first one and 20% for the other two conditions. It has to be considered that the approximations of the comparison reported in Figure 7 are the result not only of the approximations of the application of criterion for the formation of ordered morphological structures aligned along the flow direction, but they also include all approximations within the model adopted for the calculation of the molecular stretch and the mechanical work evolutions. For instance, it would be sufficient that one of the parameters (as, e.g., the ones determining the dependence of the viscosity or of the relaxation time upon pressure or temperature) is slightly either over or under estimated to determine a significant effect on the evolutions of both the molecular stretch and the mechanical work distributions and consequently on the final cross-section morphology. Moreover, it must be pointed out that the predictions of both the molecular stretch and the mechanical work evolutions and consequently the width of the oriented shear layer observed in the final cross section of the samples are sensitive to the distance from the gate of the cavity. A difference of some millimeters (a few percent of the cavity length) in the position along the flow direction could significantly reduce the discrepancy between predictions and experimental observations. Finally, it has to be pointed out that, as mentioned above, the criterion for the crystallization into fibrillar morphology, in the format proposed in this paper, is still schematic;

namely, it is an on/off model. Indeed, it does not consider intermediate morphologies when only a fraction (possibly high) of the critical mechanical work is achieved or the critical work is achieved while the stretch was not continuously above its critical value. The transition zones could be related to situations of this kind.

A further step toward a deep understanding of the morphology distribution along the sample cross section could be achieved by considering the stretch distribution along the oriented layer. In the literature [7,32–34] it was affirmed that the density of the shish increases with the increase of the shear (and thus of the stretch) experienced by the polymer. Thus, the critical stretch would be related (when and where the critical mechanical work is reached) to the minimum of shish density within the oriented layer. Such a minimum determines the minimum density of the oriented structures and therefore the maximum thickness (by effect of the lateral growth allowable) that the oriented structures can achieve. A change of the critical stretch would also induce a modification of the prediction for the width of the layer characterized by oriented structures. Furthermore, being the criterion for the fiber formation respected inside the oriented layer, the excess of stretch with respect to the critical value can be considered. This excess assumes different values along the layer, and it is non–negative. The local values of the excess of stretch should be coherent with the thickness distribution of the oriented structures and they could identify the local thickness of the oriented structures once the correlation between stretch and density of oriented and aligned structures is identified. This should be taken as the next goal.

6. Conclusions

In this paper, a criterion for the polymer crystallization into highly oriented morphological elements aligned along the flow direction was applied to the analysis of a set of isotactic polypropylene injection molded samples obtained under very different thermal boundary conditions. A formulation of the criterion in terms of critical values of both a molecular stretch parameter and the mechanical work was adopted, starting from a critical amount of mechanical work carried out when the stretch is above its critical value. In order to also take into account the fact that polymer processing operations are performed under unsteady conditions (and the injection molding process is typically unsteady), the previous formulation of the criterion was modified with the following specification: once the critical stretch is reached, if it relaxes below its critical value, the accounting of the mechanical work has to be canceled and started again.

In order to apply the criterion to the iPP injection molding samples, the evolutions of both the stretch and the mechanical work distributions were evaluated using process simulation. Taking the critical mechanical work function of temperature, as indicated by results of literature isothermal experimentations, gave rise to a satisfactory description of the morphology developed along the molding cross sections for the tests considered in this paper. In particular, widths and positions of surface (when present) and oriented layers were correctly described confirming that predictions exhibited essentially the same dependence upon the process conditions (heating power and heating time) shown by the experimental observations.

The criterion, in its actual on/off formulation, does not include the existence of intermediate transitional layers between the highly oriented and the spherulitic layers. In most of the cases, these transitions took place within very thin zones; however, under special processing conditions, the transitional zone can be wide. The possibility of transitional layers could be included in the criterion by releasing its on/off approach; namely, by considering the cases when either the mechanical work or the molecular stretch do not completely satisfy their critical conditions.

Author Contributions: Giuseppe Titomanlio and Sara Liparoti conceived and designed the experiments; Sara Liparoti performed the molding tests; Sara Liparoti, Giuseppe Titomanlio, Roberto Pantani, and Vito Speranza analyzed the acquired data; Vito Speranza and Sara Liparoti performed and analyzed the AFM tests; Vito Speranza software developing and numerical simulation; Vito Speranza: writing—original draft preparation and editing; Sara Liparoti, Giuseppe Titomanlio, Roberto Pantani, and Vito Speranza: writing—review.

Funding: This research received no external funding.

Conflicts of Interest: The authors declare no conflicts of interest.

References

1. Housmans, J.-W.; Gahleitner, M.; Peters, G.W.M.; Meijer, H.E.H. Structure–property relations in molded, nucleated isotactic polypropylene. *Polymer (Guildf)* **2009**, *50*, 2304–2319. [CrossRef]
2. Steenbakkers, R.J.A.; Peters, G.W.M. A stretch-based model for flow-enhanced nucleation of polymer melts. *J. Rheol. (N. Y. N. Y)* **2011**, *55*, 401–433. [CrossRef]
3. Gahleitner, M.; Wolfschwenger, J.; Bachner, C.; Bernreitner, K.; Neißl, W. Crystallinity and mechanical properties of PP-homopolymers as influenced by molecular structure and nucleation. *J. Appl. Polym. Sci.* **1996**, *61*, 649–657. [CrossRef]
4. Hashimoto, T.; Murase, H.; Ohta, Y. A New Scenario of Flow-Induced Shish-Kebab Formation in Entangled Polymer Solutions. *Macromolecules* **2010**, *43*, 6542–6548. [CrossRef]
5. Pogodina, N.V.; Lavrenko, V.P.; Srinivas, S.; Winter, H.H. Rheology and structure of isotactic polypropylene near the gel point: Quiescent and shear-induced crystallization. *Polymer (Guildf)* **2001**, *42*, 9031–9043. [CrossRef]
6. D'Haese, M.; Mykhaylyk, O.O.; Van Puyvelde, P. On the Onset of Oriented Structures in Flow-Induced Crystallization of Polymers: A Comparison of Experimental Techniques. *Macromolecules* **2011**, *44*, 1783–1787. [CrossRef]
7. Hsiao, B.S.; Yang, L.; Somani, R.H.; Avila-Orta, C.A.; Zhu, L. Unexpected Shish-Kebab Structure in a Sheared Polyethylene Melt. *Phys. Rev. Lett.* **2005**, *94*, 117802. [CrossRef]
8. Somani, R.H.; Yang, L.; Zhu, L.; Hsiao, B.S. Flow-induced shish-kebab precursor structures in entangled polymer melts. *Polymer (Guildf)* **2005**, *46*, 8587–8623. [CrossRef]
9. Mykhaylyk, O.O.; Chambon, P.; Graham, R.S.; Fairclough, J.P.A.; Olmsted, P.D.; Ryan, A.J. The Specific Work of Flow as a Criterion for Orientation in Polymer Crystallization. *Macromolecules* **2008**, *41*, 1901–1904. [CrossRef]
10. Mykhaylyk, O.O.; Chambon, P.; Impradice, C.; Fairclough, J.P.A.; Terrill, N.J.; Ryan, A.J. Control of Structural Morphology in Shear-Induced Crystallization of Polymers. *Macromolecules* **2010**, *43*, 2389–2405. [CrossRef]
11. An, Y.; Gu, L.; Wang, Y.; Li, Y.-M.; Yang, W.; Xie, B.-H.; Yang, M.-B. Morphologies of injection molded isotactic polypropylene/ultra high molecular weight polyethylene blends. *Mater. Des.* **2012**, *35*, 633–639. [CrossRef]
12. Yang, H.-R.; Lei, J.; Li, L.; Fu, Q.; Li, Z.-M. Formation of Interlinked Shish-Kebabs in Injection-Molded Polyethylene under the Coexistence of Lightly Cross-Linked Chain Network and Oscillation Shear Flow. *Macromolecules* **2012**, *45*, 6600–6610. [CrossRef]
13. Pantani, R.; Speranza, V.; Titomanlio, G. A Criterion for the Formation of Fibrillar Layers in Injection Molded Parts. *Int. Polym. Process.* **2018**, *33*, 355–362. [CrossRef]
14. Pantani, R.; Speranza, V.; Titomanlio, G. Effect of flow-induced crystallization on the distribution of spherulite dimensions along cross section of injection molded parts. *Eur. Polym. J.* **2017**, *97*, 220–229. [CrossRef]
15. Pantani, R.; Speranza, V.; Titomanlio, G. Thirty Years of Modeling of Injection Molding. A Brief Review of the Contribution of UNISA Code to the Field. *Int. Polym. Process.* **2016**, *31*, 655–663. [CrossRef]
16. Pantani, R.; Coccorullo, I.; Volpe, V.; Titomanlio, G. Shear-Induced Nucleation and Growth in Isotactic Polypropylene. *Macromolecules* **2010**, *43*, 9030–9038. [CrossRef]
17. Liparoti, S.; Sorrentino, A.; Titomanlio, G. Temperature and pressure evolution in fast heat cycle injection molding. *Mater. Manuf. Process.* **2018**, 1–9. [CrossRef]
18. Liparoti, S.; Landi, G.; Sorrentino, A.; Speranza, V.; Cakmak, M.; Neitzert, H.C. Flexible Poly(Amide-Imide)-Carbon Black Based Microheater with High-Temperature Capability and an Extremely Low Temperature Coefficient. *Adv. Electron. Mater.* **2016**, *2*, 1600126. [CrossRef]
19. Liparoti, S.; Sorrentino, A.; Titomanlio, G. Fast cavity surface temperature evolution in injection molding: Control of cooling stage and final morphology analysis. *RSC Adv.* **2016**, *6*, 99274–99281. [CrossRef]
20. Speranza, V.; Liparoti, S.; Pantani, R.; Titomanlio, G. Hierarchical structure of iPP during injection molding process with fast mold temperature evolution. *Materials* **2018**, *12*, 424. [CrossRef]
21. Pantani, R.; Speranza, V.; Titomanlio, G. Evolution of iPP Relaxation Spectrum during Crystallization. *Macromol. THEORY SIMULATIONS* **2014**, *23*, 300–306. [CrossRef]

22. Pantani, R.; Coccorullo, I.; Speranza, V.; Titomanlio, G. Modeling of morphology evolution in the injection molding process of thermoplastic polymers. *Prog. Polym. Sci.* **2005**, *30*, 1185–1222. [CrossRef]

23. Marrucci, G.; Ianniruberto, G. Constitutive equations for polymeric solutions close to the overlap concentration. *Chem. Eng. Sci.* **2001**, *56*, 5539–5544. [CrossRef]

24. Nakamura, K.; Katayama, K.; Amano, T. Some aspects of nonisothermal crystallization of polymers. II. Consideration of the isokinetic condition. *J. Appl. Polym. Sci.* **1973**, *17*, 1031–1041. [CrossRef]

25. Hoffman, J.D.; Davis, G.T.; Lauritzen, J.I. The Rate of Crystallization of Linear Polymers with Chain Folding. In *Treatise on Solid State Chemistry*; Hannay, N., Ed.; Springer: Boston, MA, USA, 1976; pp. 497–614.

26. Pantani, R.; De Santis, F.; Speranza, V.; Titomanlio, G. Modelling morphology evolution during solidification of IPP in processing conditions. In Proceedings of the AIP Conference Proceedings (PPS-29), Nuremberg, Germany, 15–19 July 2013; Volume 1593, pp. 636–640.

27. White, H.M.; Bassett, D.C. On row structures, secondary nucleation and continuity in α-polypropylene. *Polymer (Guildf)* **1998**, *39*, 3211–3219. [CrossRef]

28. Janeschitz-Kriegl, H.; Ratajski, E. Kinetics of polymer crystallization under processing conditions: Transformation of dormant nuclei by the action of flow. *Polymer (Guildf)* **2005**, *46*, 3856–3870. [CrossRef]

29. Jerschow, P.; Janeschitz-Kriegl, H.; Jerschow, R.; Janeschitz-Kriegl, H. On the development of oblong particles as precursors for polymer crystallization from shear flow: Origin of the so.called fine grained layers. *Rheol Acta* **1996**, *35*, 127–133. [CrossRef]

30. Janeschitz-Kriegl, H.; Ratajski, E.; Stadlbauer, M. Flow as an effective promotor of nucleation in polymer melts: A quantitative evaluation. *Rheol. Acta* **2003**, *42*, 355–364. [CrossRef]

31. Liparoti, S.; Sorrentino, A.; Guzman, G.; Cakmak, M.; Titomanlio, G. Fast mold surface temperature evolution: Relevance of asymmetric surface heating for morphology of iPP molded samples. *RSC Adv.* **2015**, *5*, 36434–36448. [CrossRef]

32. Fernandez-Ballester, L.; Thurman, D.W.; Zhou, W.; Kornfield, J.A. Effect of Long Chains on the Threshold Stresses for Flow-Induced Crystallization in iPP: Shish Kebabs vs Sausages. *Macromolecules* **2012**, *45*, 6557–6570. [CrossRef]

33. Roozemond, P.C.; Peters, G.W.M. Flow-enhanced nucleation of poly(1-butene): Model application to short-term and continuous shear and extensional flow. *J. Rheol. (N. Y. N. Y)* **2013**, *57*, 1633–1653. [CrossRef]

34. Wang, Z.; Ma, Z.; Li, L. Flow-Induced Crystallization of Polymers: Molecular and Thermodynamic Considerations. *Macromolecules* **2016**, *49*, 1505–1517. [CrossRef]

Article

Replication of Micro- and Nanofeatures in Injection Molding of Two PLA Grades with Rapid Surface-Temperature Modulation

Sara Liparoti, Vito Speranza * and Roberto Pantani

Department of Industrial Engineering, University of Salerno–via Giovanni Paolo II, 132, 84084 Fisciano (SA), Italy; sliparoti@unisa.it (S.L.); rpantani@unisa.it (R.P.)
* Correspondence: vsperanza@unisa.it; Tel.: +39-089-964145

Received: 22 July 2018; Accepted: 9 August 2018; Published: 15 August 2018

Abstract: The production by injection molding of polymeric components having micro- and nanometrical surfaces is a complex task. Generally, the accurate replication of micro- and nanometrical features on the polymeric surface during the injection-molding process is prevented by of the low mold temperature adopted to reduce cooling time. In this work, we adopt a system that allows fast heating of the cavity surface during the time the melt reaches the cavity, and fast cooling after heater deactivation. A nickel insert with micro- and nanofeatures in relief is located on the cavity surface. Replication accuracy is analyzed by Atomic Force Microscopy under different injection-molding conditions. Two grades of polylactic acid with different viscosity have been adopted. The results indicate that the higher the cavity surface temperature is, the higher the replication accuracy is. The viscosity has a significant effect only in the replication of the microfeatures, whereas its effect results are negligible in the replication of nanofeatures, thus suggesting that the interfacial phenomena are more important for replication at a nanometric scale. The evolution of the crystallinity degree on the surface also results in a key factor on the replication of nanofeatures.

Keywords: replication; microfeature; nanofeature; injection molding; polylactic acid; mold temperature

---

1. Introduction

Nanoscale technologies, advanced microfabrication, and postprocessing modification techniques support the realization of a wide range of two- and three-dimensional (2D and 3D) objects that can be adopted in several fields, from electronic to biomedical [1–5]. The micro- and nanostructured surfaces can improve cell adhesion during cell growth in tissue engineering [6,7]. Furthermore, they open the possibility to produce surfaces with super-hydrophobic characteristics without additional coating processes [8,9].

The techniques adopted for the production of micro- and nanostructured surfaces can be briefly summarized in two categories: bottom–up and top–down approaches. Bottom–up approaches are related to the construction of micro- and nanostructured materials and devices by the self-assembly of atoms or molecules. The most diffused bottom–up techniques are: the atomic layer deposition [10], sol-gel processes [11], and molecular self-assembling [12]. The top–down approach corresponds to the production of micro- and nanoscaled structures starting from larger dimensions and reducing them to the required values [13]. The most common top–down approaches are lithography-based techniques such as soft lithography, e-beam lithography, and nanoimprint lithography [14–17]. Both the mentioned approaches require high cost, and long processing times. Furthermore, there is only limited control over surface properties. This makes the application of these techniques difficult in the production of large-area micro- and nanostructured surfaces.

Replication methods can represent an excellent alternative since they couple the high dimensional accuracy of lithography, adopted for the production of the masters to be replicated, with the short processing time of techniques such as embossing and injection molding [18–21].

The injection-molding process has been adopted for the production of biocompatible and biodegradable materials; thus, this process could also be promising for the production of micro- and nanostructured surfaces made of biomaterials [8,22–25]. The injection-molding process can be divided in three main stages: The filling, during which the polymer in the molten state fills the cavity having the master to be replicated on its surface; the packing, during which additional material is forced into the cavity to compensate for the shrinkage; and the cooling. The selected mold temperature is generally low, smaller than the glass-transition temperatures for most polymers. The use of low mold temperature reduces the cooling time that represents the most processing time. However, the high cooling rates that the polymer experiences in such conditions induces the formation of a frozen layer at the polymer–mold interface that prevents the accurate replication of the micro- and nanostructures present on the master. To overcome this limitation, many authors have proposed to couple the injection-molding process with a system that allows the rapid control of mold temperature, to carry out the filling and packing with high temperature, and to carry out the cooling with low temperatures. Among these systems, induction heating [26], proximity heating [27], and infrared heating [28] have been proposed in the literature. These techniques suffer from high additional tool costs, and the processing time is significantly longer than the processing time of the conventional injection-molding process. The electrical heating of the cavity surface is undoubtedly less expensive and requires small additional tool costs [29–33]. The electrical heating of the cavity surface was efficiently applied to the injection-molding process to obtain micro- and nanostructures surfaces of polypropylene [34].

In this paper, the injection-molding process, coupled with a system that allows fast evolution of the temperature on the cavity surface, has been adopted to produce micro- and nanostructures objects made of polylactic acid (PLA), a biodegradable and biocompatible polymer. The temperature increase is achieved by the ohmic heating of thin carbon black loaded poly (imide-amide) film. The injection-molding sample has been characterized by morphological analyses and by X-ray. Two different grades of PLA have been adopted to analyze the effect of viscosity on replication ability.

2. Materials and Methods

Two grades of commercial PLA, produced by NatureWorks (Minnetonka, MN, USA), have been adopted in this work. PLA with the trade name of 4032D is characterized by about 2% of D-enantiomer, a maximum crystallinity of 45%, a molecular weight M_w = 210 kg/mole, and polydispersity index of 1.75. PLA with the trade name of 3251D is characterized by about 1% D-enantiomer content, a molecular weight M_w = 90 kg/mole, and a polydispersity index of 1.62. The rheology of both PLA grades is reported elsewhere [35].

A Negri Bossi injection-molding machine (mod. 70ton, Cologno Monzese, Italy) has been adopted to obtain molded samples of two different grades of PLA. The injection molding conditions are: $2.8 \text{ cm}^3 \cdot \text{s}^{-1}$ average flow rate, 200 °C melt temperature, 8 s holding time, and 63 MPa or 30 MPa as holding pressures. The mold is equipped with 5 pressure transducers located along the flow path, from the nozzle (position P0) to the end of the cavity (P1 in the runner, P2-P3-P4 inside the cavity). A thin gate is adopted upstream a rectangular cavity. A detailed description of the cavity and the positions of pressure transducers along the flow path are reported elsewhere [34].

The cavity surface temperature evolution has been obtained by a heating device layered below the cavity surface. The heating device is composed of a conductive layer of carbon black loaded poly (amide-imide), having 70 μm thickness. The conductive layer is electrically insulated from the mold by Kapton® layers, having thickness of 20 μm. An additional Kapton® insulating layer 140 μm thickness is located between the mold and the heater to reduce heat loss through the mold. A detailed description of the heater is reported elsewhere [36]. A steel layer of 400 μm thickness covers the heating device and protects it from the melt. Heater activation starts at the mold closing time; this time corresponds to 4 s before the melt came in contact with the cavity surface in position P2. The time that the melt reaches position P2 corresponds to t = 0 s for all pressure and temperature evolutions.

A nickel shim, a strip of 400 μm thickness, with micro- and nanocrosses in relief, is located just downstream from the gate in position P2, and replaced a part of the steel layer constituting the cavity surface. Micro- and nanocrosses were produced following a procedure reported elsewhere [37]. Each wing of the microfeatures is 5 μm in height and 20 μm in width. Each wing of the nanofeatures is 60 nm height and 500 nm width.

Table 1 summarizes all the operative conditions. The name of each test contains the constant temperature reached on the cavity surface (T_{level}), when the heater is supplied with the correct electrical power, the time that the heater is active, the adopted holding pressure, and the PLA grade (A for PLA3251D and B for PLA4032D).

Table 1. Operating conditions adopted for the injection molding experiments with the heating device used to tune the temperature on the cavity surface (T_{level} is the temperature reached on the cavity surface thanks to the activations of the heating device).

Test Name	Polylactic Acid (PLA) Grade	Holding Pressure [MPa]	Electrical Power [W/cm^2]	T_{level} [°C]	Heating Time [s]
Passive-A	3251D	63	0	30	0
Passive-B	4032D	63	0	30	0
100-1-30-A	3251D	30	5	100	1
100-8-30-A	3251D	30	5	100	8
100-13-30-A	3251D	30	5	100	13
50-1-63-A	3251D	63	2	50	1
50-8-63-A	3251D	63	2	50	8
50-13-63-A	3251D	63	2	50	13
100-1-63-A	3251D	63	5	100	1
100-8-63-A	3251D	63	5	100	8
100-13-63-A	3251D	63	5	100	13
150-1-63-A	3251D	63	10	150	1
150-8-63-A	3251D	63	10	150	8
150-13-63-A	3251D	63	10	150	13
50-1-63-B	4032D	63	2	50	1
50-8-63-B	4032D	63	2	50	8
50-13-63-B	4032D	63	2	50	13
100-1-63-B	4032D	63	5	100	1
100-8-63-B	4032D	63	5	100	8
100-13-63-B	4032D	63	5	100	13
150-1-63-B	4032D	63	10	150	1
150-8-63-B	4032D	63	10	150	8
150-13-63-B	4032D	63	10	150	13

Atomic Force Microscopy (AFM) allows analyzing the replication accuracy of the samples. The acquisitions have been carried out in the air at room temperature with Dimension 3100 coupled with a Bruker Nanoscope V controller, operating in contact mode. Commercial probe tips have been accurately selected to reduce the discrepancy between the real shape and the acquired one. The selected tip has a spring constant of 1–5 N/m, a radius of 8–12 nm, and height of 10–15 μm. Asymmetric contact angles characterize the tip: the front and the back angles are 25° and 15°, respectively. Five samples for each test were analyzed by AFM and, to reduce distortions, the samples have been rotated, and the acquisitions are referred to a distance from the cross center of 15 μm.

The height AFM patterns have been firstly derived, and then the absolute values of the first derivative have been fitted with the following Equation:

$$y = y_0 + A \frac{1}{1+e^{-\frac{x-x_c+w_1/2}{w_2}}} \left(1 - \frac{1}{1+e^{-\frac{x-x_c+w_1/2}{w_3}}} \right) \tag{1}$$

where y_0 is the offset, x_c is the center, A is the amplitude, w_1 is the width at half maximum, and w_2 and w_3 take into account the asymmetry of the curve. The Full Width calculated at the Half Maximum of the fit curve (FWHM) is an index of the channel width that is also adopted as replication accuracy index for the microfeatures [38].

The deviation of the height of the replicated nanofeatures, namely h_m, with respect to the expected value, namely h_{Ni}, which is the height of the nanofeatures located on the nickel shim, allows quantifying the replication accuracy for the nanofeatures:

$$H = \frac{h_{Ni} - h_m}{h_{Ni}} 100 \tag{2}$$

Injection-molded samples have been analyzed, in the same position where the nickel shim is located, by X-ray Diffractometry (XRD) in reflection mode by an Advance D8 Bruker (Billerica, MA, USA) diffractometer (with a continuous scan attachment and a proportional counter) with Ni-filtered Cu-K_α radiation. The full spectrum is considered as a superposition of number of reflections (5 reflections were considered: 2θ = 12.5, 14.7, 16.7, 19.1, and 22.5); each reflection being described by a combination of a Lorentzian function and a Gaussian function [39]. This analysis allows the assessment of the crystallinity degree at the surface of the sample. The same analysis has been repeated after removing 100 µm of polymer at the sample skin. To this purpose, a lapping process is performed by a Mecapol 2B lapping machine (Presi, Eybens, France) using a Mecaprex abrasive disc P1000 (Presi, Eybens, France). The lapping has been performed with a lapping pressure of 50 kPa. The repetition of the X-ray analysis provides thus the value of crystallinity degree at a distance of about 100 µm from the sample surface.

3. Results

3.1. Temperature and Pressure Evolutions

Pressure and temperature evolutions give an indication of the thermomechanical history experienced by the polymer during the process. The thermomechanical history influences the replication accuracy. Figure 1 shows the temperature evolutions recorded during the experiments on PLA4032D with different electrical powers, which means different T_{level} (see Table 1).

Temperature on the cavity surface increases from the bulk temperature of the mold, 30 °C, to the temperature T_{level} before the melt reaches the cavity. The contact of the melt with the cavity induces an additional and sharp temperature increase at t = 0 s; after that, the melt starts to cool down to T_{level}, and this value is kept constant as long as the heater is active. At heater deactivation, the complete cooling of the polymer takes place. Obviously, when the heater is not activated, the polymer starts to cool down soon after the first contact with the cavity surface. The temperature evolutions recorded during the experiments on the PLA3251D show similar trends.

Figure 2 shows the pressure evolutions along the flow path, recorded during the experiments with two PLA grades and with different T_{level}.

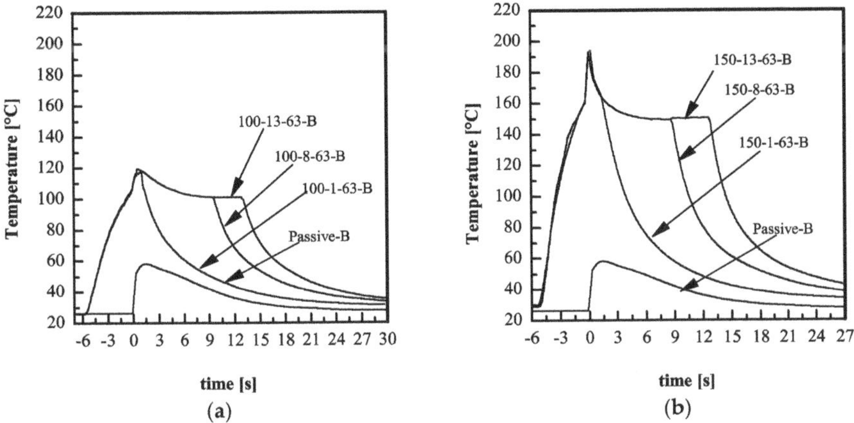

Figure 1. Temperature evolutions recorded during the experiments performed on PLA4032D (**a**) with T_{level} = 100 °C, and (**b**) T_{level} = 150 °C; the temperature evolution of the test Passive-B is also reported for comparison.

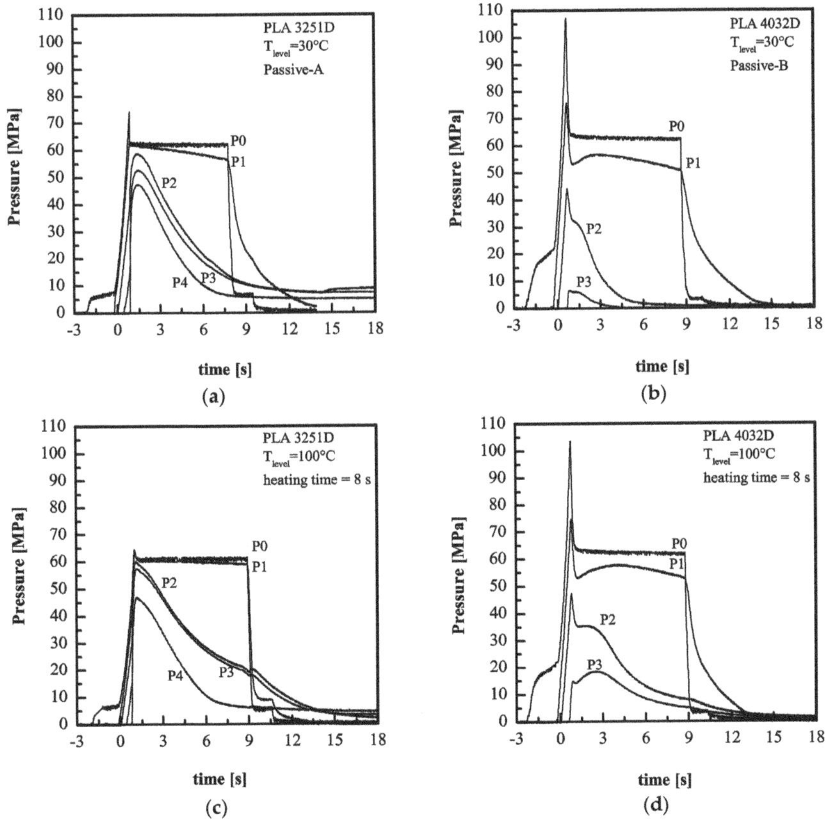

Figure 2. *Cont.*

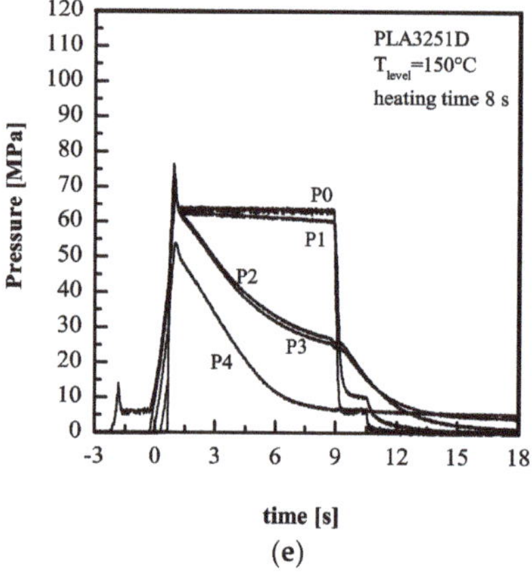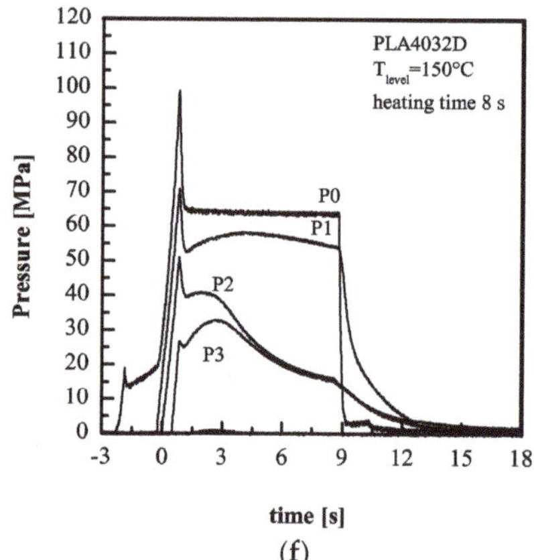

Figure 2. Pressure evolutions in different positions along the flow path recorded during the experiments (**a**) Passive-A, (**b**) Passive-B, (**c**) 100-8-63-A, (**d**) 100-8-63-B, (**e**) 150-8-63-A, and (**f**) 150-8-63-B. In all the figures, t = 0 s corresponds to the time that the melt came in contact with the cavity surface in the position P2.

The comparison between the temperature evolutions of Figure 1a,b with the pressure evolutions of Figure 2c,f, demonstrates that the temperature on the cavity surface reaches the value T_{level} during the time the melt fills the sprue, and the runner.

The pressure evolutions (see Figure 2) allow identifying the main steps of the injection-molding process: Filling, packing, and cooling. During the filling step, pressure increases in all the positions along the flow path. The pressure peak in position P0 (namely in the nozzle) is due to the filling end and its value decreases with the polymer viscosity; thus, the pressure peak is smaller in the Passive-A case than in the Passive-B case. After the filling, the pressure in position P0 decreases down to the value selected for the packing stage, 63 MPa in the cases shown in the Figure 2. During the packing, additional flow is forced into the cavity to compensate for the shrinkage due to the polymer solidification. This process ends when the gate is sealed. Gate solidification can be detected when the pressure evolution in position P2 presents an inflection point, changing the concavity from downward to upward (Figure 2) [40]. It takes place at about 3 s for all the materials and molding conditions; after this time, the solidification of the polymer takes place along the whole thickness, and pressures in the cavity decrease down to zero for the passive cases (Figure 2a,b). When the heater is active, the cooling takes place in two steps [30] (see Figure 2c–f). The pressure reached after the gate sealing, during the time the heater was active, increased with T_{level}. The main difference among the pressure evolutions during the packing for the two PLA grades was related to the pressure drop between P0 and P3, which was significantly higher for PLA4032D than for PLA3251D due to the high viscosity.

A further consideration concerns the pressure evolutions in positions P2 and P3. When PLA3251D was adopted, the pressure in position P3 overlapped the pressure evolutions in position P2; when PLA4032D was adopted, the overlap was not present. The reason for this behavior was due to the polymer viscosity. When PLA3251D was adopted, between positions P2 and P3 the polymer was still in the molten state and, being confined between two sealed parts, the pressure drop between P2 and P3 became negligible [34]. PLA4032D has a higher viscosity and the pressure drop between positions P2 and P3 was still significant; the pressure drop tended to disappear with the increase of T_{level}. Additionally, the high viscosity prevented the complete filling of the cavity, and pressure in position P4 was not recorded.

3.2. Replication of Microfeatures

Figure 3 shows the micro-feature acquired by AFM on the nickel shim and the micro-features acquired on the injection molded samples, with T_{level} = 30 °C (i.e., Passive-A) and with T_{level} = 150 °C kept for 13 s heating time.

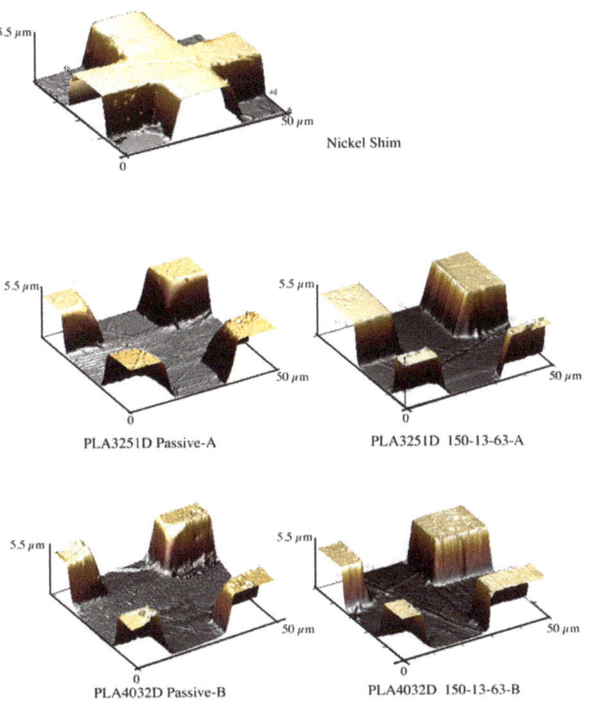

Figure 3. AFM height images of the micro-features on the nickel shim, injection molded samples of PLA3251D and PLA4032D obtained in different conditions of cavity surface temperature.

The samples Passive-A and Passive-B do not show an accurate replication of the micro-feature; indeed, no sharp edge can be detected. The micro-features acquired on the samples 150-13-63-A and 150-13-63-B, obtained with T_{level} = 150 °C, and 13 s heating time, show an accurate replication, as confirmed by the sharp edges. PLA3251D and PLA4032D show a similar dependence of the replicability on the T_{level}: Accurate replication is achieved only with high T_{level}. For both the PLA grades, the heights of the micro–features are close to the height of the nickel shim with a small deviation, i.e., ±0.05 µm. Figure 4 shows the replication accuracy measured on different injection molded samples in term of FWHM. The values of the FWHM related to the micro-feature on the nickel shim is also reported for comparison.

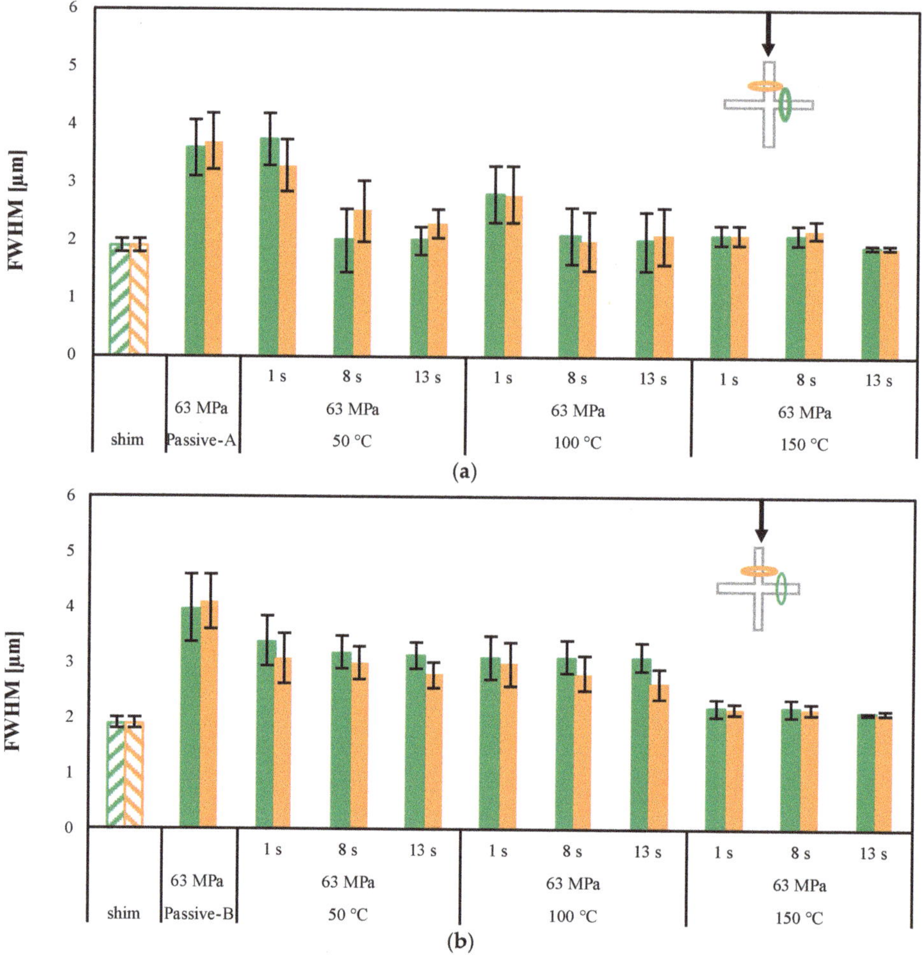

Figure 4. FWHM index of micro-feature on PLA3251D (**a**) and PLA4032D (**b**) samples obtained as indicated in Table 1. A sketch of the area where the AFM patterns have been acquired is also reported.

For the passive samples, Passive-A and Passive-B, the FWHM is far from the optimum value, 1.97 μm measured on the nickel shim. Figure 4 suggests that the FWHM depends on both the T_{level}, and the heating time: the FWHM decreases, namely the replication accuracy increases, as the T_{level} and the heating time increase. When the PLA3251D is adopted, the dependence of FWHM on the cavity surface temperature evolution is significant: with T_{level} = 50 °C and 100 °C, a significant reduction of FWHM can be observed increasing the heating time from 1 s to 8 s. With T_{level} = 150 °C the FWHMs approach values close to that one measured on the nickel shim already with 1 s heating time, and additional decrease with heating time cannot be observed. When PLA4032D is adopted (Figure 4b), the values of FWHM show a smaller dependence on the heating times in all the explored T_{level} ranges. For both the considered materials, the alignment between the wings of the micro-features and the direction of flow does not influence the FWHM values. The micro-features replicated on the PLA4032D show values of FWHM higher than those calculated on PLA3251D. Figure 5 summarizes the differences in replication accuracy, in the direction of the flow front, among PLA3251D and PLA4032D.

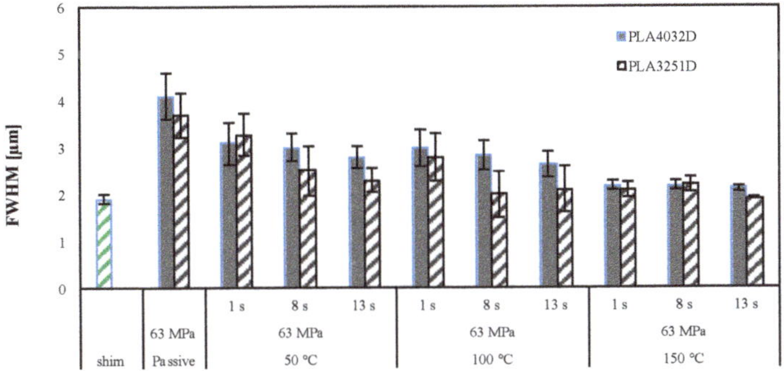

Figure 5. Comparison between the FWHM index, measured in the direction of the flow front, of micro-feature on PLA4032D and FWHM index of micro-features on PLA3251D samples obtained as indicated in Table 1.

Additional experiments have been carried out also with smaller holding pressure, 30 MPa, to highlight the effect of holding pressure on the replication accuracy. Figure 6a shows the results in term of FWHM for the experiments carried out on PLA3251D sample with T_{level} = 100 °C. Figure 6b shows the pressure evolutions recorded in position P2, where the nickel shim is located, for the experiments 100-1-63-A and 100-1-30-A, carried out with different holding pressure.

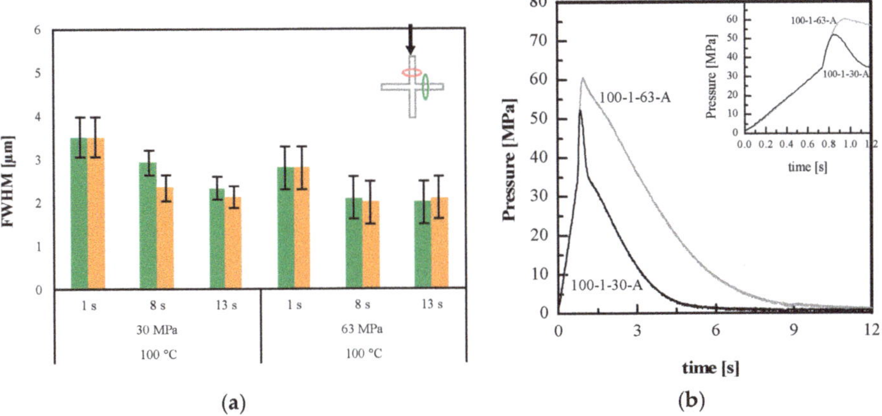

Figure 6. (a) FWHM index of micro-feature on PLA3251D samples obtained with different holding pressures and with T_{level} = 100 °C, as indicated in Table 1. A sketch of the area where the AFM patterns have been acquired is also reported. (b) Pressure evolutions in position P2 obtained for the experiments 100-1-63-A and 100-1-30-A.

Figure 6a shows that the FWHM decreases with the heating time for both the considered holding pressures. Generally, an increase in holding pressure, giving higher pressure levels during the whole process (see Figure 6b), improves replication accuracy. However, Figure 6a shows that the increase of the holding pressure induces only a slightly decrease of the FWHM index. This means that most of the replication process is achieved during the macroscopic cavity filling stage (which does not depend on the holding pressure adopted) that ends at t = 0.8 s. The replication process is completed at the beginning of the packing stage, so that an increase of the packing pressure improves, even if marginally, the replication process.

The replication process can be considered as the filling of micro and nano-cavity by a pressure driven flow [34,41]. During such a flow, the pressure exerted on the polymer has to overcome the pressure drop due to the filling of micro-cavity, the pressure due to the trapped air, and the pressure due to the surface tension [42]. During the filling of micro-cavity, the volume between the feature to be replicated and the polymer undergoes to a reduction. The air that is partially trapped in this volume is compressed. The pressure due to the trapped air, becoming comparable to the pressure exerted on the polymer, delays the filling of the micro-cavity [34]. The pressure due to the surface tension becomes more significant as the dimension of the feature decreases. The pressure due to the surface tension also delays the filling of micro-cavity. Furthermore, during such a pressure driven flow, the polymer undergoes a solidification and a frozen layer is formed at the polymer-air interface. The frozen layer has to elastically deform to allow the replication process to proceed [43]. All these phenomena have to be taken into account in the analysis of the replication accuracy.

To obtain an order of magnitude of the forces involved in the replication process, the melt front has been considered as a cylinder that wets the cavity surface with a contact angle of 180° on both the perpendicular surfaces of the micro or nano-feature (see Figure 7b). The radius of the cylinder (R) is initially equal to the height of the feature to be replicated (i.e., 5 µm for the micro-feature), and then the radius decreases with the ongoing of replication process. The distance between the melt front and the corner of the feature is equal to $d = R\left(\sqrt{2}-1\right)$ and the accuracy of the replication increases as d decreases. Assuming isothermal conditions, considering air as an ideal gas, the pressure due to trapped air, P_{air}, is:

$$P_{air} = \frac{V_0}{V}P_0 = \frac{R_0^2}{R^2}P_0 = \frac{d_0^2}{d^2}P_0 \tag{3}$$

where V_0 and P_0 are the initial volume and pressure, respectively, and V is the current volume.

The pressure due to surface tension is:

$$P_s = \frac{\sigma}{R} = \frac{\sigma\left(\sqrt{2}-1\right)}{d} \tag{4}$$

where σ is the surface tension (reported in the literature for PLA is 21 mN/m [44]).

Once the two pressures are related to the distance d, their value can be calculated as reported in the Figure 7a. Obviously, P_S depends only on d, whereas P_{air} also depends on the initial volume occupied by the gas, in which a pressure $P_0 = 1$ bar is assumed.

As clear from Figure 7, surface tension should not play any role in the replication of micro-features. When the features have dimensions in the nano-scale, surface tension becomes significant, although the pressure due to trapped air is dominant, if air cannot escape the volume to be replicated.

The pressure exerted on the melt during the filling of the micro-cavity is essentially the pressure recorded in the position P2, since in this position, the nickel shim containing the features to be replicated is located. The pressure exerted on the melt depends on the viscosity. The viscosity, on its turn, increases with the molecular weight. Figure 8 shows the influence of the viscosity on the pressure evolution in position P2, related to the two PLA grades adopted in this work, with $T_{level} = 100$ °C. It can be noticed that the pressures reached during the cavity filling (t = 0–0.8 s), reach a value of about 50 MPa for both materials, high enough to overcome the resistance due to P_{air} and P_s for both micro- and nanofeatures until a reduction of d of about one order of magnitude is attained.

The PLA4032D experiences higher pressure during the filling and smaller pressure during the packing than PLA3251D. During the filling (t = 0–0.8 s), the pressure is higher for PLA4032D, than PLA3251D. This is due to the fact that the higher is the viscosity, the higher the pressure required to fill the macroscopic cavity. During the packing stage, the pressure is smaller for PLA4032D, than PLA3251D, since the pressure drop between the injection point and the position inside cavity is higher because the viscosity is higher. Figure 5 shows that the FWHM obtained with 1 s heating time is similar for the two PLA grades, at all the adopted T_{level}. This could be due to the fact that the high-pressure levels (certainly higher than those recorded with PLA3251D) recorded with PLA4032D compensates for the pressure required to fill the micro-cavity. With long heating times, the pressure exerted on the polymer during the micro-cavity filling is higher adopting PLA3251D than PLA4032D, thus the replication accuracy is lower for the latter polymer. This also confirms that the replication process completes during the early packing stage.

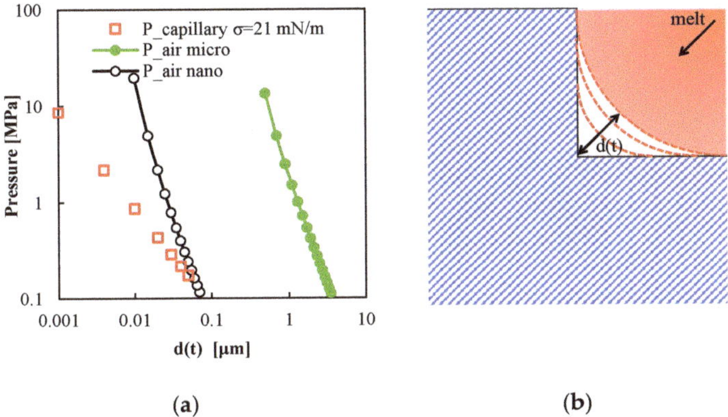

Figure 7. (**a**) Pressure due to the trapped air and to the surface tension as function of the unfilled distance d(t). (**b**) Sketch of the replication process viewed as the filling of a micro-cavity.

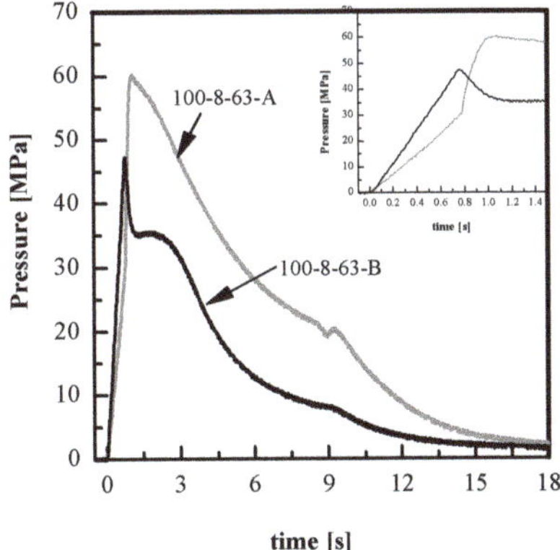

Figure 8. Pressure evolutions in position P2 obtained for the experiments 100-8-63-A and 100-8-63-B.

3.3. Replication of Nanofeatures

The replication ability has been also analyzed in the cases of nanofeatures replication. It is important to highlight that nanofeatures have not been found on the passive samples for both PLA grades. Figure 9 shows the AFM acquisitions of the nanofeatures of the nickel shim and the nanofeatures produced during the injection-molding process with two T_{level}, 50 °C and 150 °C, and 13 s heating time.

Figure 10a,b shows that the replication accuracy does not depend on the direction of nanofeatures (any difference is within the confidence range of the measurement) with respect the direction of the flow front for both the PLA grades. Furthermore, the higher T_{level} is, the smaller the H index is, for both

PLA grades. Figure 11 summarizes the differences in replication accuracy in terms of H index, in the direction of the flow front, between PLA3251D and PLA4032D.

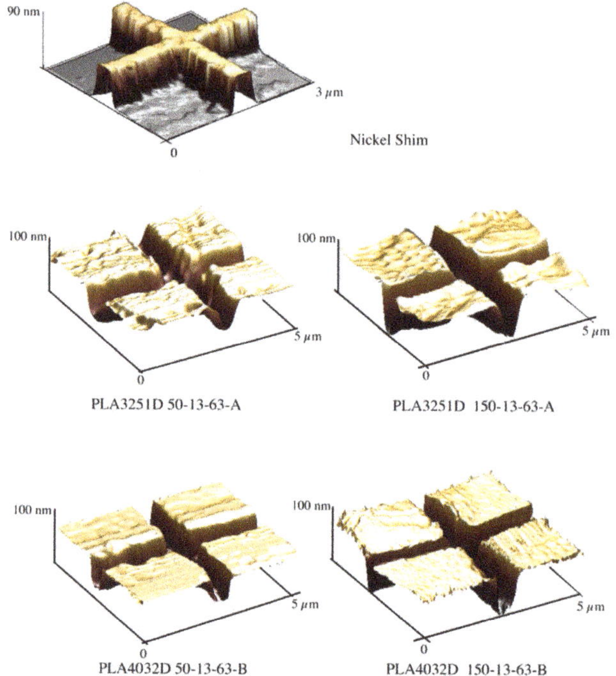

Figure 9. AFM height images of the nanofeatures on the nickel shim, injection-molded samples of PLA3251D and PLA4032D obtained in different conditions of cavity surface temperature.

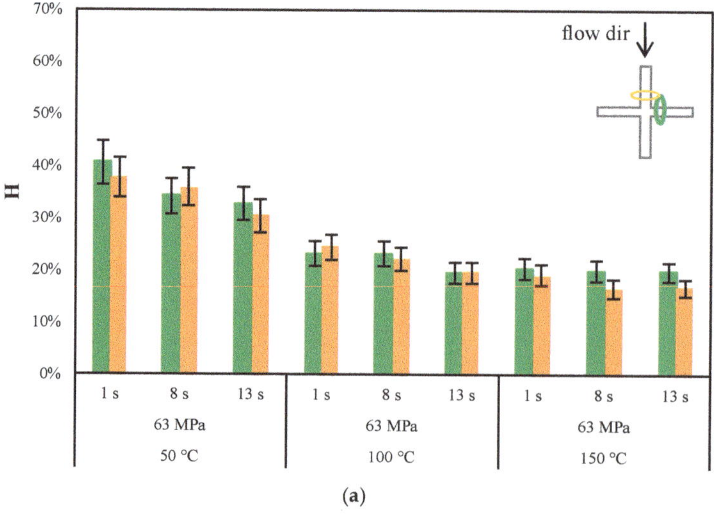

(a)

Figure 10. *Cont.*

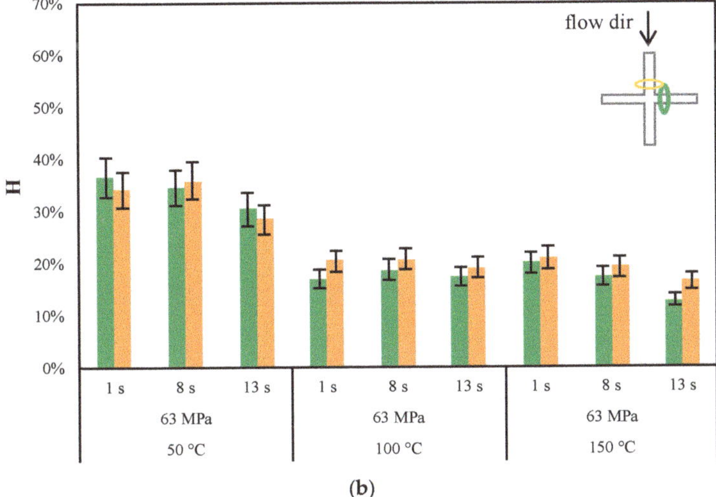

(b)

Figure 10. H index of nanofeatures on (**a**) PLA3251D samples and (**b**) PLA4032D samples obtained as indicated in Table 1. A sketch of the area where the AFM patterns have been acquired is also reported.

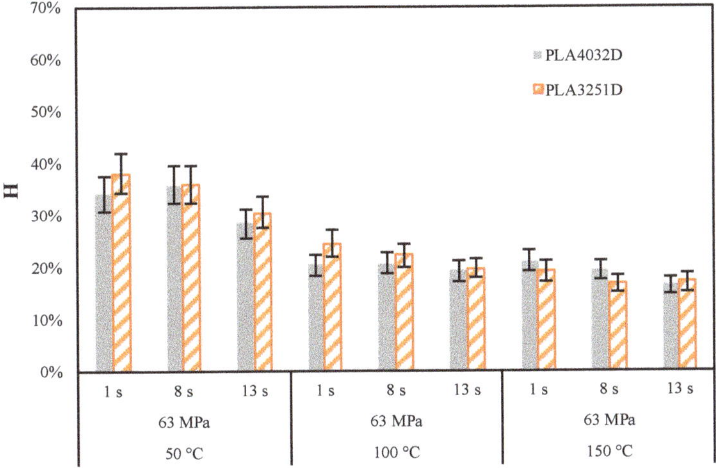

Figure 11. Comparison between the H index of nanofeatures on PLA3251D and H index of nanofeatures on PLA4032D samples obtained as indicated in Table 1.

The AFM acquisitions reported in Figure 9 show that replication accuracy was poor with T_{level} = 50 °C. The height of the nanofeature replicated on the polymer is significantly smaller than that on the nickel shim. The higher is the T_{level}, the closer is the height of the replicated nanofeature to the height measured for the nanofeature on the nickel shim. A quantitative analysis of the replication accuracy is given in terms of H index in the Figure 10, for both the adopted PLA grades.

First of all, Figure 11 shown that good replication has been reached for both PLA grades: the value of the H index is 16 ± 2% (i.e., the height of the replicated nanofeature is 50 ± 1 nm) with T_{level} = 100 °C.

The heating time seems to have had only a slight effect on nanofeature-replication accuracy. Furthermore, replication accuracy seems to be not dependent on the PLA grade. These observations suggest that the replication mechanism is less dependent on viscosity with respect to the cases of the microfeatures. The phenomena relating to the interaction between the polymer and the cavity surface, the nickel shim in this work, become more important as the dimension of the feature to be replicated decreases down to a nanometrical level.

3.4. Analysis of Morphology

Morphological analysis has been also performed on the molding surface in the areas where nanofeatures have been found. Figure 12 shows the height and amplitude error AFM acquisitions related to the samples 100-13-63-B and 150-13-63-B. The cross sections of the height maps were also reported for each sample.

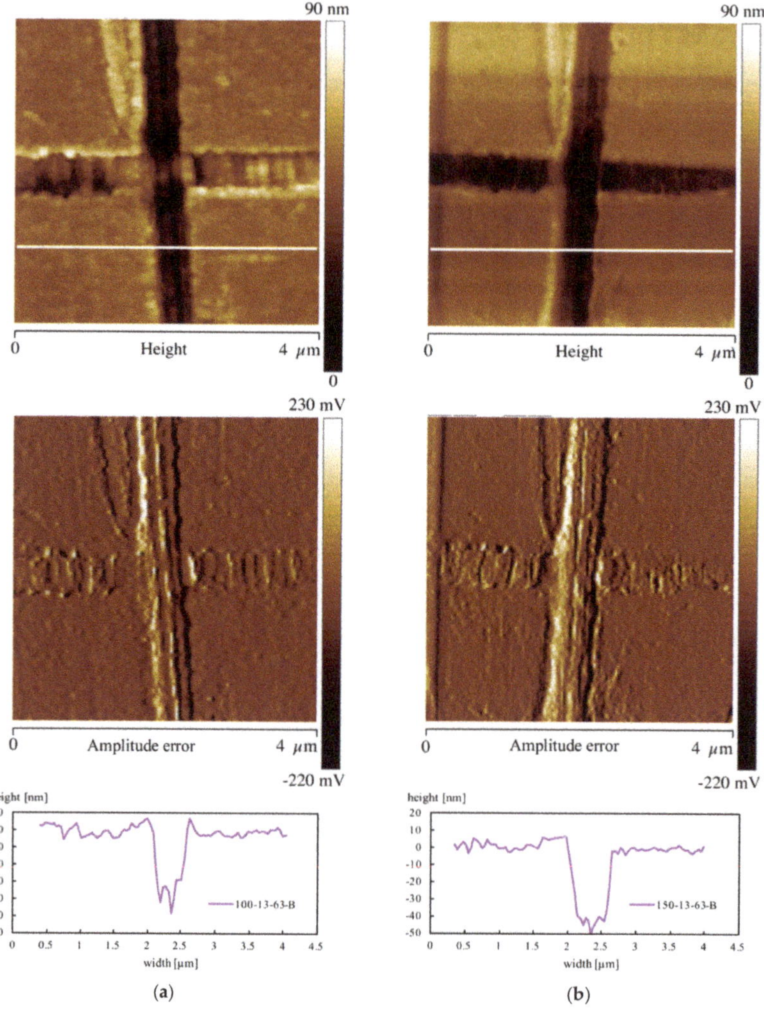

Figure 12. Height and amplitude error AFM maps of the samples (**a**) 100-13-63-B and (**b**) 150-13-63-B. The pattern related to the white line reported in the height map is also shown.

The AFM maps show that the samples produced activating the heating device are characterized by the presence of structures aligned with the flow direction. The pattern of the cross sections reported in the bottom of the figure shows that the height of the structures, 10 ± 5 nm, is comparable with the height of the features to be replicated. Similar patterns have been also obtained adopting PLA3251D with the same T_{level} and heating time. Thus, the H value obtained at 100 °C and 150 °C T_{level} cannot further decrease at values smaller than 16% because of the presence of the structures aligned with the flow front.

The XRD analysis performed on the surface of the injection-molded samples is reported in the Figure 13.

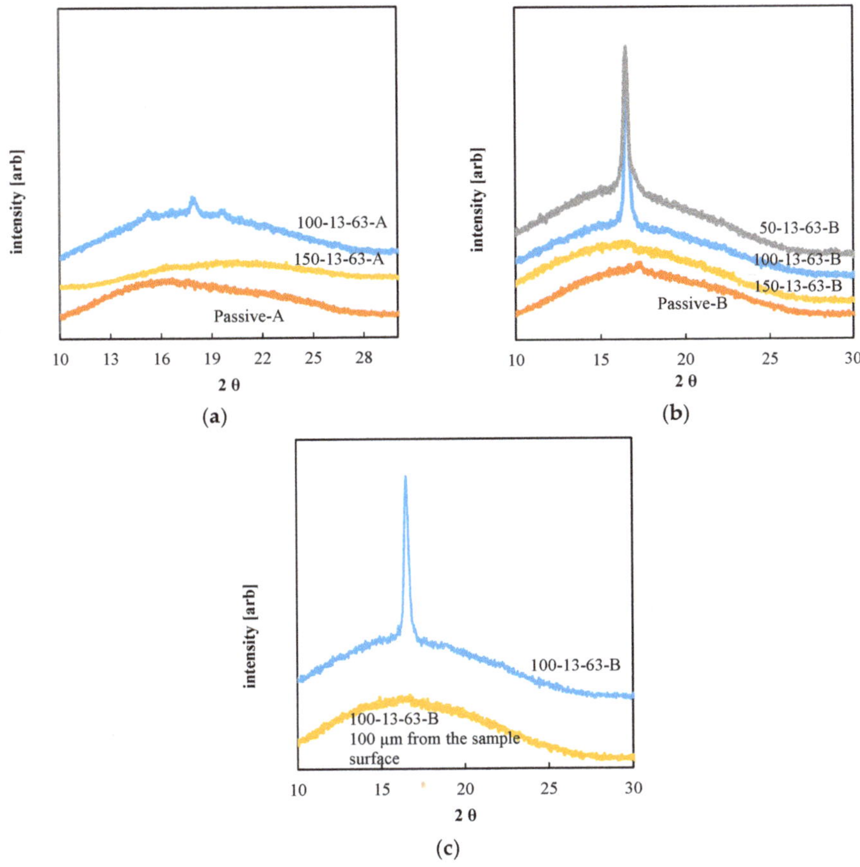

Figure 13. X-ray Diffractometry (XRD) spectra of the PLA injection-molding samples (**a**) PLA3251D, and (**b**) PLA4032D. (**c**) XRD spectra along the sample thickness for the sample 100-13-63-B.

The samples obtained with low cavity-surface temperature, Passive-A and Passive-B, show a low crystalline degree due to the high cooling rates that the polymer experiences during the process; the samples obtained with T_{level} = 150 °C also show a low crystallinity degree. The samples obtained with T_{level} = 100 °C show the highest crystallinity degree (see the peak at $2\theta \cong 16.7°$), $5 \pm 3\%$ for PLA3251D, and $10 \pm 3\%$ for PLA4032D. Such crystalline degrees are significant for PLA, since the maximum value of the crystallinity degree of these thermoplastic materials is about 40% [45]. Figure 13b shows that there was a shift of the peak $2\theta \cong 16.7°$ (attributed to the α-phase [46,47]) toward high values of 2θ with the decrease of T_{level}. This shift is due to the different cooling and flow conditions that the polymer experiences during the solidification that took place with different T_{level} and heating times [47–49].

The crystalline degree found for both the PLA grades was also consistent with the presence of structures aligned with the flow front and shown the Figure 12. Figure 13c shows the XRD spectra of the sample 100-13-63-B on the sample surface and 100 μm distance from the sample surface. The comparison between the two spectra demonstrates that the high crystallinity was limited to a thin layer close to the sample surface. This suggests that the formation of the structures aligned with the flow front was also limited to a narrow layer close to the sample surface.

It is worth mentioning that the PLA grades adopted in this work, and generally all the commercial grades of PLA, had crystallization times (in quiescent conditions) much longer than the processing times of injection molding. This means that the crystalline structures detected in our samples had to be due to flow-induced crystallization, and their fibrillary morphology confirms this interpretation [50].

4. Discussion

In this work, the replication of micro- and nanofeatures on two different grades of PLA during the injection-molding process has been experimentally analyzed. The system adopted to modulate the temperature on the cavity surface during the injection-molding process has been demonstrated as efficient in the enhancing of replication accuracy of both micro- and nanofeatures.

The replication accuracy of microfeatures increases with both T_{level} and heating time. An increase of packing pressure only marginally improves the replication. The replication is more accurate when the polymer with low viscosity is adopted. The results suggest that the replication process has to take place during both the filling (referred to the macrocavity), and the early packing. When the heating time is comparable with the filling time of the macroscopic cavity, the high-pressure levels recorded with PLA4032D, the high viscosity polymer compensates for the pressure required to fill the microcavity. The frozen layer, which starts to form immediately after heater deactivation, prevents any additional filling of the microcavity during the packing stage. As a result, the differences observed in the replication accuracy with 1 s heating time are within the confidence range of the measurement (see Figure 5). Adopting heating times longer than the filling time, significant pressure is exerted on the polymer for longer, and the formation of the frozen layer, because the high T_{level} adopted, is significantly delayed. As a consequence, an increase in replication accuracy with heating time is observed. Additionally, thanks to the smaller viscosity, which reduces the pressure drop at the flow front, PLA3251D shows higher replication accuracy than PLA4032D when heating times longer than filling times are adopted. The interpretation of the replication process described above implies that for the microcavities the replication process can be considered as a viscous filling, for which the interfacial phenomena can be neglected.

In the cases of nanofeatures, the replication accuracy is poorly dependent on the heating time and on the polymer grade. This suggests that the replication mechanism has to be different from the one hypothesized for the replication of microfeatures. In particular, interfacial phenomena become more significant the smaller the dimension of the feature to be replicated is. Interestingly, the AFM maps show the presence of structures aligned along the flow direction. The presence of these structures is consistent with the crystallization degree found on the sample surface. The mean dimension of the structures is in the range of few nanometers and their presence certainly prevents a further increase of the replication accuracy.

Author Contributions: Investigation, S.L. and V.S.; Supervision, R.P.; Writing—review & editing, S.L., V.S. and R.P.

Funding: This research received no external funding

Acknowledgments: The authors acknowledge Giovanni Marmora for giving precious support to this research.

Conflicts of Interest: The authors declare no conflict of interest.

References

1. Cesca, F.; Limongi, T.; Accardo, A.; Rocchi, A.; Orlando, M.; Shalabaeva, V.; Di Fabrizio, E.; Benfenati, F. Fabrication of biocompatible free-standing nanopatterned films for primary neuronal cultures. *RSC Adv.* **2014**, *4*, 45696–45702. [CrossRef]
2. Limongi, T.; Schipani, R.; Di Vito, A.; Giugni, A.; Francardi, M.; Torre, B.; Allione, M.; Miele, E.; Malara, N.; Alrasheed, S.; et al. Photolithography and micromolding techniques for the realization of 3D polycaprolactone scaffolds for tissue engineering applications. *Microelectron. Eng.* **2015**, *141*, 135–139. [CrossRef]
3. Rahim, K.; Mian, A. A review on laser processing in electronic and MEMS packaging. *J. Electron. Packag.* **2017**, *139*, 030801. [CrossRef]
4. Nihtianov, S.; Luque, A. *Smart Sensors and MEMS: Intelligent Devices and Microsystems for Industrial Applications*; Woodhead Publishing Limited: Cambridge, UK, 2014.
5. Nuxoll, E. BioMEMS in drug delivery. *Adv. Drug Deliv. Rev.* **2013**, *65*, 1611–1625. [CrossRef] [PubMed]
6. Ranella, A.; Barberoglou, M.; Bakogianni, S.; Fotakis, C.; Stratakis, E. Tuning cell adhesion by controlling the roughness and wettability of 3D micro/nano silicon structures. *Acta Biomater.* **2010**, *6*, 2711–2720. [CrossRef] [PubMed]
7. Sorkin, J.A.; Hughes, S.; Soares, P.; Popat, K.C. Titania nanotube arrays as interfaces for neural prostheses. *Mater. Sci. Eng. C* **2015**, *49*, 735–745. [CrossRef] [PubMed]
8. Weng, C.; Wang, F.; Zhou, M.; Yang, D.; Jiang, B. Fabrication of hierarchical polymer surfaces with superhydrophobicity by injection molding from nature and function-oriented design. *Appl. Surf. Sci.* **2018**, *436*, 224–233. [CrossRef]
9. Liparoti, S.; Pantani, R.; Sorrentino, A.; Speranza, V.; Titomanlio, G. Hydrophobicity tuning by the fast evolution of mold temperature during injection molding. *Polymers* **2018**, *10*, 322. [CrossRef]
10. Eigenfeld, N.T.; Gray, J.M.; Brown, J.J.; Skidmore, G.D.; George, S.M.; Bright, V.M. Ultra-thin 3D nano-devices from atomic layer deposition on polyimide. *Adv. Mater.* **2014**, *26*, 3962–3967. [CrossRef] [PubMed]
11. Kustra, J.; Martin, E.; Chateau, D.; Lerouge, F.; Monnereau, C.; Andraud, C.; Sitarz, M.; Baldeck, P.L.; Parola, S. Two-photon controlled sol–gel condensation for the microfabrication of silica based microstructures. The role of photoacids and photobases. *RSC Adv.* **2017**, *7*, 46615–46620. [CrossRef]
12. Borges, J.; Mano, J.F. Molecular interactions driving the layer-by-layer assembly of multilayers. *Chem. Rev.* **2014**, *114*, 8883–8942. [CrossRef] [PubMed]
13. Rogers, J.A.; Lee, H.H. *Unconventional Nanopatterning Techniques and Applications*; John Wiley & Sons, Inc.: New York, NY, USA, 2008.
14. Mogi, K.; Sugii, Y.; Yamamoto, T.; Fujii, T. Rapid fabrication technique of nano/microfluidic device with high mechanical stability utilizing two-step soft lithography. *Sens. Actuators B Chem.* **2014**, *201*, 407–412. [CrossRef]
15. Pinto, E.; Faustino, V.; Rodrigues, R.; Pinho, D.; Garcia, V.; Miranda, J.M.; Lima, R. A rapid and low-cost nonlithographic method to fabricate biomedical microdevices for blood flow analysis. *Micromachines* **2014**, *6*, 121–135. [CrossRef]
16. Tan, G.; Inoue, N.; Funabasama, T.; Mita, M.; Okuda, N.; Mori, J.; Koyama, K.; Kaneko, S.; Nakagawa, M.; Matsuda, A.; et al. Formation of 0.3-nm-high stepped polymer surface by thermal nanoimprinting. *Appl. Phys. Express* **2014**, *7*, 055202. [CrossRef]
17. Kooy, N.; Mohamed, K.; Pin, L.T.; Guan, O.S. A review of roll-to-roll nanoimprint lithography. *Nanoscale Res. Lett.* **2014**, *9*, 320. [CrossRef] [PubMed]
18. Mitra, J.; Tripathi, G.; Sharma, A.; Basu, B. Scaffolds for bone tissue engineering: Role of surface patterning on osteoblast response. *RSC Adv.* **2013**, *3*, 11073–11094. [CrossRef]
19. Masato, D.; Sorgato, M.; Lucchetta, G. Analysis of the influence of part thickness on the replication of micro-structured surfaces by injection molding. *Mater. Des.* **2016**, *95*, 219–224. [CrossRef]
20. Lucchetta, G.; Fiorotto, M.; Bariani, P.F. Influence of rapid mold temperature variation on surface topography replication and appearance of injection-molded parts. *CIRP Ann. Manuf. Technol.* **2012**, *61*, 539–542. [CrossRef]
21. Kuo, C.-C.; Wang, Y.-J. Development of a micro-hot-embossing mold with high replication fidelity using surface modification. *Mater. Manuf. Process.* **2014**, *29*, 1101–1110. [CrossRef]

22. Pantani, R.; Volpe, V.; Titomanlio, G. Foam injection molding of poly(lactic acid) with environmentally friendly physical blowing agents. *J. Mater. Process. Technol.* **2014**, *214*, 3098–3107. [CrossRef]

23. Feldmann, M.; Fuchs, J. *Specialized Injection Molding Techniques*; Heim, H.P., Ed.; William Andrew Publishing: Oxford, UK, 2016.

24. Pan, Y.; Liu, X.; Shi, S.; Liu, C.; Dai, K.; Yin, R.; Schubert, D.W.; Zheng, G.; Shen, C. Annealing induced mechanical reinforcement of injection molded iPP parts. *Macromol. Mater. Eng.* **2016**, *301*, 1468–1472. [CrossRef]

25. Jiang, J.; Liu, X.; Lian, M.; Pan, Y.; Chen, Q.; Liu, H.; Zheng, G.; Guo, Z.; Schubert, D.W.; Shen, C.; et al. Self-reinforcing and toughening isotactic polypropylene via melt sequential injection molding. *Polym. Test.* **2018**, *67*, 183–189. [CrossRef]

26. Guerrier, P.; Tosello, G.; Nielsen, K.K.; Hattel, J.H. Three-dimensional numerical modeling of an induction heated injection molding tool with flow visualization. *Int. J. Adv. Manuf. Technol.* **2016**, *85*, 643–660. [CrossRef]

27. Yao, D.; Kimerling, T.E.; Kim, B. High-frequency proximity heating for injection molding applications. *Polym. Eng. Sci.* **2006**, *46*, 938–945. [CrossRef]

28. Yu, M.C.; Young, W.B.; Hsu, P.M. Micro-injection molding with the infrared assisted mold heating system. *Mater. Sci. Eng. A* **2007**, *460–461*, 288–295. [CrossRef]

29. De Santis, F.; Pantani, R. Development of a rapid surface temperature variation system and application to micro-injection molding. *J. Mater. Process. Technol.* **2016**, *237*, 1–11. [CrossRef]

30. Liparoti, S.; Sorrentino, A.; Titomanlio, G. Fast cavity surface temperature evolution in injection molding: Control of cooling stage and final morphology analysis. *RSC Adv.* **2016**, *6*, 99274–99281. [CrossRef]

31. Liparoti, S.; Sorrentino, A.; Guzman, G.; Cakmak, M.; Titomanlio, G. Fast mold surface temperature evolution: Relevance of asymmetric surface heating for morphology of iPP molded samples. *RSC Adv.* **2015**, *5*, 36434–36448. [CrossRef]

32. Liparoti, S.; Titomanlio, G.; Sorrentino, A. Analysis of asymmetric morphology evolutions in iPP molded samples induced by uneven temperature field. *AIChE J.* **2016**, *62*, 2699–2712. [CrossRef]

33. Jansen, K.M.B.; Flaman, A.A.M. Construction of fast-response heating elements for injection molding applications. *Polym. Eng. Sci.* **1994**, *34*, 894–897. [CrossRef]

34. Speranza, V.; Liparoti, S.; Calaon, M.; Tosello, G.; Pantani, R.; Titomanlio, G. Replication of micro and nano-features on iPP by injection molding with fast cavity surface temperature evolution. *Mater. Des.* **2017**, *133*, 559–569. [CrossRef]

35. Volpe, V.; De Filitto, M.; Klofacova, V.; De Santis, F.; Pantani, R. Effect of mold opening on the properties of PLA samples obtained by foam injection molding. *Polym. Eng. Sci.* **2017**, *58*, 475–484. [CrossRef]

36. Liparoti, S.; Landi, G.; Sorrentino, A.; Speranza, V.; Cakmak, M.; Neitzert, H.C. Flexible poly(amide-imide)-carbon black based microheater with high-temperature capability and an extremely low temperature coefficient. *Adv. Electron. Mater.* **2016**, *2*, 1600126. [CrossRef]

37. Calaon, M.; Hansen, H.N.; Tosello, G.; Garnaes, J.; Nørregaard, J.; Li, W. Microfluidic chip designs process optimization and dimensional quality control. *Microsyst. Technol.* **2013**, *21*, 561–570. [CrossRef]

38. Hansen, H.N.; Hocken, R.J.; Tosello, G. Replication of micro and nano surface geometries. *CIRP Ann. Manuf. Technol.* **2011**, *60*, 695–714. [CrossRef]

39. Pantani, R.; Coccorullo, I.; Speranza, V.; Titomanlio, G. Modeling of morphology evolution in the injection molding process of thermoplastic Polymers. *Prog. Polym. Sci.* **2005**, *30*, 1185–1222. [CrossRef]

40. Pantani, R.; De Santis, F.; Brucato, V.; Titomanlio, G. Analysis of gate freeze-off time in injection molding. *Polym. Eng. Sci.* **2004**, *44*, 1–17. [CrossRef]

41. Zhang, N.; Browne, D.J.; Gilchrist, M.D. Characterization of micro injection molding process for the replication of micro/nano features using bulk metallic glass insert. *Int. J. Eng. Technol.* **2013**, 198–201. [CrossRef]

42. Lu, Z.; Zhang, K.F. Morphology and mechanical properties of polypropylene micro-arrays by micro-injection molding. *Int. J. Adv. Manuf. Technol.* **2009**, *40*, 490–496. [CrossRef]

43. Theilade, U.A.; Hansen, H.N. Surface microstructure replication in injection molding. *Int. J. Adv. Manuf. Technol.* **2007**, *33*, 157–166. [CrossRef]

44. Sarikhani, K.; Jeddi, K.; Thompson, R.B.; Park, C.B.; Chen, P. Effect of pressure and temperature on interfacial tension of poly lactic acid melt in supercritical carbon dioxide. *Thermochim. Acta* **2015**, *609*, 1–6. [CrossRef]

45. Refaa, Z.; Boutaous, M.; Xin, S.; Siginer, D.A. Thermophysical analysis and modeling of the crystallization and melting behavior of PLA with talc: Kinetics and crystalline structures. *J. Therm. Anal. Calorim.* **2017**, *128*, 687–698. [CrossRef]

46. Tsuji, H.; Nakano, M.; Hashimoto, M.; Takashima, K.; Katsura, S.; Mizuno, A. Electrospinning of poly(lactic acid) stereocomplex nanofibers. *Biomacromolecules* **2006**, *7*, 3316–3320. [CrossRef] [PubMed]

47. Huang, S.; Li, H.; Jiang, S.; Chen, X.; An, L. Crystal structure and morphology influenced by shear effect of poly(l-lactide) and its melting behavior revealed by WAXD, DSC and in-situ POM. *Polymer* **2011**, *52*, 3478–3487. [CrossRef]

48. De Santis, F.; Pantani, R.; Titomanlio, G. Nucleation and crystallization kinetics of poly(lactic acid). *Thermochim. Acta* **2011**, *522*, 128–134. [CrossRef]

49. Jalali, A.; Shahbikian, S.; Huneault, M.A.; Elkoun, S. Effect of molecular weight on the shear-induced crystallization of poly(lactic acid). *Polymer* **2017**, *112*, 393–401. [CrossRef]

50. De Meo, A.; De Santis, F.; Pantani, R. Dynamic local temperature control in micro-injection molding: Effects on poly(lactic acid) morphology. *Polym. Eng. Sci.* **2018**, *58*, 586–591. [CrossRef]

Article

Multi-Objective Optimizations for Microinjection Molding Process Parameters of Biodegradable Polymer Stent

Hongxia Li [1], Kui Liu [1], Danyang Zhao [1], Minjie Wang [1], Qian Li [2,*] and Jianhua Hou [2]

1   School of Mechanical Engineering, Dalian University of Technology, Dalian 116023, China;
     hxli@dlut.edu.cn (H.L.); Lkui@mail.dlut.edu.cn (K.L.); danyangz@163.com (D.Z.);
     mjwang@dlut.edu.cn (M.W.)
2   National Center for International Joint Research of Micro-Nano Molding Technology, Zhengzhou University,
     Zhengzhou 450000, China; houjianhua@zzu.edu.cn
*   Correspondence: qianli@zzu.edu.cn; Tel.: +86-371-6778-1101

Received: 18 October 2018; Accepted: 15 November 2018; Published: 19 November 2018

Abstract: Microinjection molding technology for degradable polymer stents has good development potential. However, there is a very complicated relationship between molding quality and process parameters of microinjection, and it is hard to determine the best combination of process parameters to optimize the molding quality of polymer stent. In this study, an adaptive optimization method based on the kriging surrogate model is proposed to reduce the residual stress and warpage of stent during its injection molding. Integrating design of experiment (DOE) methods with the kriging surrogate model can approximate the functional relationship between design goals and design variables, replacing the expensive reanalysis of the stent residual stress and warpage during the optimization process. In this proposed optimization algorithm, expected improvement (EI) is used to balance local and global search. The finite element method (FEM) is used to simulate the micro-injection molding process of polymer stent. As an example, a typical polymer vascular stent ART18Z was studied, where four key process parameters are selected to be the design variables. Numerical results demonstrate that the proposed adaptive optimization method can effectively decrease the residual stress and warpage during the stent injection molding process.

Keywords: polymeric stent; injection molding; residual stress; warpage; kriging surrogate model; design optimization

1. Introduction

Polymeric stents are a promising prospect [1]. However, the thin-walled tubular surface of polymeric stents has a discontinuous mesh structure. Furthermore, the stent has a tiny entirety, big length-to-diameter ratio, partial precision and unique structure, which greatly limits the machining and application of polymeric stents. The traditional machining method for stents is laser cutting. However, there are some uncontrolled machining problems, such as cutting seam width, surface roughness, taper of cut, back injury, slag, recast layer, etc. Fortunately, Clarke et al. presented an innovative technique for manufacturing polymer stents through the injection molding process [2]. This method provides a new line of thought for the high efficiency machining of stents. The injection molding stents have many remarkable advantages such as its strong reproducibility, high molding efficiency, good surface quality, good material condensation orientation, good molding consistency, etc.

At present, the injection molding method is not widely used in stent manufacturing. An important cause is that the parameters greatly influence molding quality of the polymeric stent during the injection molding process. Moreover, it is hard to determine a reasonable process parameter combination for the injection molding process, and this can easily result in some problems. For example, the flow of the melt is difficult to manage or the melt flows unevenly, it is hard to fill the mold cavity, the relatively large residual stress in the stent's cavity results in difficulty in demolding, and the stent undergoes a warpage phenomenon after demolding. Therefore, there is a need to investigate the effect of some important process parameters, such as mold temperature, melt temperature, flow rate and packing pressure, on stent quality. A reasonable process parameter combination is able to ensure the molding quality of the polymeric stent. Some scholars have conducted studies on the effect of injection molding process parameters on product quality. Liu et al. studied how the mold temperature affected the shrinkage of rapid heat cycle injection molded parts [3]. Duo performed the Taguchi experiment to investigate the effect of the injection molding process parameters, such as packing time, molding temperature and cooling time, on the warpage of flat tiny devices [4]. Jiang et al. revealed the great influence of mold temperature on the structure and mechanical properties of microinjection molding PP products [5]. In the meantime, Singh et al., Annicchiarico et al. and Mohan et al. conducted a detailed study of the effect of injection molding process parameters on the quality of injection molding products, providing a reference value for the stent injection molding process [6–8]. The above-mentioned studies were based on the effect between process parameters and product quality indicators, and some of these scholars were able to establish a function relationship between process parameters and quality indicators from the perspective of the optimal design. Based on the combination of the artificial neural network and genetic algorithm, Shen et al. optimized the injection molding process parameters, and improved the volume shrinkage of parts [9]. Based on the software tools, Florin et al. optimized the microinjection molding process of polymer medical apparatuses and instruments [10]. Kitayama et al. realized the multi-objective optimization of volume shrinkage and clamping force in the plastic injection molding process through the sequential approximate optimization method [11]. Based on Taguchi, ANOVA and the artificial neural network method, Oliaei et al. investigated the warpage and shrinkage optimization of biodegradable polymer injection molding plastic spoon parts [12]. Dang proposed a general framework for optimizing injection molding process parameters [13]. Kashyap et al. presented a detailed summary of related methods for the optimization of injection molding process parameters [14]. Kitayama et al. performed a multi-objective optimization of process parameters in plastic injection molding for simultaneously minimizing the warpage, cycle time and clamping force using radial basis function [15]. These studies have laid a solid foundation for the process parameter control of the stent injection molding process.

However, the polymeric vascular stent has a thin-walled tubular network structure with micro-scale geometry (strut width and thickness of 0.1 mm), large ratio of length to diameter and local precision, which make molding quality very sensitive to changes in process parameters. Furthermore, the relationship between process parameters and molding quality is very complicated, non-linear, implicit and multimodal. In addition, the number of functional assessments is extremely time consuming. In this case, if using a conventional optimization method (such as a gradient-based algorithm), it will be difficult to get a global optimization design for the process parameters. Here, it is worth recommending that the surrogate modeling, which mainly consisting of the Kriging method, can effectively solve the engineering problems mentioned above [16]. It is well known that the microscopic properties of polymer materials affect macroscopic properties, but the relationship is complex, and the computational process is highly time-consuming [17]. Fortunately, one of the great benefits is that the surrogate model actually calculates material properties at the micro level, which can save the computation time. The surrogate model can establish a suitable approximation functional relationship between process parameters (input) and molding quality (output), thus taking the place of the complicated engineering calculation process at lower computational cost. In this respect, Gao et al. [18], Li et al. [19] have done a lot of related work. Gao et al. used Surrogate-based process optimization for reducing warpage in macro-injection molding. Li et al. have developed a multi-objective optimization based on the Kriging method. The method was applied to optimize the geometry of the biodegradable polymer stent, which finally showed that the stent expansion performance could be successfully improved.

In the present paper, the multi-objective process optimization about process parameters and molding quality was researched by constructing the adaptive optimization algorithm based on Kriging surrogate modeling. The Kriging model was applied to establish an approximate functional relationship between stent molding quality (residual stress and warpage) and injection molding parameters, thus making it possible to avoid the time-consuming finite element reanalysis in microinjection molding process optimization. The optimization iterative process is based on approximate functional relationships to decrease the computation time. Meanwhile, the 40 trial sample points were obtained by using the Optimal Latin Hypercube Sampling Method (Optimal LHS) [20]. Also, the expected improvement (EI) function plays a huge role when performing the adaptive optimization process. Even in the case of small sample sizes, the local and global searches can also be balanced, making it more likely to find global optimization designs. The finite element method based on MOLDFLOW (Autodesk Moldflow Insight 2012, Autodesk Inc., San Rafael, CA, USA) was ran simultaneously to model the injection molding process of the stent to obtain the response of molding quality under the specific process parameter combination. The combination of the numerical results and the design optimization method of the study facilitates further optimization studies and development of a method for further manufacturing a polymer stent by a microinjection molding process.

2. Materials and Methods

2.1. PLA Stent Material

Biodegradable polymeric stents are mainly made of polylactic acid (PLA) and modified materials. In the present study, as an example, the biodegradable polymeric stent was made of semi-crystalline PLA (Manufacturer: Kao Corporation, Trademark: ECOLA S-1010, Tokyo, Japan), with a solid particle density of 1.6416 g/cm^3 and a melt density of 1.4836 g/cm^3.

The melt rheology curve of PLA is presented in Figure 1. On one hand, the flow process of the PLA melt under constant temperature exhibits a significant shear thinning behavior. In the actual injection molding process, and within the allowable shear rate of the material, this rheology should be fully used in selecting an appropriate melt flow shear rate, adjusting the melt viscosity, and improving flowability, thereby reaching a better flow and filling ability. On the other hand, the shear viscosity of the PLA melt was correlated to temperature. Within a certain range of shear rate, as a whole, viscosity decreases as temperature increases. Furthermore, the effect of temperature on viscosity can basically be ignored when the melt flow exceeds a certain shear rate. As a result, overall consideration should be given to the effect of temperature and shear rate on melt flowability for full utilization.

The PVT curve of PLA plastic is presented in Figure 2 to investigate the relationship between pressure, volume and temperature. Under specific pressure conditions, the volume ratio of PLA increases as temperature increases. When pressure is relatively small, and as the temperature increases, PLA is converted from a solid particle status into a melt status, and the relationship between the volume and temperature of PLA tends to undergo a staged linear change. Furthermore, there will be a small upward slope near 130 °C. This is because PLA is a semi-crystalline material, and 130 °C is the crystalline transition temperature. The molecular chain of PLA is converted from a partially ordered arrangement status into an overall disordered status, which results in a relatively big volume expansion. The temperature drop led to significant volume shrinkage. In the injection molding process, consideration should be given to the effect of the material's crystalline. When the pressure is very high, and the temperature increases, a few upward slopes under the crystalline transition temperature appear. This is due to the very high pressure at this moment, which greatly inhibits the movement of the molecular chain and the transition of material volume, resulting in the single linear change of the whole process. Therefore, it is necessary to flexibly control the injection pressure and the injection temperature due to the characteristics of PLA as a semi-crystalline material when performing the actual injection molding process.

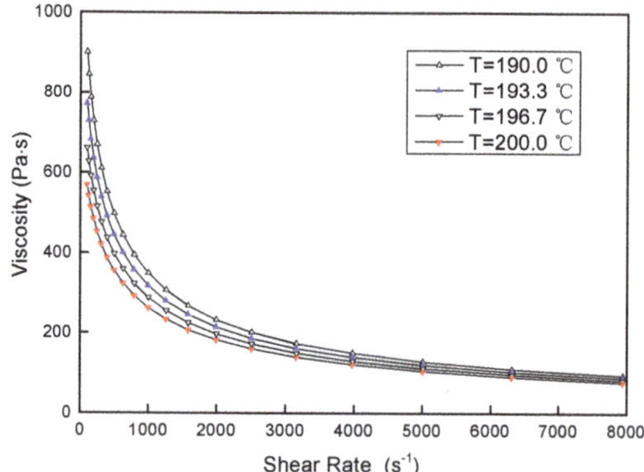

Figure 1. The rheological properties of polylactic acid (PLA).

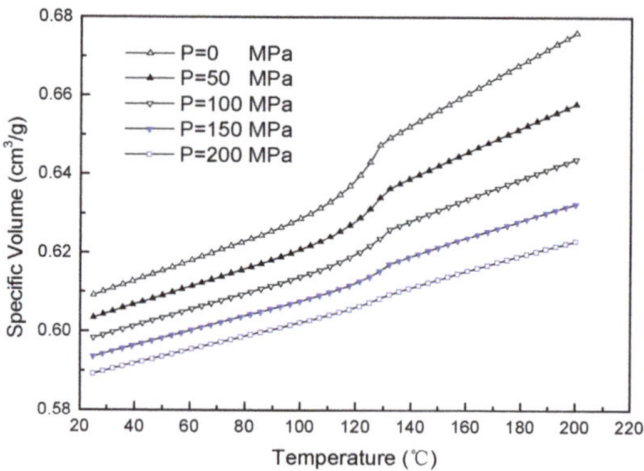

Figure 2. The PVT properties of polylactic acid (PLA).

2.2. Biodegradable Polymeric Stent Structure

The biodegradable polymeric stent made of PLA has good biocompatibility and a relatively short biodegradation period. The structure of this stent is presented in Figure 3: overall length, 13.75 mm; thickness, 0.17 mm; outside diameter, 3.36 mm.

It can be observed that it has a typical thin-walled structure and a large number of discontinuous mesh structures. Full consideration should be given to demolding properties and post-demolding warpage.

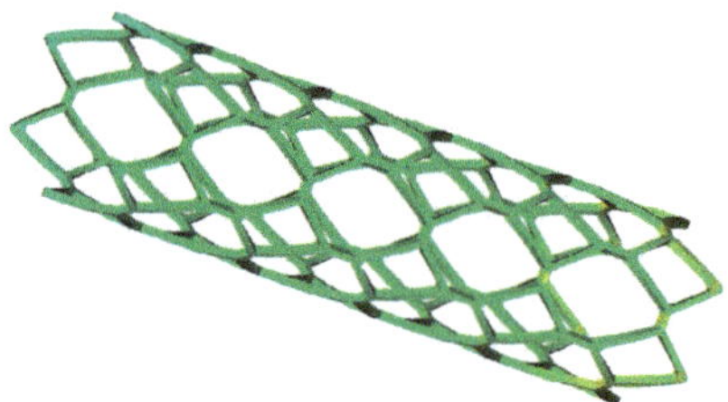

Figure 3. Generic polymeric stent with straight bridges.

2.3. Microinjection Molding Optimization of Biodegradable Polymeric Stents

In the PLA stent injection molding process, due to its thin-wall and mesh structure feature, its common phenomena, such as demolding difficulty and post-demolding warpage, greatly influence the efficiency and molding quality of the stent injection molding process. Its demolding process is mainly influenced by the residual stress of the stent in the orientation direction, indicating the stress of parts before ejection, instead of after ejection. An excessively large residual stress in the cavity would cause the stent in the cavity to maintain a relatively high-tension status, which is not good for the demolding process. Warpage is a size deformation caused by the unbalanced change of inner stress. Serious warpage can even influence the normal use of stents. The residual stress and warpage of stents are influenced by processes, such as mold temperature, melt temperature, flow rate and packing pressure. Hence, a smaller residual stress and warpage would be better. Therefore, the four process parameters, that is, mold temperature, melt temperature, flow rate and packing pressure, should be viewed as design variables, while the two evaluation indicators, that is, residual stress of the stent in the orientation direction in the cavity (affecting the stent demolding process) and warpage (affecting the use properties of the stent), should be viewed as objective variables. Consequently, the multi-objective optimization design for the molding quality optimization problem during the stent injection molding process can be defined as follows:

$$\begin{aligned} \min \quad & \text{residual stress(RS), warpage(W)} \\ \text{s.t.} \quad & x_1 \leq x \leq x_2 \end{aligned} \tag{1}$$

Residual stress refers to the residual stress of the stent in the cavity in the first principal direction. Warpage refers to the total warpage of the stent, where x represents the optimization variables, and x_1 and x_2 represent the upper and lower limits of these optimization variables.

Residual stress and warpage are the two main indicators for stent injection molding quality. However, the RS and W of a polymeric stent varies with significantly different scales, making it difficult to define a suitable weight. Naturally, if both of these are scaled to (0, 1), the weight can be assigned an intermediate value of 0.5. In this case, the optimization problem mentioned above can be written as:

$$\begin{aligned} \min \quad & f(x) = 0.5 \frac{RS - RS_{\min}}{RS_{\max} - RS_{\min}} + 0.5 \frac{W - W_{\min}}{W_{\max} - W_{\min}} \\ \text{s.t.} \quad & x = \begin{bmatrix} T_{mold} & T_{melt} & v_{flow} & P_{pack} \end{bmatrix}^{\mathrm{T}} \\ & 80\,^{\circ}\mathrm{C} \leq T_{mold} \leq 90\,^{\circ}\mathrm{C} \\ & 190\,^{\circ}\mathrm{C} \leq T_{melt} \leq 205\,^{\circ}\mathrm{C} \\ & 0.13\ \mathrm{cm}^3/\mathrm{s} \leq v_{flow} \leq 0.25\ \mathrm{cm}^3/\mathrm{s} \\ & 75\% \leq P_{pack} \leq 90\% \end{aligned} \tag{2}$$

where RS represents the residual stress of the stent in the cavity in the first principal direction and W is the total warpage. T_{mold} is the temperature of the mold which can cool the melt, T_{melt} is the melt temperature and has a large influence on the melt flow viscosity, v_{flow} is the volume flow rate affecting the molding cycle, and P_{pack} is the packing pressure.

2.4. Finite Element Method of the PLA Stent Injection Molding Process

The stent microinjection molding process can be researched by using the finite element method (FEM). The numerical simulation platform based on the MOLDFLOW injection molding process was designed with a stent injection molding runner system, which includes six injecting gates (Figure 4). The runner system uses the cylinder units to divide the grids, which includes 501 cylinder units and 502 nodes, with an injecting system volume of 0.109295 cm^3. The stent structure has a uniform wall thickness. Hence, it uses the double-layer grid division, which includes 80,670 triangular units and 40,227 nodes, with a volume of 0.00333709 cm^3 and a grid matching ratio of 92.8%.

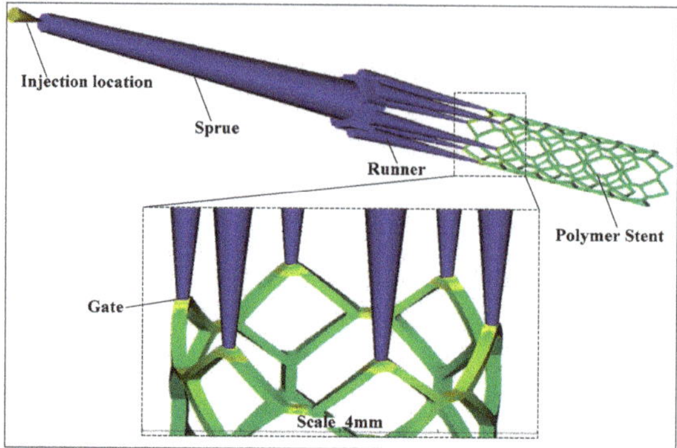

Figure 4. The injection molding process of stent based on the MOLDFLOW.

The injection molding runner system was adopted to conduct the numerical simulation of the stent injection molding process. The melt filling process is presented in Figure 5. The polymer melt is injected slowly into the cavity from the left end shown in Figure 5 and eventually fills the entire cavity. Compared to the eight filling stages, the filling process along the longitudinal direction is uniform.

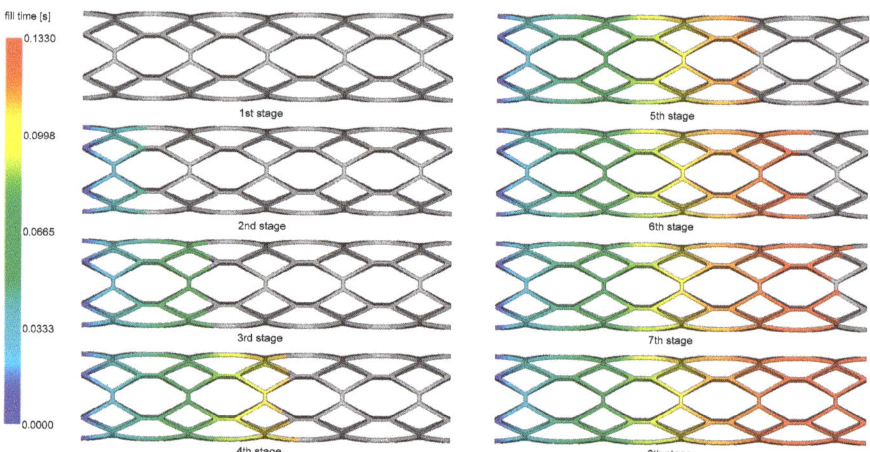

Figure 5. The injection molding process of a stent.

2.5. Optimization Algorithm

2.5.1. Approximate Method

The kriging model is an approximation technique that consists of a regression and a non-parametric part, including a polynomial and a random distribution:

$$\hat{y}(\mathbf{x}^i) = F(\beta, \mathbf{x}^i) + z(\mathbf{x}^i) = \mathbf{f}^T(\mathbf{x}^i)\beta + z(\mathbf{x}^i),$$ (3)

in which, β is the regression coefficient; $f(\mathbf{x}^i)$ is a polynomial that provides a global approximation of the simulation; and $z(\mathbf{x}^i)$ is a randomly distributed error that provides an approximation of the simulated local deviation and has the following statistical properties:

$$E[z(\mathbf{x})] = 0$$
$$\text{var}[z(\mathbf{x})] = \sigma_z^2$$
$$\text{corr}\left[z(\mathbf{x}^i), z(\mathbf{x}^j)\right] = R(\theta, \mathbf{x}^i, \mathbf{x}^j) = \prod_{l=1}^{m} \exp\left[-\theta(x_l^i - x_l^j)^2\right] \tag{4}$$

where $\mathbf{x}^i$ and $\mathbf{x}^j$ are any two points in the training samples; $R(\theta, \mathbf{x}^i, \mathbf{x}^j)$ is the correlation function with parameter θ, characterizing the spatial correlation between the points of training samples, and in this paper, Gaussian correlation function was used, which can be described as:

$$R(\theta, \mathbf{x}^i, \mathbf{x}^j) = \exp\left(-\sum_{l-1}^{n} \theta_k \left| x_l^i - x_l^j \right|^2\right) \tag{5}$$

where n is the number of design variables; x_l^i and x_l^j are the l-th component of the training sample.

2.5.2. Predictor

Given the training samples $\mathbf{S} = [x_1, x_2, \ldots, x_n]$ and corresponding response $\mathbf{Y} = [y_1, y_2, \ldots y_n]$, the response at a new point $\mathbf{x}^*$ can be expressed by a linear combination of $\mathbf{Y}$

$$\hat{y}(\mathbf{x}^*) = \mathbf{c}^T \mathbf{Y}, \tag{6}$$

The error is

$$\hat{y}(\mathbf{x}^*) - y(\mathbf{x}^*) = \mathbf{c}^T \mathbf{Y} - y(\mathbf{x}^*), \tag{7}$$

Substituting Equation (3) into Equation (7) gives

$$\begin{aligned}
\hat{y}(\mathbf{x}^*) - y(\mathbf{x}^*) &= \mathbf{c}^T (F\beta + \mathbf{Z}) - \left(f(\mathbf{x}^*)^T \beta + z\right) \\
&= \mathbf{c}^T \mathbf{Z} - z + \left(F^T \mathbf{c} - f(\mathbf{x}^*)\right)^T \beta
\end{aligned} \tag{8}$$

where $\mathbf{Z} = [z_1, z_2, \ldots, z_n]$, $\mathbf{F} = [f_1, f_2, \ldots, f_n]$. To get the unbiased predictor for $\mathbf{x}^*$, the mean error at this point should be zero, i.e.,

$$E(\hat{y}(\mathbf{x}^*) - y(\mathbf{x}^*)) = 0, \tag{9}$$

And then, $F^T \mathbf{c}(\mathbf{x}^*) - f(\mathbf{x}^*) = 0$. Therefore, we have the mean squared error of the predictor (8),

$$\begin{aligned}
\varphi(\mathbf{x}^*) &= E\left[(\hat{y}(\mathbf{x}^*) - y(\mathbf{x}^*))^2\right] \\
&= \sigma^2(1 + \mathbf{c}^T \mathbf{R}\mathbf{c} - 2\mathbf{c}^T \mathbf{r})
\end{aligned} \tag{10}$$

where

$$\mathbf{r}(\mathbf{x}^*) = [R(\theta, \mathbf{x}_1, \mathbf{x}^*), \ldots R(\theta, \mathbf{x}_n, \mathbf{x}^*)], \tag{11}$$

It represents the spatial correlation between new sample $\mathbf{x}^*$ and training samples. $\mathbf{c}(\mathbf{x}^*)$ can be obtained by minimizing $\varphi(\mathbf{x}^*)$, i.e.,

$$\begin{aligned}
\mathbf{c} &= \mathbf{R}^{-1}\left(\mathbf{r} - F\tilde{\lambda}\right) \\
\tilde{\lambda} &= (F^T \mathbf{R}^{-1} F)^{-1} (F^T \mathbf{R}^{-1} \mathbf{r} - f)
\end{aligned} \tag{12}$$

then

$$\hat{y}(\mathbf{x}^*) = f(\mathbf{x}^*)\hat{\beta} + \mathbf{r}(\mathbf{x}^*)^T \gamma, \tag{13}$$

where

$$\gamma = \mathbf{R}^{-1}(\mathbf{Y} - F\hat{\beta}), \tag{14}$$

Thus, we can predict the function value $\hat{y}(\mathbf{x}^*)$ at every new point $\mathbf{x}^*$ by using Equation (13).

2.5.3. Expected Improvement (EI)

Maximizing expectation improvement is an aspect of the method that adds new points that consider the prediction value and the prediction variance. Expected improvement (EI) calculates the probability of a target's response improvement at a given point. $Y(\mathbf{x})$ at some point $\mathbf{x}$ is unknown before sampling at this point. However, kriging can predict its mean $\hat{y}(\mathbf{x})$ and variance σ^2. Assuming that the response value of the current optimal design is Y_{min}, the improvement of the response of this point is $I = Y_{min} - Y(x)$ with a normal distribution. Thus, its probability density function is

$$\frac{1}{\sqrt{2\pi}\sigma(\mathbf{x})} \exp\left[-\frac{(Y_{min} - I - \hat{y}(\mathbf{x}))^2}{2\sigma^2(\mathbf{x})}\right], \tag{15}$$

Therefore, the expected improvement of the response value is

$$E[I(\mathbf{x})] = \int_{I=0}^{I=\infty} I\left\{\frac{1}{\sqrt{2\pi}\sigma(\mathbf{x})} \exp\left[-\frac{(Y_{min} - I - \hat{y}(\mathbf{x}))^2}{2\sigma^2(\mathbf{x})}\right]\right\} dI, \tag{16}$$

Integrating Equation (16) by parts, it gives

$$E[I(\mathbf{x})] = \sigma(\mathbf{x})[u\Phi(u) + \phi(u)], \tag{17}$$

where

$$u = \frac{Y_{min} - \hat{y}(\mathbf{x})}{\sigma(\mathbf{x})} \tag{18}$$

and where Φ and ϕ are the normal cumulative distribution and density functions, respectively.

2.5.4. The Convergence Criterion

The convergence condition of EI is

$$\frac{EI(\mathbf{x})}{Y_{max} - Y_{min}} < \varepsilon_1, \tag{19}$$

where ε_1 is the convergence tolerance. Y_{max} and Y_{min} are the maximal and minimal function values in samples, respectively.

In addition, the predicted value converges to the numerical result to characterize the accuracy of kriging model,

$$|f(\mathbf{x}_l) - \hat{y}_l| < \varepsilon_2 \tag{20}$$

where $f(\mathbf{x}_l)$ is the numerical result; $\hat{y}_l$ is the predicted value based on kriging; ε_2 is the convergence tolerance.

Moreover, the optimization results of the last two iterations should be similar,

$$|f(\mathbf{x}_l) - f(\mathbf{x}_{l-1})| < \varepsilon_3 \tag{21}$$

where $f(\mathbf{x}_l)$ and $f(\mathbf{x}_{l-1})$ are the response values of the last two iterations, respectively. Figure 6 shows the complete process of the optimization algorithm based on the Kriging surrogate model, making this process easier to understand.

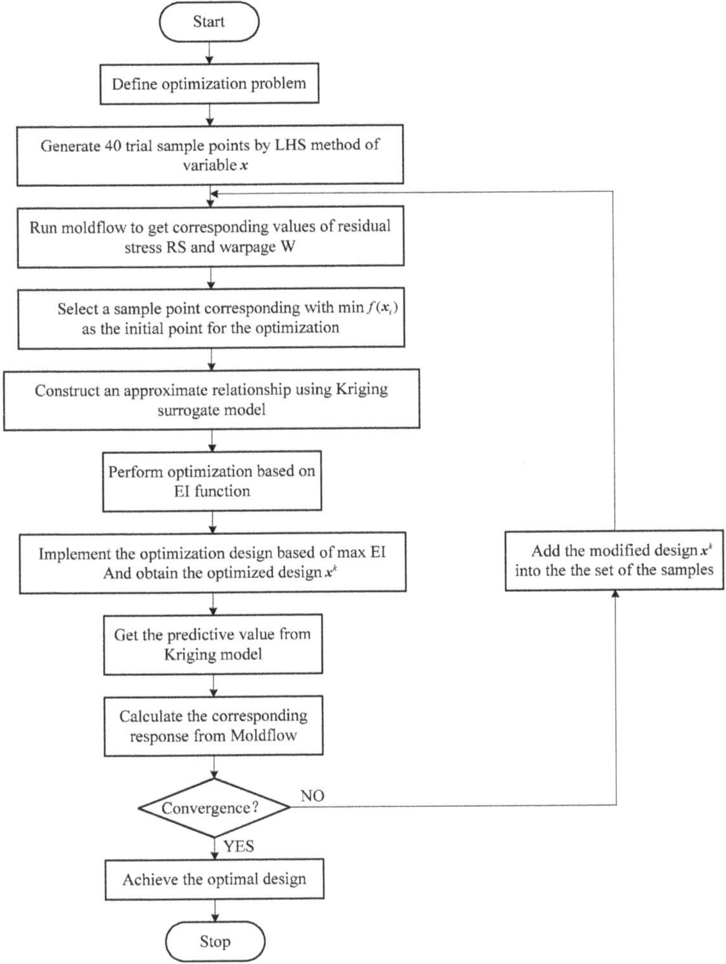

Figure 6. Complete process of the optimization algorithm based on the Kriging surrogate model.

3. Results

3.1. Results of Stent Injection Molding Process

The boundary values of the four design variables were respectively given. Then, 40 initial training sample points were generated using the orthogonal LHS method to build the Kriging surrogate model, with the aim of establishing the approximate relationship between the two objective variables of the stent (residual stress and warpage) and the four design variables.

The design variables included mold temperature T_{mold}, melt temperature T_{melt}, flow rate v_{flow}, and packing pressure P_{pack} and the range of the four process parameters were shown in Table 1.

Table 1. The range of the process parameter.

Stent	T_{mold} (°C)	T_{melt} (°C)	v_{flow} (cm³/s)	P_{pack} (%)
Lower limit	80	190	0.13	75
Upper limit	90	205	0.25	90

The Kriging surrogate model optimization process needs 20 iterations to obtain the optimal solution corresponding to the minimum values of objective functions. The optimization iterative process is shown in Figure 7.

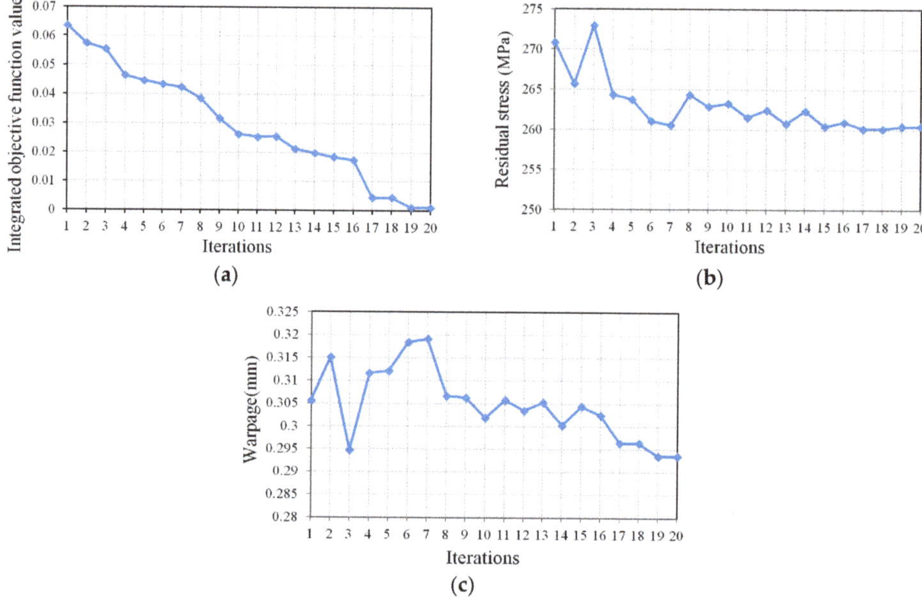

Figure 7. Optimization iterative process. (**a**) Relationship between integrated objective function value and iterations; (**b**) Relationship between residual stress and iterations; (**c**) Relationship between warpage and iterations.

The comparison between the optimal design result and comparable design result is presented in Table 2. The comparable design results, which adopt the intermediate values of the boundary values of the four design variables recommended by MOLDFLOW, correspond to the residual stress and warpage results.

Table 2. Comparison of the two design results.

Stent	T_{mold} (°C)	T_{melt} (°C)	v_{flow} (cm³/s)	P_{pack} (%)	RS	W
Comparable design	85	197.5	0.19	82.5	326.9	0.5304
Optimal design	87.5	205	0.13	75	260.5	0.2936

It can be observed that merely the design recommended by MOLDFLOW could not effectively reduce the residual stress and warpage of the stent, while the optimal design based on the iteration process can significantly reduce residual stress and warpage. Compared to the results of the comparable design, residual stress decreased by 20.3% and warpage decreased by 44.6%. Apparently, this was attributable to the enhancement of the molding quality of the stent.

Figures 8 and 9 show the corresponding residual stress and warpage of these two kinds of designs, and the results of the optimal design obviously had smaller residual stress and warpage.

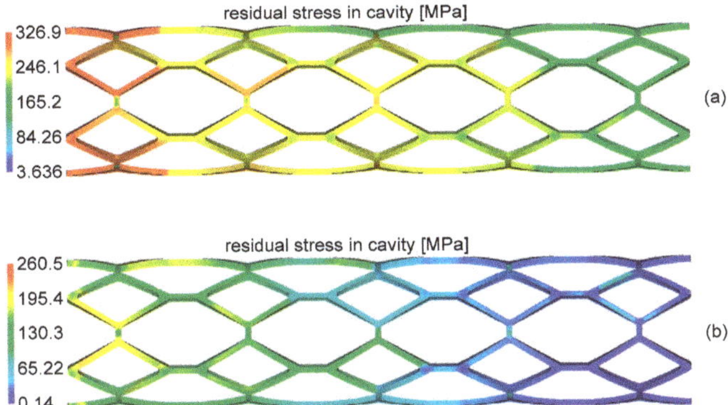

Figure 8. The distribution of residual stress of the comparable design and optimal design. (**a**) The distribution of residual stress of the comparable design; (**b**) The distribution of residual stress of the optimal design.

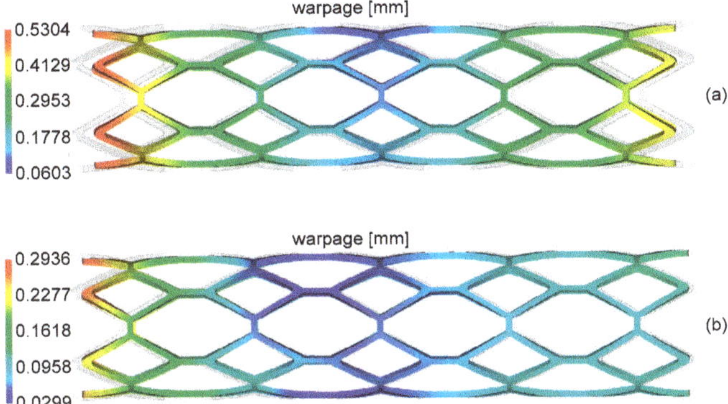

Figure 9. The distribution of warpage of the comparable design and optimal design. (**a**) The distribution of warpage of the comparable design; (**b**) The distribution of warpage of the optimal design.

3.2. Results of Single Factor Analysis

In order to investigate the influences of these four important parameters on the residual stress and warpage of the stent during the injection molding process, it is necessary to carry out a single factor analysis. In the optimal process combination, the influence of the remaining factor on the result was analyzed by fixing three factors, and the results are shown as follows. It revealed that the effects and mechanisms of these four parameters on the residual stress and warpage of the stent are significantly different.

The effect of mold temperature on the residual stress and warpage of the stent is presented in Figure 10. As illustrated by the curve change trend, mold temperature has a relatively significant influence on the residual stress and warpage of a stent. Raising the mold temperature could decrease the warpage. As for the PLA semi-crystalline material, high mold temperatures can reduce the flow viscosity of the melt, causing the melt to easily flow and enhance mold-filling efficiency. A relatively high mold temperature can decrease the temperature difference of the melt, cool the melt uniformly, reduce shrinkage deformation, and prevent defects, such as depression, deformation, etc. Excessively high mold temperatures and being close to the boundary value expands the volume of the melt, causes warpage, and prolongs the cooling time, thereby reducing production efficiency. Raising the mold temperature is attributable to improving surface roughness, reducing inner stress and the degree of orientation, enhancing the strength of the weld line and product density, which thereby reduces the impact strength of the streamline direction, and reduces residual stress in the cavity.

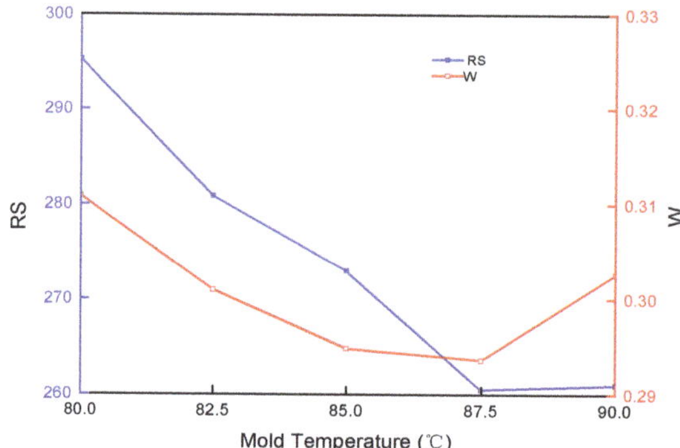

Figure 10. The effect of mold temperature on the residual stress and warpage.

The effect of melt temperature on the residual stress and warpage of a stent is presented in Figure 11. As for the PLA semi-crystalline material, with its small crystalline tendency, the melt temperature would mainly affect the viscosity, flowability and molecular orientation of the melt. On one hand, enhancing the melt temperature would expand the volume of the melt, reducing the amount of material into the cavity. Furthermore, this would cause the loose status of the polymer in the orientation direction, and enhance the molecular chain orientation ability, resulting in an increase in warpage. On the other hand, enhancing the melt temperature can reduce the viscosity of the melt, and enhance the shear rate of the melt shear rate, when the injection pressure and packing pressure remain unchanged. This is attributable to the transfer of the pressure to the cavity, reducing the warpage. In general, as for the latter, viscosity has a relatively significant influence on temperature-sensitive materials. As for the former, viscosity has a relatively significant influence on non-temperature-sensitive material. PLA melt viscosity is relatively sensitive to temperature. Therefore, enhancing the melt temperature, as a whole, can reduce warpage. Furthermore, enhancing the melt temperature can effectively reduce melt viscosity. This would cause filling to be more uniform and reduce residual stress in the cavity. The stent has a number of uniformly-distributed injecting gates, which are contributive to the transfer of pressure, increasing the material supplement, reducing the shrinkage of the flat plate, and decreasing the warpage. Therefore, the effect of melt temperature on residual stress and warpage is the comprehensive result of these above-mentioned factors.

The influence of flow rate on the residual stress and warpage of the PLA stent is presented in Figure 12, showing a non-linear relationship obviously. The flow rate of the melt in the cavity is basically controlled by the flow rate of the injection. If the cavity structure is the same as other conditions, a relatively small flow rate would cause the melt to flow slowly in the cavity, and the melt compressed at the injecting gate can be fully loosened during the slow flow process. Furthermore, the melt has a relatively small uniformly-distributed residual stress in the follow-up packing and cooling process, which makes the warpage very small. With the increase in flow rate, the quicker the melt flows in the cavity, the bigger the flow and shear rate of the melt is, and the higher its shear stress becomes. Typically, the above-mentioned facts result in the higher residual stress of a stent, and even serious warpage. With the further increase in flow rate, the flow and shear rate of the melt increases. At the same time, melt viscosity also decreases to make the melt easily flow, and allow the cavity to be filled well, which can reduce residual stress and warpage. Apparently, the flow rate can affect the looseness and viscosity of the melt; weighing the proportion of these two by adjusting the flow rate can effectively reduce residual stress and warpage. Indeed, a relatively small flow rate correlates to longer injection molding time, while a relatively large flow rate correlates to greater injection pressure. In the actual molding process, equipment and production efficiency should be considered. Furthermore, as for various types of materials, the effect of the flow rate on residual stress and warpage is relatively complex, because the mold-filling rate of the melt depends on the flow rate, while the effect of the mold-filling rate on plastic product properties is also relatively complex, which is correlated to the type of material.

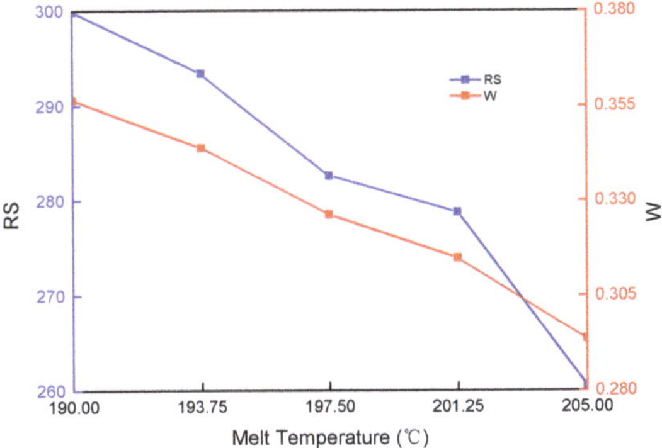

Figure 11. The effect of melt temperature on the residual stress and warpage.

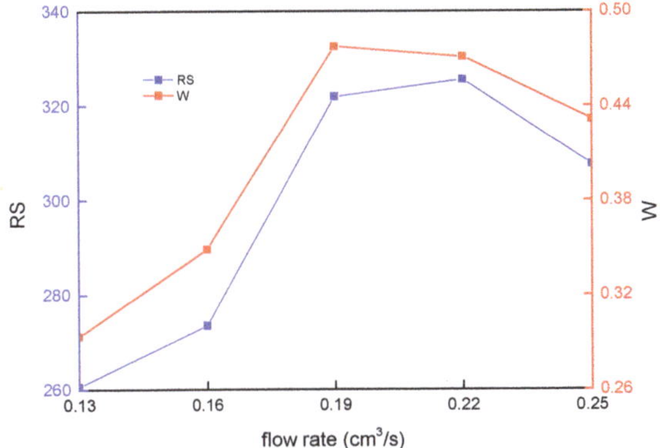

Figure 12. The effect of packing pressure on the residual stress and warpage.

The effect of packing pressure on the residual stress and warpage of a stent is presented in Figure 13. As for the thin-wall stent structure of the PLA material, merely relatively low packing pressure can ensure the relatively small warpage and residual stress. As for thin-walled parts, a relatively high packing pressure increases residual stress, causing non-uniform shrinkage, thereby resulting in a relatively larger warpage. Indeed, plastics and products with a relatively large viscosity, and have relatively high requirements in small wall thickness, a long process and high accuracy, needs a relatively high packing pressure. The present study considers these from the perspective of the optimal design.

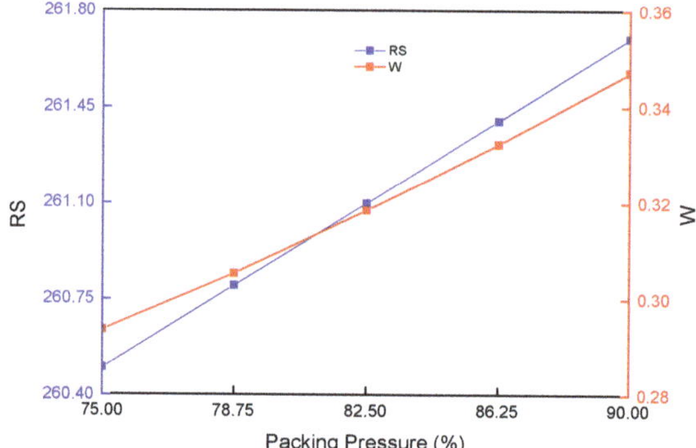

Figure 13. The effect of packing pressure on the residual stress and warpage.

4. Discussion

As for the biodegradable PLA polymeric stent, the shortage of an accurate and effective manufacturing method is a potential limit to the development of polymeric stents. A PLA stent, which has low mechanical strength and an extremely tiny volume, is a typical precision thin-walled product. Due to its complex stent structure and tiny local size, the microinjection molding stent is greatly different from products manufactured by conventional microinjection molds. The stent product has a large length-to-diameter ratio, and a relatively low surface continuity, which influences the flowability of the polymeric melt in the cavity. Therefore, it would be difficult to completely fill the cavity during the experiment process. This is one of main difficulties of microinjection molding. In addition, the stent can easily form weld marks on the surface, and the number of the weld marks is expected to be more than that of conventional injection molding products.

In the design process, injection molding process parameters should be repeatedly adjusted to meet the quality requirements of plastic parts. The unreasonable parameters in the injection molding process, such as mold temperature, melt temperature, flow rate, and packing pressure, can result in serious residual stress and warpage of the stent. A relatively large residual stress in the cavity can result in difficulties during the demolding process, and even damage the stent during the demolding process. Furthermore, a relatively large warpage can greatly influence the product's use properties, and even its normal use.

In the present study, in order to reduce residual stress and warpage in the polymeric stent injection molding process, the study on the effect of process parameters, such as mold temperature, melt temperature, flow rate and packing pressure, can provide guidance in selecting the process parameter combination during the injection molding process. The scope of the studied process parameters are as follows: mold temperature (80, 90), melt temperature (190, 205), flow rate (0.13, 0.25), and packing pressure (75%, 90%). Therefore, the purpose of the optimal design in the stent injection molding process was to maximize the residual stress and warpage of the stent, thereby reducing the difficulties of demolding and the use properties of the product. Compared to the results of the comparable design, residual stress decreased by 20.3%, while warpage decreased by 44.6%. These apparently contributed to enhancing the molding quality of the stent. It should be acknowledged that the present study came to this conclusion based on the PLC material, without a setting of a cooling waterway, and without considering the effect of the actual cooling process. When the stent materials were different, or the mold structure was different from that in the present study, this would result in the optimal results of the stent's different process parameters. Of course, when investigating polymer stent of different materials or different structures, the optimization results of the injection molding process will be different. This research highlights how the surrogate model can be established and used to optimize the stent injection molding process in order to demonstrate that the process can be designed and improved through this important approach.

5. Conclusions

The present study presents a multi-objective optimization method combined with the Kriging surrogate model to improve PLA stent injection molding process parameters, with the aim of enhancing the injection molding quality of micro absorbable polymeric stents. The Kriging surrogate model coupled with DOE methods was adopted to establish the approximate relationship between objective function and design variables. These results show that the proposed optimization method could be effectively used for the microinjection molding process. These results indicate that it is very useful based on the proposed optimization method to improve the molding quality during the microinjection molding process. This exhibits a novel methodology for the injection molding process design of polymer stents, representing a new research direction. The combination of this optimization method and experimental verification can play an important role in the process design of injection molding stents.

Author Contributions: H.L., D.Z., M.W., Q.L. and J.H. were responsible for the design, data collection and overall investigation, and established the optimization model. H.L. and Q.L. were responsible for the optimization method. H.L. and K.L. were responsible for the finite element analysis and the writing of the paper.

Funding: This research was funded by the National Natural Science Foundation of China (grant no.11502044 and 51675079), the Fundamental Research Funds for the Central Universities (grant no.DUT17RC(3)104) and National Center for International Research of Micro-nano Molding Technology & Key Laboratory for Micro Molding Technology of Henan Province Joint Open Fund (grant no. MMT2017-03).

Conflicts of Interest: The authors declare no conflict of interest.

References

1. Onuma, Y.; Serruys, P.W. Bioresorbable scaffold: the advent of a new era in percutaneous coronary and peripheral revascularization? *Circulation* **2011**, *123*, 779–797. [CrossRef] [PubMed]
2. Clarke, G.; Mulvihill, H.; Duffy, A. Bioresorbable Stent and Method of Making. U.S. Patent 20,080,234,831, 25 September 2008.
3. Liu, D.L.; Shen, C.Y.; Liu, C.T.; Xin, Y.; Sun, L. Investigation of mold temperature affecting on shrinkage of rapid heat cycle molding plastic part. *Adv. Mater. Res.* **2011**, *189*, 2477–2481. [CrossRef]
4. Duo, Y. Effect of micro-injection molding process parameters for warpage in Micro Plate Devices. *Adv. Mater. Res.* **2014**, 265–268.
5. Jiang, J.; Wang, S.; Sun, B.; Ma, S.J.; Zhang, J.M.; Li, Q.; Hu, G.H. Effect of mold temperature on the structures and mechanical properties of micro-injection molded polypropylene. *Mater. Des.* **2015**, *88*, 245–251. [CrossRef]
6. Singh, G.; Verma, A. A Brief Review on injection moulding manufacturing process. *Mater. Today* **2017**, *2*, 1423–1433. [CrossRef]
7. Annicchiarico, D.; Alcock, J.R. Review of factors that affect shrinkage of molded part in injection molding. *Mater. Manuf. Process.* **2014**, *29*, 662–682. [CrossRef]
8. Mohan, M.; Ansari, M.N.M.; Shanks, R.A. Review on the effects of process parameters on strength, shrinkage, and warpage of injection molding plastic component. *Polym. Plast. Technol. Eng.* **2017**, *56*, 1–12. [CrossRef]
9. Shen, C.; Wang, L.; Li, Q. Optimization of injection molding process parameters using combination of artificial neural network and genetic algorithm method. *J. Mater. Process. Technol.* **2007**, *183*, 412–418. [CrossRef]
10. Gheorghe, O.C.; Florin, T.D.; Vlad, G.T.; Gabriel, D.T. Optimization of micro injection molding of polymeric medical devices using software tools. *Procedia Eng.* **2014**, *69*, 340–346. [CrossRef]
11. Kitayama, S.; Natsume, S. Multi-objective optimization of volume shrinkage and clamping force for plastic injection molding via sequential approximate optimization. *Simul. Model. Pract. Theory* **2014**, *48*, 35–44. [CrossRef]

12. Oliaei, E.; Heidari, B.S.; Davachi, S.M.; Bahrami, M.; Davoodi, S.; Hejazi, I.; Seyfi, J. Warpage and shrinkage optimization of injection-molded plastic spoon parts for biodegradable polymers using taguchi, ANOVA and artificial neural network methods. *J. Mater. Sci. Technol.* **2016**, *32*, 710–720. [CrossRef]

13. Dang, X.P. General frameworks for optimization of plastic injection molding process parameters. *Simul. Model. Pract. Theory* **2014**, *41*, 15–27. [CrossRef]

14. Kashyap, S.; Datta, D. Process parameter optimization of plastic injection molding: A review. *Int. J. Plast. Technol.* **2015**, *19*, 1–18. [CrossRef]

15. Kitayama, S.; Yamazaki, Y.; Takano, M.; Aiba, S. Numerical and experimental investigation of process parameters optimization in plastic injection molding using multi-criteria decision making. *Simul. Model. Pract. Theory* **2018**, *85*, 95–105. [CrossRef]

16. Krige, D.G. A statistical approach to some basic mine valuation problems on the Witwatersrand. *J. South. Afr. Inst. Min. Metall.* **1951**, *52*, 119–139.

17. Spina, R.; Spekowius, M.; Hopmann, C. Multiphysics simulation of thermoplastic polymer crystallization. *Mater. Des.* **2016**, *95*, 455–469. [CrossRef]

18. Gao, Y.H.; Wang, X.C. Surrogate-based process optimization for reducing warpage in injection molding. *J. Mater. Process. Technol.* **2009**, *209*, 1302–1309. [CrossRef]

19. Li, H.X.; Wang, X.Y.; Wei, Y.B.; Liu, T.; Gu, J.F.; Li, Z.; Wang, M.J.; Zhao, D.Y.; Qiao, A.; Liu, Y.H. Multi-Objective Optimizations of biodegradable polymer stent structure and stent microinjection molding process. *Polymers* **2017**, *9*, 20. [CrossRef]

20. Joseph, V.R.; Hung, Y. Orthogonal-maximin Latin hypercube designs. *Stat. Sin.* **2008**, 171–186.

Article

Laser-Assisted Thermal Imprinting of Microlens Arrays—Effects of Pressing Pressure and Pattern Size

Keisuke Nagato *, Yuki Yajima and Masayuki Nakao

Department of Mechanical Engineering, Graduate School of Engineering, The University of Tokyo, Tokyo 113-8656, Japan; yajima@hnl.t.u-tokyo.ac.jp (Y.Y.); nakao@hnl.t.u-tokyo.ac.jp (M.N.)
* Correspondence: nagato@hnl.t.u-tokyo.ac.jp; Tel.: +81-3-5841-6361

Received: 30 January 2019; Accepted: 22 February 2019; Published: 25 February 2019

Abstract: Polymer films with nano- or microstructured surfaces have been widely applied to optical devices, bioplates, and printed electronics. Laser-assisted thermal imprinting (LATI), in which a laser directly heats the surfaces of a mold and a thermoplastic polymer, is one of the high-throughput methods of replicating nano- or microstructures on polymer films. Only the surfaces of the mold and polymer film are heated and cooled rapidly, therefore it is possible to replicate nano- or microstructures on polymer films more rapidly than by using conventional thermal nanoimprinting. In this study, microlens arrays (MLAs) were replicated on polymethylmethacrylate (PMMA) films using LATI, and the effects of the pressing pressure (10−50 MPa) and the pattern size (33- and 5-μm pitch) of the MLA on the filling ratio were investigated by analyzing a microlens replicated using different laser-irradiation times (0.1−2 ms). The filling ratio increased with increasing pressing pressure and laser-irradiation time in the replication of MLAs with varying sizes, while the flow of the PMMA varied with the pressing pressure and laser-irradiation time. It was found that during filling, the shape of the polymer cross-sectional surface demonstrated a double and single peak in the 33- and 5-μm-pitch patterns, respectively. This was because the depth of the heated area in the 33-μm-pitch pattern was smaller than the pattern size, whereas that of the 5-μm-pitch pattern was comparable to (or larger) than the pattern size.

Keywords: laser-assisted thermal imprinting; pressure; pattern size; thermoplastic polymer; microlens array

---

1. Introduction

Nano- or microstructured polymer surfaces perform various functions, such as hydrophilic or hydrophobic interactions with liquids, reduction of friction or sticking to solid surfaces, demonstrating an anchoring effect for adhesives, and optical or photonic behaviors. In terms of their use in optical or photonic capacities, antireflection structures have been fabricated to reduce light reflection and to increase the incident light transmittance in solar cells, flat panel displays and so forth [1]. Microlens arrays (MLAs) are fabricated on OLED devices, or OLEDs are fabricated on polymer films with photonic crystal surfaces [2–6] to improve the light extraction efficiency of organic light-emitting diodes (OLEDs). Furthermore, various nanostructures have been fabricated on polymer films for organic photovoltaic cells [7] and nanopillars have been fabricated for plasmonic biosensing [8]. As described above, nano- or microstructures are expected to become more commonly applied to various optical and photonic devices.

For these applications, it is necessary to fabricate nano- or microstructures on polymer films with a high throughput and low cost. Nanoimprint lithography (NIL), which was developed by Chou et al., is one of the highest-throughput and most inexpensive methods of fabricating nano- or microstructures [9]. To obtain nano- or microstructured surfaces on polymer films, three main kinds of nanoimprint method have been proposed and developed: thermal nanoimprinting [9], ultraviolet (UV) nanoimprinting [10], and soft lithography [11]. UV lithography involves the use of a UV-curable polymer which is coated and molded and then cured using UV light. Soft lithography involves the use of thermocurable polymer coatings, which are cured post heating. These lithography techniques possess advantages associated with the use of soft molds and the use of low applied pressures because of the low viscosity of polymers. Flexible substrates can thus be coated and highly-functional materials, such as those possessing ferroelectric functionality, can be used [12,13]. On the other hand, thermal nanoimprinting is a direct imprinting method used to fabricate nano- and microstructures on polymer films. Therefore, it has the advantage of simplicity in terms of the manufacturing process because of the absence of a coating process [14]. In thermal nanoimprinting, a mold with a nano- or microstructured surface is heated and pressed directly onto a thermoplastic polymer film. The polymer is heated to a temperature above its glass transition temperature (T_g), causing it to flow and fill the nano- or microstructures of the mold. The mold and polymer are then cooled and demolded. Nano- or microstructures can be replicated on polymer films using these simple procedures. This simplicity has increased the popularity of thermal nanoimprinting among researchers as both a promising and convenient method [15].

The mold and polymer require heating and cooling in the thermal nanoimprinting process. The reduction in cycle time is thus limited by the thermal conductivities and specific heat capacities of the mold and polymer. On the other hand, in laser-assisted nanoimprinting, which was developed by Chou et al., nano- or microstructures are replicated by heating only the surfaces of a mold and a substrate using a laser [16]. Xia et al. and Grigalinūnas et al. replicated the nanostructures of a glass mold on a thermoplastic polymer film [17,18]. Grigalinūnas et al. also demonstrated nanoimprint lithography by using an Si mold and a CO_2 laser [19]. Nagato et al. fabricated nanostructures of diamond-like carbon on a glass mold and replicated them on thermoplastic polymer films in a predefined area by scanning with a laser. They also demonstrated that laser heating reduced the cycle time for thermal imprinting below that associated with conventional thermal nanoimprinting [20].

However, although it has already been demonstrated that various nanostructures could be replicated on polymer films using laser-assisted thermal imprinting (LATI), this has only been achieved in a few instances. When microstructures are replicated on a polymer film using LATI, the polymer cannot effectively fill the microstructures of the mold, and the microstructures can therefore not be completely replicated. This is because it takes a finite amount of time for the polymer to flow and fill the microstructures of the mold. The flow of the polymer is caused by the applied pressure generated by pressing the mold onto the polymer film and the decrease in viscosity of the polymer is caused by laser heating. Therefore, the pressing pressure and laser-irradiation time both have an important effect on the filling ratio of the polymer (Figure 1).

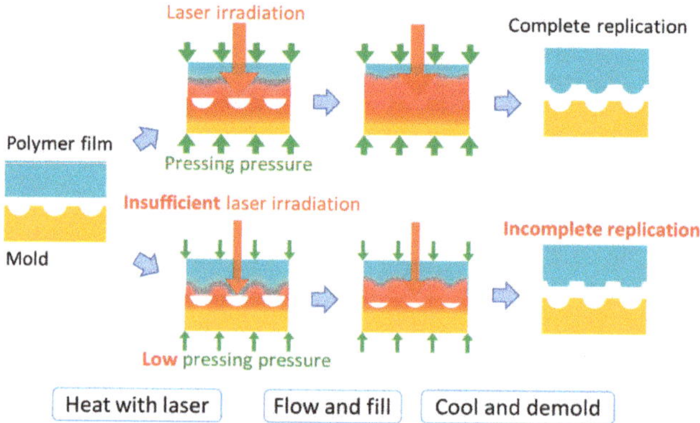

Figure 1. Schematic showing complete and incomplete replication.

In this study, microstructures were replicated on polymer films using LATI and the phenomenon of polymer filling was investigated. MLAs with different pattern sizes were used as microstructures and the effects of the pressing pressure, and the pattern size of the MLA, were investigated and the filling ratio was measured by analyzing a microlens replicated using different laser-irradiation times.

2. Experimental

2.1. MLA Molds

Figure 2 shows scanning electron microscope (SEM) (semi-inlens field-emission type, SU-8010, Hitachi High-Technologies Corporation, Tokyo, Japan) secondary-electron images of the Ni concave molds with MLAs used in this study. It is possible to fabricate MLAs with various pattern sizes, which are used to improve the light extraction efficiency of OLEDs [21]. One of the MLAs used in this study had a lens diameter of 30 µm, a lens depth of 14 µm, and a pitch of 33 µm, and the other had a spherical lens diameter of 4.2 µm, a lens depth of 2.1 µm, and a pitch of 5 µm. The microlenses were arranged in a regular triangular lattice on both MLAs. The Ni concave molds for the MLAs were constructed using electroforming on convex master molds. Each Ni concave mold had a size of 10×10 mm^2 and a thickness of 200 µm. The convex master molds for the electroforming process were prepared using photolithography and etching on a glass substrate. The mold surfaces were treated with a lubricant for the nanoimprinting mold, which was a hydrophobic coating material (DURASURF, Harves Co. Ltd., Saitama, Japan). A polymethylmethacrylate (PMMA) film was used as a thermoplastic film. The PMMA film had a thickness of 75 µm and T_g of approximately 89 °C (HBA002P, ACRYPLEN, Mitsubishi Chemical Corporation, Tokyo, Japan).

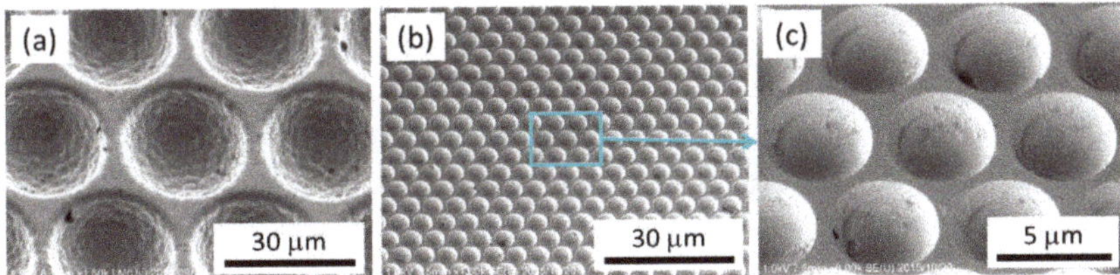

Figure 2. Scanning electron microscope (SEM) images of the microlens arrays (MLAs) in the Ni concave molds all shown as 30°-tilted views. (**a**) 33-µm-pitch MLA, (**b**) 5-µm-pitch MLA, and (**c**) magnified view of 5-µm-pitch MLA.

2.2. Experimental Setup and Conditions

Figure 3 shows the experimental setup for the LATI process. A pressure-sensitive paper (Prescale MS R270 10M, of range: 10–50 MPa and accuracy: $<\pm10\%$, Fujifilm Corporation, Tokyo, Japan) was used to measure the pressing pressure. The pressure was applied using an air cylinder (Misumi, Tokyo, Japan). The pressure-sensitive paper was placed on the Ni concave mold, and the paper and mold were pressed onto a glass plate with a thickness of 15 mm. The pressure applied to the pressure sensitive paper was then measured. In this study, the pressure measured using the pressure-sensitive paper was defined as the pressing pressure. The pressing pressure was set to 10, 20, 30, 40, or 50 MPa by adjusting the air pressure in the cylinder. The above procedure was conducted for each Ni concave mold [11].

After the measurement of the pressing pressure, the pressure-sensitive paper was removed and a PMMA film was placed on the Ni concave mold. The PMMA film and mold were then pressed onto the glass plate with pressing pressures of 10, 20, 30, 40, or 50 MPa. A continuous-wave single-mode fiber laser (SPI Lasers, Southampton, UK) with a wavelength of 1070 nm, 100 W power, and 500 µm diameter was used in this study to irradiate the surface of the Ni concave mold. After the laser was scanned once along a line of length 10 mm, the PMMA film was switched to an unimprinted one and the laser-irradiation time was varied by varying the scanning speed. The scanning speeds were 240,

320, and 500 mm/s for the 33-μm-pitch MLA and 1000, 2000, and 5000 mm/s for the 5-μm-pitch MLA. The laser irradiation-time was defined as the laser-irradiation diameter divided by the scanning speed. The laser-irradiation times were calculated to be 1.0, 1.6, and 2.1 ms for the 33-μm-pitch MLA and 0.10, 0.25, and 0.50 ms for the 5-μm-pitch MLA. Then, for each condition, the MLA replication on the PMMA film was analyzed using a laser microscope (OLS4100, Olympus Corporation, Tokyo, Japan) and the filling ratio of the PMMA was calculated. The height within each cell of area 0.125×0.125 μm^2 was obtained using the laser microscope. The product of the height and area of each cell was then summed for each replicated microlens, and the sum was defined as the volume of the replicated microlens. The volume of the microlens was calculated for each concave master mold. The filling ratio was then defined as the volume of the replicated microlens as a percentage of the volume of the microlens on the concave master mold.

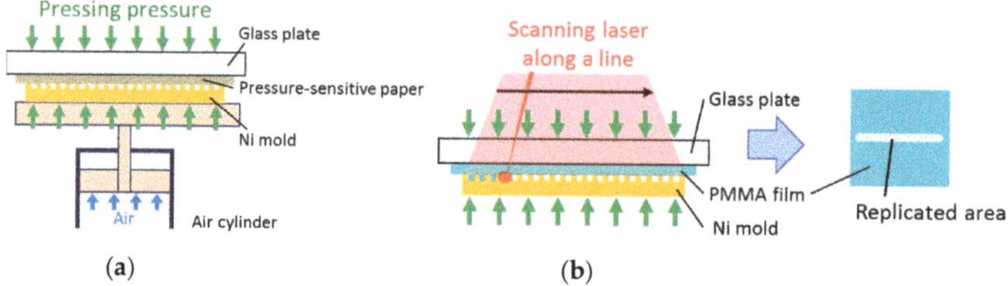

Figure 3. Schematics of (**a**) measurement of the pressing pressure and (**b**) replication by scanning the laser along a line.

3. Results and Discussion

3.1. Replication of 33-μm-Pitch MLA

Figure 4 shows the relationship between the filling ratio and the pressing pressure for each laser-irradiation time in the replication of the 33-μm-pitch MLA. In addition, Figure 4 shows laser microscopic images taken at pressing pressures of 10, 30, and 50 MPa for the laser-irradiation time of 1.6 ms, an image taken at a pressing pressure of 40 MPa and laser-irradiation time of 1.0 ms, and an image taken at a pressing pressure of 20 MPa and laser-irradiation time of 2.1 ms. The filling ratio increased with increasing pressing pressure and laser-irradiation time. This tendency showed that the higher the pressing pressure, the faster the PMMA flowed and filled the mold. When the pressing pressure was increased, the thermal contact resistance between the PMMA film and the Ni mold decreased. The temperature of the PMMA then increased because of the increase in heat input, and the viscosity of the PMMA decreased. This decrease in the viscosity allowed the PMMA to flow more easily. At a pressing pressure of 50 MPa, although the filling ratio was approximately 100% for the laser-irradiation times of 1.6 and 2.1 ms, it was only approximately 60% for the laser-irradiation time of 1.0 ms. This was due to the laser irradiation being insufficient for the laser-irradiation time of 1.0 ms. When the laser irradiation was insufficient, the depth of the PMMA experiencing a temperature exceeding T_g was smaller than the lens depth of the 33-μm-pitch MLA. The filling ratio was almost identical when the pressing pressure and laser-irradiation time were 40 MPa and 1.0 ms, 30 MPa and 1.6 ms, and 20 MPa and 2.1 ms, respectively. In the laser microscopy images obtained under these conditions, the height difference between the highest area and the center of the replicated microlens decreased with increasing laser-irradiation time. This tendency showed that the temperature of the PMMA exceeded T_g not only in the mold contact area but also in the noncontact area when sufficient laser irradiation was provided, causing the PMMA to flow in both the contact area and the noncontact area. On the other hand, when the laser irradiation was insufficient, the temperature of the PMMA only exceeded T_g in the contact area and the PMMA flowed only in the contact area. As can be seen in the result associated with an irradiation time of 2.1 ms and pressing pressure of 50 MPa, the filling was almost perfect. The LATI process possibly did not reach 100% filling because of shrinkage due to rapid cooling after the laser irradiation. However, the filling was successful using the various pressing pressures applied in this study.

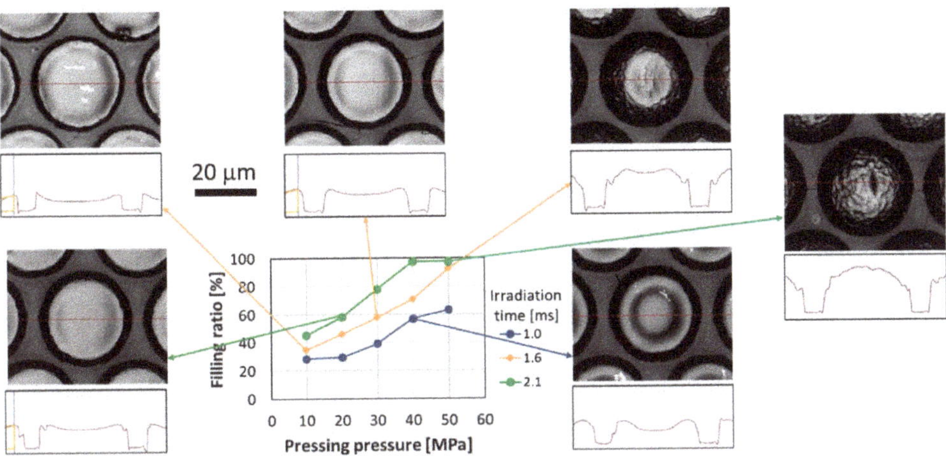

Figure 4. Filling ratio as a function of pressing pressure for each irradiation time in the replication of the 33-μm-pitch MLA.

3.2. Replication of 5-μm-Pitch MLA

Figure 5 shows the relationship between the filling ratio and the pressing pressure for each laser-irradiation time in the replication of the 5-μm-pitch MLA. In addition, Figure 5 shows laser microscopic images taken at pressing pressures of 10, 30, and 50 MPa for the laser-irradiation time of 0.25 ms, and an image taken at a pressing pressure of 50 MPa and laser-irradiation time of 0.10 ms. For the laser-irradiation times of 0.10 and 0.25 ms, the filling ratio increased with increasing pressing pressure and laser-irradiation time. For the laser-irradiation times of 0.25 and 0.50 ms, the filling ratio decreased for some pressing pressures. These decreases in the filling ratio may have been due to calculation errors because the PMMA filled the mold almost completely according to the laser microscopy image obtained at a pressing pressure of 50 MPa and laser-irradiation time of 0.25 ms. The filling ratio was approximately 60% at a pressing pressure of 50 MPa and laser-irradiation time of 0.10 ms. This is because the laser irradiation was insufficient for the laser-irradiation time of 0.10 ms. The filling ratio was almost identical when the pressing pressures and laser-irradiation times were 50 MPa and 0.10 ms, and 10 MPa and 0.25 ms, respectively. The laser microscopy images obtained under these conditions showed that the height difference between the highest area and the center of a replicated microlens was almost identical. This indicated that the temperature of the PMMA exceeded T_g not only in the mold contact area, but also in the noncontact area, causing the PMMA to flow from both the contact area and the noncontact area. For an irradiation time of 0.50 ms, the filling ratio was unstable up to approximately 5% throughout the range of pressing pressures. One of the reasons for this is the limitation in accuracy of the laser microscope. A 5% reduction corresponds to a 2% or 3% error in measured length, which equates to 100 or 150 nm error over 5 μm. To investigate this stability, more precise measurements, such as might be attained using an atomic force microscope (AFM), are necessary. The other reason for this 5% disparity is that the heated and low-viscosity area was deeper than shorter irradiation time and the effect of supplying the polymer from underneath and that of escaping the pressure to unirradiated area in planar direction.

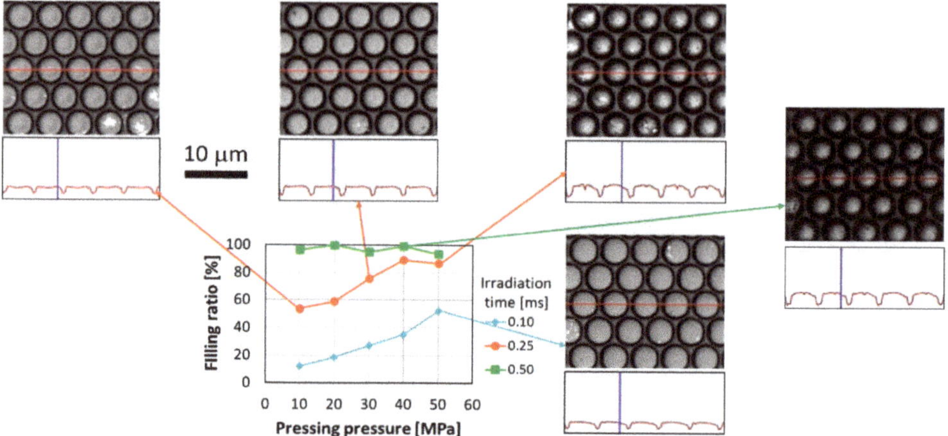

Figure 5. Filling ratio as a function of pressing pressure for each irradiation time in the replication of the 5-µm-pitch MLA.

3.3. Comparison of Surface Shapes during Imprinting

This section discusses the differences in the shape of the polymer surfaces after undergoing the LATI process. As shown in Sections 3.1 and 3.2, the leading surface of the polymer in the cavity of the 33- and 5-µm-pitch MLAs showed double and single peaks in the cross-section view after imprinting, respectively. Figure 6 shows the schematic of the heat conduction and polymer flow for the 33- and 5-µm-pitch MLAs. In the larger cavity, the depth of the heated area was smaller than the scale of the cavity and the polymer near the contact surface preferentially decreased in viscosity and a resultant double peak was formed, as shown in Figure 6a. In the smaller cavity, the depth of the heated area was comparable to, or larger than, the scale of the cavity and all the polymer around the contact surface was heated and the resultant polymer flow caused a single peak as shown in Figure 6b. To more comprehensively analyze the polymer flow phenomena seen in Figure 6a,b, experiments should be conducted using MLA with other sizes ranging from between 33 and 5 µm pitch. Applying simulations that incorporate the modelling of viscoelastic body behavior [22–24] to the MLA imprinting process would also be useful. Furthermore, other parameters such as laser power, spot diameter, and offset temperature are important and further investigation of these parameters would expand the discussion.

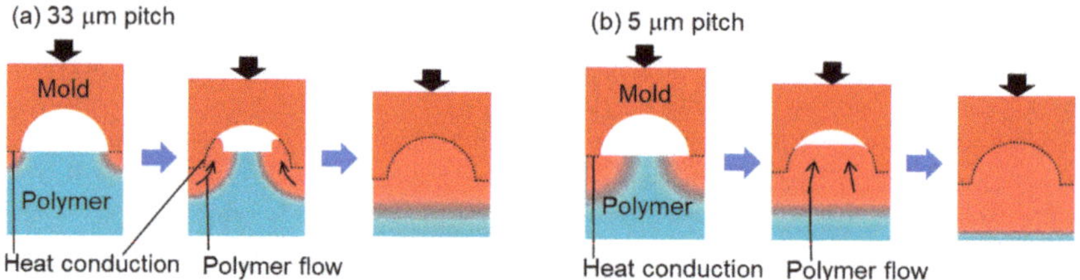

Figure 6. Schematics showing heat conduction and polymer flow during the laser-assisted thermal imprinting (LATI) process applied to (**a**) 33-µm-pitch and (**b**) 5-µm-pitch MLAs.

4. Conclusions

33- and 5-µm-pitch MLAs were replicated on PMMA films using LATI. The effects of the laser-irradiation time, the pressing pressure, and the pattern size of the MLA were investigated by analyzing the replicated microlens and measuring the filling ratio. In the replication of the 33-µm-pitch MLA, the filling ratio increased with increasing pressing pressure and laser-irradiation time. However, for a laser-irradiation time of 1.0 ms, the PMMA did not fill the mold completely because of insufficient laser-irradiation even though the pressing pressure was increased to 50 MPa, which was the highest pressing pressure applied in this study. Furthermore, the area of the PMMA in which the temperature exceeded T_g and the flow of the PMMA varied with laser-irradiation time. In the replication of the 5-μm-pitch MLA, the filling ratio also increased with increasing pressing pressure and laser-irradiation time. By comparing the surface shapes during filling, it was found that the cross-sectional surface shape possessed a double and single peak in the 33- and 5-μm-pitch pattern, respectively. The depth of heated area of the 33-μm-pitch pattern was smaller than the pattern size, whereas that of the 5-μm-pitch pattern was comparable to or larger than the pattern size. Future experiments will need to be conducted with patterns of other sizes along with simulation modelling of heat conduction and polymer flow to expand on the current discussion. MLA performance characteristics, such as light-extraction efficiency, will also be verified as part of future work.

Author Contributions: Conceptualization, K.N.; methodology, K.N. and Y.Y.; validation, Y.Y.; data curation, Y.Y.; writing—original draft preparation, Y.Y.; writing—review and editing, K.N.; project administration and funding acquisition, M.N. and K.N.

Funding: This study was founded by JSPS KAKENHI Grant Numbers 24246027 and 26709004 from the Ministry of education, Culture, Sports, Science and Technology (MEXT), Japan.

Conflicts of Interest: The authors declare no conflict of interest.

References

1. Yao, L.; He, J. Recent progress in antireflection and self-cleaning technology—From surface engineering to functional surfaces. *Prog. Mater. Sci.* **2014**, *61*, 94–143. [CrossRef]
2. Galeotti, F.; Mróz, W.; Scavia, G.; Botta, C. Microlens arrays for light extraction enhancement in organic light-emitting diodes; A facile approach. *Org. Electron.* **2013**, *14*, 212–218. [CrossRef]
3. Takahashi, K.; Nagato, K.; Hamaguchi, T.; Nakao, M. High-speed replication of light-extraction surface with thermal roller nanoimprinting. *Microelectron. Eng.* **2015**, *141*, 285–288. [CrossRef]
4. Zhou, L.; Dong, X.X.; Lv, G.C.; Chen, J.; Shen, S. Fabrication of concave microlens diffuser films with a soft transparent mold of UV-curable polymer. *Opt. Commun.* **2015**, *342*, 167–172. [CrossRef]
5. Altun, A.O.; Jeon, S.; Shim, J.; Jeong, J.H.; Choi, D.G.; Kim, K.D.; Choi, J.H.; Lee, S.W.; Lee, E.S.; Park, H.D.; et al. Corrugated organic light emitting diodes for enhanced light extraction. *Org. Electron.* **2010**, *11*, 711–716. [CrossRef]
6. Reboud, V.; Kehagias, N.; Kehoe, T.; Leveque, G.; Mavidis, C.; Kafesaki, M.; Torres, C.M.S. Nanoimprinted plasmonic crystals for light extraction applications. *Microelectron. Eng.* **2010**, *87*, 1367–1369. [CrossRef]
7. Yang, Y.; Mielczarek, K.; Aryal, M.; Zakhidov, A. Nanoimprinted polymer solar cell. *ACS Nano.* **2012**, *6*, 2877–2892. [CrossRef] [PubMed]
8. Saito, M.; Kitamura, A.; Murahashi, M.; Yamanaka, K.; Hoa, L.Q.; Yamaguchi, Y.; Tamiya, E. Novel gold-capped nanopillars imprinted on a polymer film for highly sensitive plasmonic biosensing. *Anal. Chem.* **2012**, *84*, 5494–5500. [CrossRef] [PubMed]
9. Chou, S.Y.; Krauss, P.R.; Renstron, P.J. Imprint of sub-25 nm vias and trenches in polymers. *Appl. Phys. Lett.* **1995**, *67*, 3114–3116. [CrossRef]
10. Haisma, J.; Verheijen, M.; van den Heuvel, K. Mold-assisted nanolithography: A process for reliable pattern replication. *J. Vac. Sci. Technol. B* **1996**, *14*, 4124–4126. [CrossRef]
11. Qin, D.; Xia, Y.; Whitesides, G.M. Soft lithography for micro- and nanoscale patterning. *Nature Protoc.* **2010**, *5*, 491–502. [CrossRef] [PubMed]
12. Song, J.; Lu, H.; Li, S.; Tan, L.; Gruverman, A.; Ducharme, S. Fabrication of ferroelectric polymer nanostructures on flexible substrates by soft-mold reverse nanoimprint lithography. *Nanotechnol.* **2016**, *27*, 015302. [CrossRef] [PubMed]
13. Song, J.; Lu, H.; Foreman, K.; Li, S.; Tan, L.; Adenwalla, S.; Gruverman, A.; Ducharme, S. Ferroelectric polymer nanopillar arrays on flexible substrates by reverse nanoimprint lithography. *J. Mater. Chem. C* **2016**, *4*, 5914–5921. [CrossRef]

14. Schift, H.; David, C.; Gabriel, M.; Gobrecht, J.; Heyderman, L.J.; Kaiser, W.; Köppel, S.; Scandella, L. Nanoreplication in polymers using hot embossing and injection molding. *Microelectron. Eng.* **2000**, *53*, 171–174. [CrossRef]

15. Guo, L.J. Recent progress in nanoimprint technology and its applications. *J. Phys. D Appl. Phys.* **2004**, *37*, 123–141. [CrossRef]

16. Chou, S.Y.; Keimel, C.; Gu, J. Ultrafast and direct imprint of nanostructures in silicon. *Nature* **2002**, *417*, 835–837. [CrossRef] [PubMed]

17. Xia, Q.; Keimel, C.; Ge, H.; Yu, Z.; Wu, W.; Chou, S.Y. Ultrafast patterning of nanostructures in polymers using laser assisted nanoimprint lithography. *Appl. Phys. Lett.* **2003**, *83*, 4417–4419. [CrossRef]

18. Grigalinūnas, V.; Tamulevičius, S.; Tomašiūnas, R.; Kopustinskas, V.; Guobienė, A.; Jucius, D. Laser pulse assisted nanoimprint lithography. *Thin Solid Films* **2004**, *453–454*, 13–15.

19. Grigalinūnas, V.; Tamulevičius, S.; Muehlberger, M.; Jucius, D.; Guobienė, A.; Kopustinskas, V.; Gudonytė, A. Nanoimprint lithography using IR laser irradiation. *Appl. Surf. Sci.* **2006**, *253*, 646–650. [CrossRef]

20. Nagato, K.; Takahashi, K.; Sato, T.; Choi, J.; Hamaguchi, T.; Nakao, M. Laser-assisted replication of large area nanostructures. *J. Mater. Proc. Technol.* **2014**, *214*, 2444–2449. [CrossRef]

21. Saxena, K.; Jain, V.K.; Mehta, D.S. A review on the light extraction techniques in organic electroluminescent devices. *Opt. Mater.* **2009**, *32*, 221–233. [CrossRef]

22. Rowland, H.D.; Sun, A.C.; Schunk, P.R.; King, W.P. Impact of polymer film thickness and cavity size on polymer flow during embossing; Toward process design rules for nanoimprint lithography. *J. Micromech. Microeng.* **2005**, *15*, 2414–2425. [CrossRef]

23. Mooney, M. A theory of large elastic deformation. *J. Appl. Phys.* **1940**, *11*, 582–592. [CrossRef]

24. Gingold, R.A.; Monaghan, J.J. Kernel estimates as a basis for general particle methods in hydrodynamics. *J. Comput. Phys.* **1982**, *46*, 429–453. [CrossRef]

Article

Preparation Nano-Structure Polytetrafluoroethylene (PTFE) Functional Film on the Cellulose Insulation Polymer and Its Effect on the Breakdown Voltage and Hydrophobicity Properties

Jian Hao [1,2,*], Cong Liu [1], Yanqing Li [1], Ruijin Liao [1], Qiang Liao [3] and Chao Tang [4]

1 The State Key Laboratory of Power Transmission Equipment & System Security and New Technology, Chongqing University, Chongqing 400044, China; 20171101003z@cqu.edu.cn (C.L.); cqu0926@126.com (Y.L.); rjliao@cqu.edu.cn (R.L.)
2 Postdoctoral Station of Power Engineering and Engineering Thermophysics, Chongqing University, Chongqing 400044, China
3 College of Power Engineering, Chongqing University, Chongqing 400044, China; lqzx@cqu.edu.cn
4 College of Engineering and Technology, Southwest University, Chongqing 400715, China; tangchao_1981@163.com
* Correspondence: haojian2016@cqu.edu.cn; Tel.: +86-182-2301-0926

Received: 21 April 2018; Accepted: 16 May 2018; Published: 21 May 2018

Abstract: Cellulose insulation polymer is an important component of oil-paper insulation, which is widely used in power transformer. The weight of the cellulose insulation polymer materials is as high as tens of tons in the larger converter transformer. Excellent performance of oil-paper insulation is very important for ensuring the safe operation of larger converter transformer. An effective way to improve the insulation and the physicochemical property of the oil impregnated insulation pressboard/paper is currently a popular research topic. In this paper, the polytetrafluoroethylene (PTFE) functional film was coated on the cellulose insulation pressboard by radio frequency (RF) magnetron sputtering to improve its breakdown voltage and the hydrophobicity properties. X-ray photoelectron spectroscopy (XPS) results show that the nano-structure PTFE functional film was successfully fabricated on the cellulose insulation pressboard surface. The scanning electron microscopy (SEM) and X-ray diffraction (XRD) present that the nanoscale size PTFE particles were attached to the pressboard surface and it exists in the amorphous form. Atomic force microscopy (AFM) shows that the sputtered pressboard surface is still rough. The rough PTFE functional film and the reduction of the hydrophilic hydroxyl of the surface due to the shielding effect of PTFE improve the breakdown and the hydrophobicity properties of the cellulose insulation pressboard obviously. This paper provides an innovative way to improve the performance of the cellulose insulation polymer.

Keywords: cellulose insulation pressboard; magnetron sputtering; polytetrafluoroethylene; nano structure; breakdown; hydrophobicity

---

1. Introduction

Converter transformer is one of the key equipment in HVDC (High Voltage Direct Current) transmission network. The main insulation of the valve side for converter transformer usually has a high failure rate because of the complicated electric field in the operating condition, including alternating current (AC), direct current (DC), AC, and DC compound electric field [1,2]. Oil-paper insulation is the main insulation form of the valve side in converter transformer. Its insulation reliability is closely related to the safe operation of a converter transformer. The insulation performance of the insulation paper/pressboard under DC condition is an important factor to decide the operation reliability of the converter transformer.

It has been reported that polymer nanocomposites with metal oxide nanoparticle fillers can exhibit enhanced electrical breakdown strength [3,4]. In order to improve the breakdown and the mechanical properties of insulation paper, Liao Ruijin et al. developed the nano-Al_2O_3 doped cellulose insulation paper [5–7]. Results show that the nano-Al_2O_3 doped insulation paper possesses the better dielectric properties and AC breakdown strength. Chi Minghe et al. also studied the breakdown behavior of the Al_2O_3 modified pressboard [8]. Results show the breakdown strength of modified pressboard firstly increases and then decreases with the increase of nano-doping content, and reaches the peak value at 2.5% concentration. Liao Ruijin and Chi Minghe et al. also investigated the dielectric characteristics of nano-montmorillonite (MMT) modification insulation pressboard [9,10]. It is found that the breakdown strength of modified pressboard firstly increases and it then decreases with the growth of nano-doping content [5–10]. However, nanoparticle fillers research indicated that the difficulty of nano-doping is the agglomeration of nanoparticles [4–10]. In addition to adding nano-fillers to the material, it is worth investigating the fabrication of a special functional nano-structure film on the surface of the insulating material, which could provide an effective function for enhancing the electrical or the physicochemical properties of the oil impregnated insulation pressboard/paper that is used in the power transformer.

Moreover, moisture plays a detrimental role in the oil-paper insulation lifetime by reducing the thermal resistance and electrical breakdown strength and is regarded as "the first enemy" after temperature [11–13]. The production of moisture is inevitable in the service of a transformer. Field experience shows that the moisture content of a transformer is usually <0.5% in the initial stage of its operation, and it may increase to 2~4% at the last stage of its life [12]. The more moisture the insulation contains and the higher the temperature of the insulation system, the faster the oil-paper insulation material degrades [14,15]. Therefore, improving the hydrophobicity of insulating paper is very important for ensuring the performance of insulation paper.

Effective ways to improve the breakdown and hydrophobicity property of the oil impregnated insulation pressboard/paper used in power transformer are currently a popular research topic. PTFE has excellent insulation and hydrophobicity performance [16,17]. In this paper, the PTFE functional film on the cellulose insulation polymer was prepared by RF magnetron sputtering. Firstly, the structure and the existence form of the PTFE functional film were characterized. Then, the effect of the PTFE functional film on the breakdown and the hydrophobicity properties of the cellulose polymer were analyzed.

2. Materials and Methods

2.1. Materials and Sample Preparation

The cellulose insulation pressboard with a thickness 0.45 mm was used at here. The JPGF-480 reactive RF magnetron sputtering device at 13.56 MHz (Beijing Instrument Factory, Beijing, China) was used for pressboard coating. For PTFE film deposition, the insulation pressboard substrates were cut into 15 cm × 10 cm pieces, and a PTFE target (diameter 61.5 mm, thickness 5 mm) was sputtered. The distance between the target and the substrate sample was 10 cm. The vacuum chamber was pumped down to a base pressure of 4.0×10^{-3} Pa before sputtering. Deposition was conducted by using forward power 100 W. Argon was used as the working gas with a constant pressure of 1.5 Pa. The deposition mode was static, double-sided coating, and the deposition time was 10 min and 20 min at 28 °C. The sketch map for the RF magnetron sputtering PTFE functional film on the cellulose pressboard surface is shown in Figure 1. The sample composition is shown in Table 1.

Table 1. Sample composition.

Sample	Abbreviation
new pressboard	NP
new pressboard deposited PTFE for 10 min	NP-PTFE10
new pressboard deposited PTFE for 20 min	NP-PTFE20

Figure 1. RF magnetron sputtering polytetrafluoroethylene (PTFE) functional film on the cellulose pressboard surface.

2.2. Characterization Methods and Sample Treatment

X-ray photoelectron spectroscopy (XPS) with Al Kα X-ray source (XPS, Thermo escalab 250Xi, Waltham, MA, USA) was used to characterize the chemical binding state of the deposition film. The XPS spectra without argon etching were recorded in the fixed analyzer transmission mode with pass energy of 20 eV and a resolution of 0.1 eV. The deviation that is caused by the charging effect was calibrated using adventitious carbon referencing (C 1s, 284.6 eV). The scanning electron microscopy (SEM) (JSM-7800F, JEOL, Tokyo, Japan) and atomic force microscopy (AFM, Bruker Daltonics Inc., Billerica, MA, USA) were used to investigate the surface morphology of the coated surface. The X-ray diffractometer using Cu Ka (λ = 0.154 nm) radiation at a fixed incident angle of 2° was used to obtain the X-ray diffraction (XRD) (PANalytical Empyrea, Almelo, The Netherlands) patterns of the samples.

Before the DC (direct current) breakdown experiment, firstly, all of the samples were dried at 90 °C for 24 h in a vacuum box (1000 Pa). Then, new mineral oil was infused into the vacuum box and the temperature of the vacuum box was adjusted to 40 °C. The pressboard was impregnated at 40 °C for 48 h. The parameter of the oil used for impregnation is shown in Table 2. The measured moisture content of the oil-impregnated pressboard using Karl Fischer titration method was 0.95% after impregnation, and being cooled to room temperature. The DC breakdown voltage was measured according to Figure 2a. The pre-pressure DC voltage (15 kV/mm) was applied for 5 min. Then, the voltage was increased at 1 kV/s until sample breakdown. Five breakdown voltages were recorded for each sample. The breakdown test electrode setup is shown in Figure 2b. The test temperature is 28 °C.

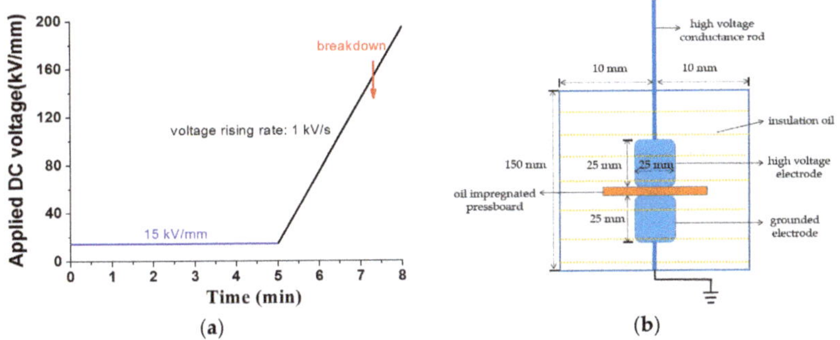

Figure 2. Direct current (DC) breakdown test process and the DC breakdown test electrode setup. (**a**) DC breakdown test process; (**b**) DC breakdown test electrode setup.

Table 2. The parameter of the oil used for impregnation.

Property	Mineral Oil
Kinematic viscosity 40 °C (mm^2/s)	9.7
Acidity (mg KOH/g)	0.008
Breakdown voltage (2.5 mm gap electrodes) (kV)	47.0
Relative permittivity (50 Hz)	2.2
Moisture content (ppm)	9

At last, the contact angle was measured with a Kyowa contact angle meter. Three measurements on different sample spots were made for each specimen. An average of the measurements was used for analysis. For XPS, SEM, AFM, XRD and contact angle test, non-impregnated pressboard samples were used. For the DC breakdown voltage test, the oil impregnated insulation pressboard was used.

3. Results and Discussions

3.1. XPS Analysis

Figure 3 shows the XPS survey spectra of the new pressboard, new pressboard surface as-prepared PTFE film for 10 min and 20 min. Cellulose insulation pressboard consists of linear, polymeric chains of cyclic, β-D-glucopyranose units, which are composed of C, H, and O element [18]. Therefore, there is only C 1s and O 1s peak, and extremely weak O 2s peak shown in Figure 3a. The molecular formula of PTFE is (C$_2$F$_4$)n. As shown in Figure 3b,c, it is obvious that the F 1s peak appears on the pressboard surface as-prepared PTFE film for 10 min and 20 min. With the increase of the coating time, the F 1s peak is obviously enhanced [19]. While the O 1s peak becomes weaker and weaker because of the coverage of the PTFE film on the surface of the cellulose pressboard. The O 1s peak almost disappeared for the pressboard surface deposited PTFE 20 min.

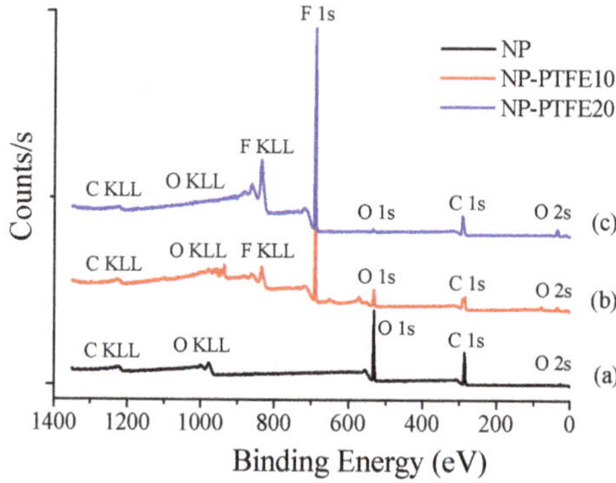

Figure 3. The X-ray photoelectron spectroscopy (XPS) spectra of the new pressboard (**a**), new pressboard surface as-prepared. PTFE film for 10 min (**b**) and 20 min (**c**).

Figures 4–6 show the C 1s, O 1s, and F 1s peak fitting in the XPS narrow scan spectra, respectively. The peak fitting can be used to make identify the chemical components. The identified chemical components with different binding energies and its concentration by C 1s, O 1s, and F 1s peak fitting are shown in Tables 3–5, respectively. The C1, C6, and C7 peaks that are presented in Figure 4a are attributed to C–C/C=C, C–O, and O–C=O, respectively [19,20]. It is particularly noteworthy from Figure 4b,c that new C2, C3, C4, C5 peaks appear for the sample NP-PTFE10 and NP-PTFE20. The C2, C3, C4, C5 peaks are attributed to O–C–CF$_3$/CF$_2$, CF, CF$_2$, and CF$_3$, respectively [19,20]. From the O 1s high resolution spectra shown in Figure 5b,c, it can be seen that new O3 peak appears for the coated samples. The O1 and O2 are attributed to O=C–O and O–C. The O3 is attributed to

O–C–CF$_3$/CF$_2$ [19,20]. Figure 6b,c show that there are also new F1 peak attributed to F–C appears for the pressboard surface deposited PTFE for 10 min and 20 min. With the coating time increase, due to the covering PTFE film, Figure 4b,c show that the intensity of the new C3, C4, C5 peaks becomes stronger, while the intensity of C1, C6, and C7 peaks becomes weaker. The intensity of new F 1 peaks for the pressboard surface as-prepared PTFE film for 20 min is significantly stronger than that of the sample as-prepared PTFE film for 10 min. From Figures 5 and 6, Tables 3–5, it could be deduced that the PTFE has been successfully fabricated on the cellulose insulation pressboard surface.

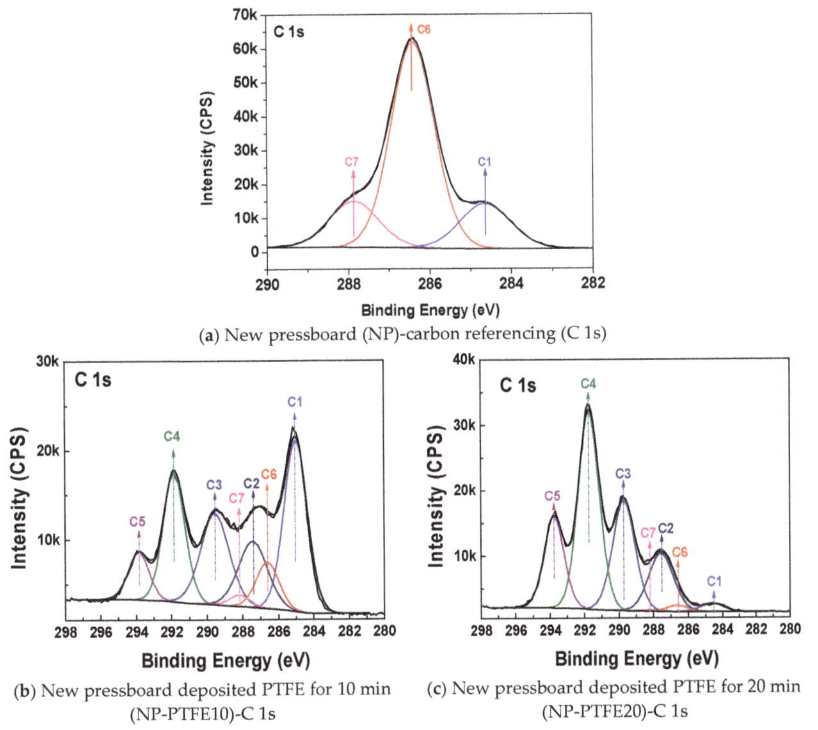

(a) New pressboard (NP)-carbon referencing (C 1s)

(b) New pressboard deposited PTFE for 10 min (NP-PTFE10)-C 1s

(c) New pressboard deposited PTFE for 20 min (NP-PTFE20)-C 1s

Figure 4. C 1s peak fitting for the new pressboard (**a**), new pressboard coated PTFE film for 10 min (**b**), and 20 min (**c**).

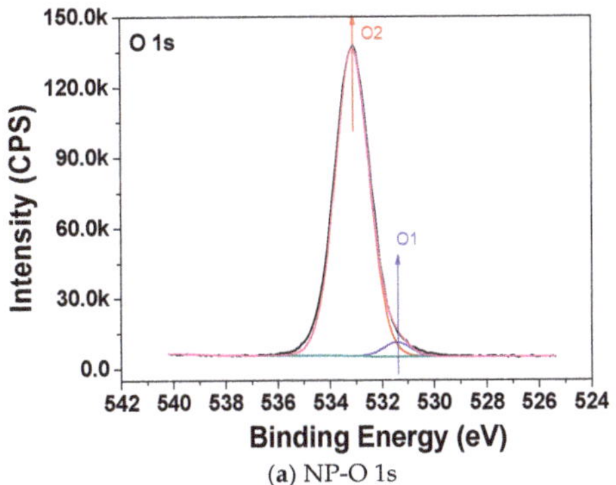

(a) NP-O 1s

Figure 5. *Cont.*

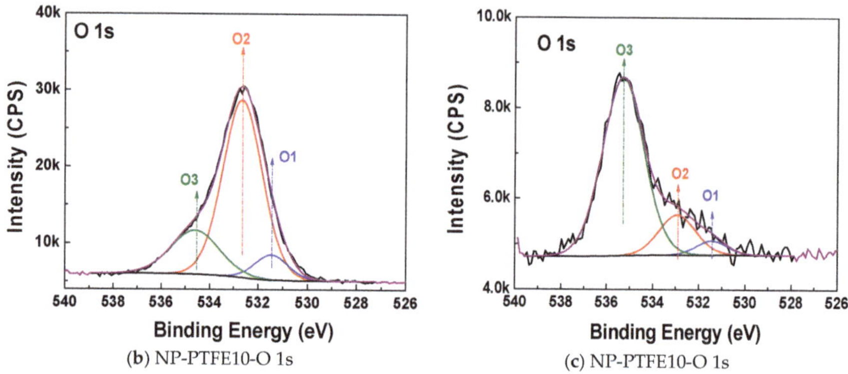

(b) NP-PTFE10-O 1s

(c) NP-PTFE10-O 1s

Figure 5. O 1s peak of the new pressboard, pressboard surface coated PTFE film for 10 min and 20 min.

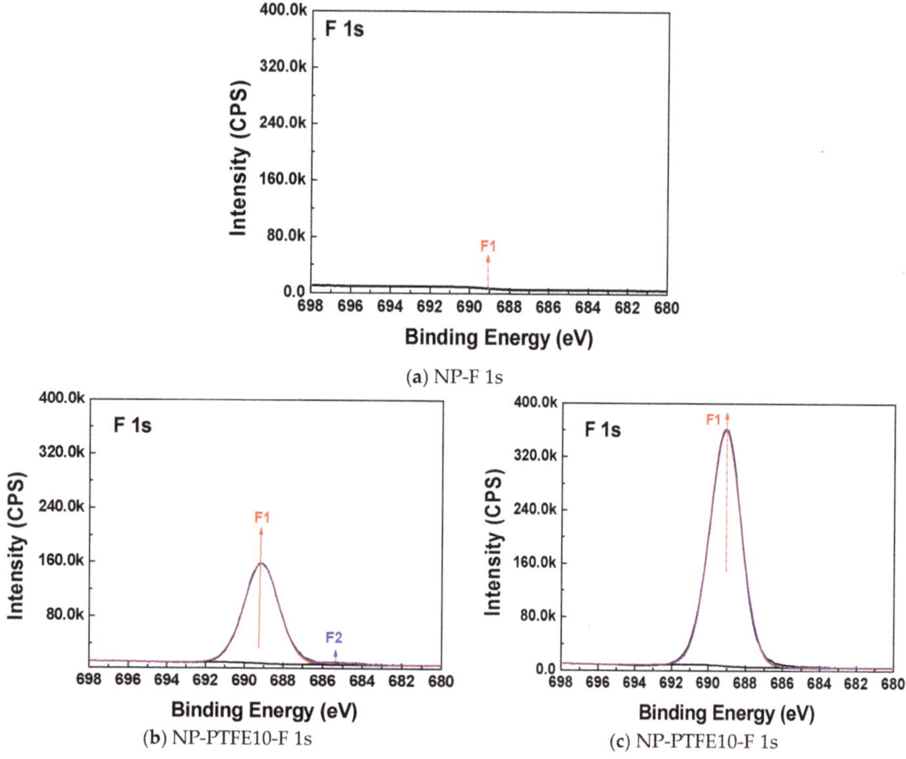

(a) NP-F 1s

(b) NP-PTFE10-F 1s

(c) NP-PTFE10-F 1s

Figure 6. F 1s peak of the new pressboard, pressboard surface coated PTFE film for 10 min and 20 min.

Table 3. C 1s peak fitting result for NP, NP-PTFE10, and NP-PTFE20.

Sample	Data Set	Name	Position	% Conc.
	C–C/C=C	C1	284.6	16.9
NP	C–O	C6	286.4	66.2
	O–C=O	C7	287.9	16.9

Table 3. *Cont.*

Sample	Data Set	Name	Position	% Conc.
	C–C/C=C	C1	285.0	29.4
	O–C–CF$_3$/CF$_2$	C2	287.4	14.0
	C–F	C3	289.5	19.2
NP-PTFE10	C–F$_2$	C4	291.8	20.5
	C–F$_3$	C5	293.9	7.3
	C–O	C6	286.6	7.7
	O–C=O	C7	288.2	1.8
	C–C/C=C	C1	284.6	1.9
	O–C–CF$_3$/CF$_2$	C2	287.6	14.0
	C–F	C3	289.7	24.2
NP-PTFE20	C–F$_2$	C4	291.7	40.3
	C–F$_3$	C5	293.8	18.0
	C–O	C6	286.6	1.2
	O–C=O	C7	288.2	0.3

Table 4. O 1s peak fitting result for NP, NP-PTFE10, and NP-PTFE20.

Sample	Data Set	Name	Position	% Conc.
NP	O=C–O	O1	531.4	3.6
	O–C	O2	533.1	96.4
	O=C	O1	531.4	8.4
NP-PTFE10	O–C	O2	532.6	69.9
	O–C–CF$_3$/CF$_2$	O3	534.6	21.7
	O=C	O1	531.4	4.8
NP-PTFE20	O–C	O2	532.9	16.2
	O–C–CF$_3$/CF$_2$	O3	535.3	79.0

Table 5. F 1s peak fitting result for NP, NP-PTFE10, and NP-PTFE20.

Sample	Data Set	Name	Position	% Conc.
NP	/	/	/	/
NP-PTFE10	F$_2$–C	F1	689.1	97.5
	F–C	F2	685.4	2.5
NP-PTFE20	F$_2$–C	F1	689.1	100.0

3.2. Surface Topography Analysis

The SEM micrographs of the untreated pressboard and the coated pressboard are shown in Figure 7. We can observe that the cellulose fibers of untreated pressboard (Figure 7a) intersect each other and its surface is relatively rough. There are some cracks where the fibers intersect. The pressboard surface with magnetron sputtering treatment for 10 min (Figure 7b) is more smooth and dense. There are many very small PTFE particles with nanometer covered on the surface. The PTFE particles filled the cracks between the fibers and were distributed uniformly on the surface. However, for the pressboard surface sputtered for 20 min, as shown in Figure 6c, the cracks also are be filled. Besides, PTFE is present in larger particles due to the agglomeration of particles. The PTFE particles are about a few dozen nanometers in size.

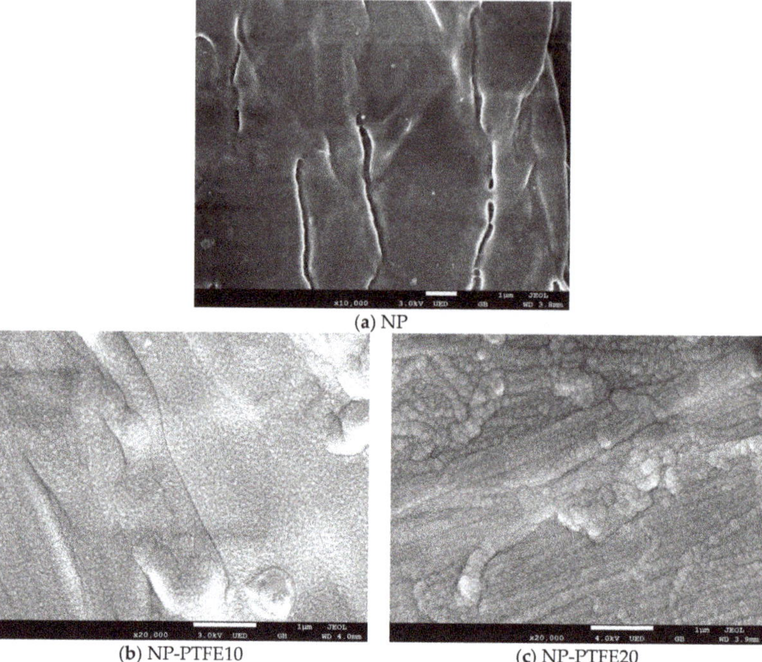

(a) NP

(b) NP-PTFE10 (c) NP-PTFE20

Figure 7. The scanning electron microscopy (SEM) of the new pressboard, pressboard surface coated PTFE film for 10 min and 20 min.

The microscopic appearance of insulating pressboard specimen before and after magnetron sputtering was measured by AFM (Figure 8). The AFM image shows a greater longitudinal undulating and some sharp protrusions, in good agreement with the SEM micrograph revealing a relatively rough surface. After specimen treatment, there are some obvious changes that have taken place. By comparing the fresh pressboard, we can find that the pressboard surface coated PTFE for 10 min (Figure 8b) is smoother than that of new pressboard, as well as the raised part is granular and relatively flat. Figure 8c shows the surface topography of sample that is modified by magnetron sputtering for 20 min, and as the sputtering time increases, the sample surface becomes slightly rougher again.

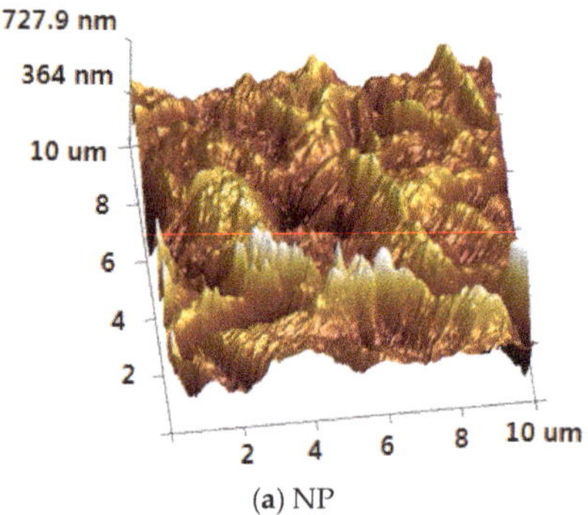

(a) NP

Figure 8. *Cont.*

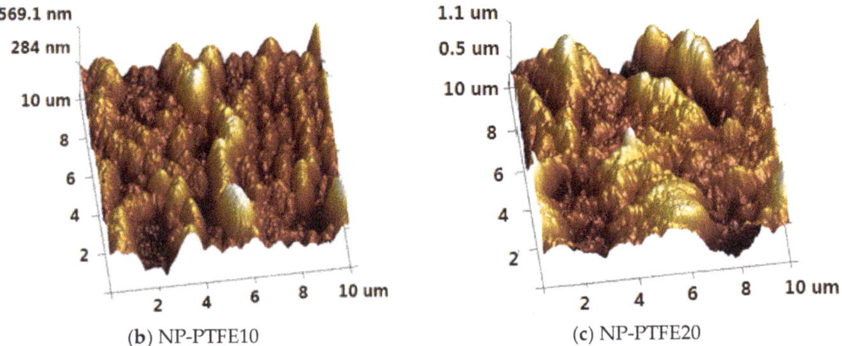

Figure 8. The atomic force microscopy (AFM) of the new pressboard, pressboard surface coated PTFE film for 10 min and 20 min.

3.3. XRD Analysis

Figure 9 shows the XRD spectrum of new pressboard and the PTFE coated pressboard. Diffraction pattern for the pressboard has three broad peaks at $2\theta = 15°$, $2\theta = 22°$, and $2\theta = 34°$, corresponding to (101), (002), and (040) diffraction peaks of cellulose, respectively [21]. In the diffraction pattern of the new pressboard, there is a sharp peak and some dispersive diffraction peaks, which means that the cellulose has a mixed structure of crystallization and amorphous phase. The diffraction peak of PTFE is at $17°$, $30°$, and $35°$ [22]. We can notice that there is no peak at $2\theta = 17°$, $2\theta = 30°$, and $2\theta = 35°$ in the XRD results of coated pressboard, which proves that the PTFE film exists on the surface of insulation pressboard surface in the amorphous form.

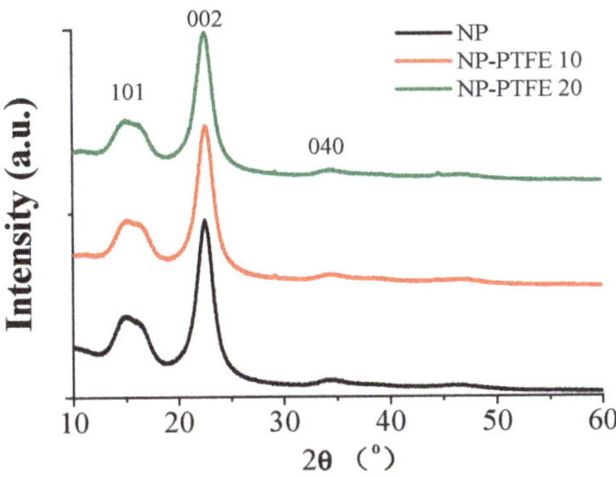

Figure 9. X-ray diffraction (XRD) of the the new pressboard, pressboard surface coated PTFE film for 10 min and 20 min.

3.4. DC Breakdown Analysis

The DC pre-pressure breakdown strength of the new pressboard (NP), new pressboard deposited PTFE for 10 min and 20 min (NP-PTFE10, NP-PTFE20) is shown in Figure 10. The "pre-pressure breakdown strength" means the breakdown voltage obtained through the test process shown in the Figure 2. The average DC pre-pressure breakdown voltage for the NP, NP-PTFE10 and NP-PTFE20 is 136.37 kV/mm, 142.53 kV/mm, and 151.77 kV/mm, respectively. When compared with the new pressboard, the DC pre-pressure breakdown enhancement is 5% and 11% for NP-PTFE10 and NP-PTFE20, respectively. The PTFE functional film improves the breakdown property of the coated insulation pressboard, especially for the sample NP-PTFE20. This mainly because the nano PTFE particles filled the surface defects (Figure 7) and improved the breakdown performance.

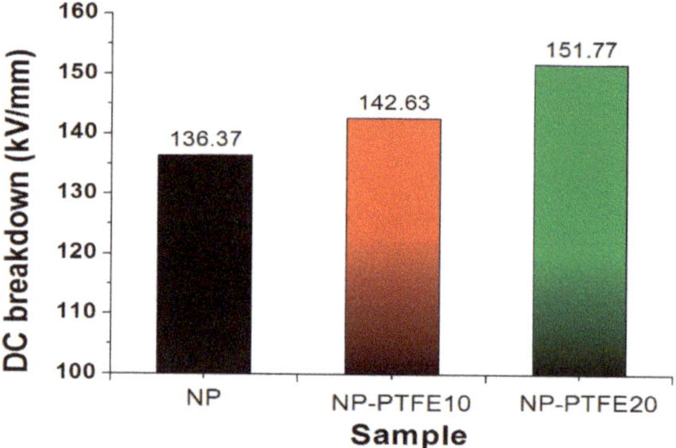

Figure 10. DC pre-pressure breakdown strength of the new pressboard, new pressboard deposited. PTFE for 10 min and 20 min.

3.5. Hydrophobicity and Hygroscopicity Analysis

The insulation paper was developed from natural fiber. The structural characteristics of the fiber determines that it absorbs water very easily. However, the hygroscopicity of the insulation paper is a very bad feature when it is being used for insulation in the transformer. Therefore, if the insulation paper has better hydrophobicity, its performance is not easily destroyed by moisture. Figure 11 shows the detailed dynamic process of liquid droplets that are dripping on the surface of each sample. The contact angle is about 0° that it is impossible to measure, indicating that water droplet penetrated through the surface of pressboard due to the hydrophilicity of cellulose. However, the pressboard surface coated by PTFE shows hydrophobicity. Water droplets can last a long time on the PTFE surface. As reported in [23–25], the surface hydrophobicity should increase in the order $-CH_2 < -CH_3 < -CF_2 < -CF_2H < -CF_3$. The pressboard surface coated PTFE has much C–F groups which improve its hydrophobicity. At the beginning, the contact angle of pressboard deposited PTFE for 10 min and 20 min is 118.2° and 116.6°, respectively. As time increases, the contact angle decreases gradually. Before 45 min, both of the samples have the same change, and both are greater than 90°. Then, the sample contact angle of pressboard surface as-prepared PTFE film for 10 min decreased faster than the PTFE film coated for 20 min.

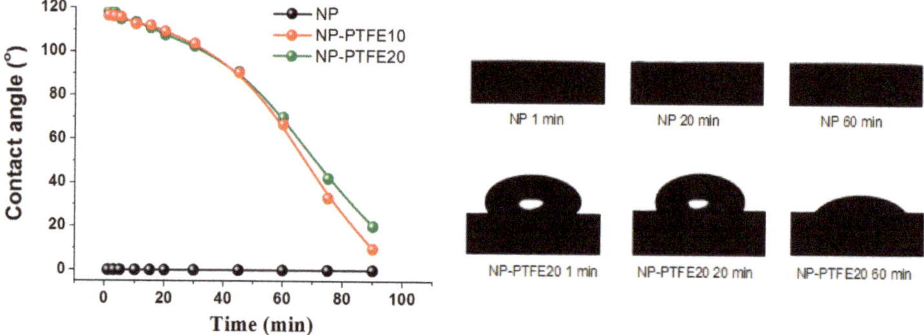

Figure 11. The contact angle of the new pressboard, new pressboard deposited PTFE for 10 min and 20 min.

PTFE is a non-polar polymer with symmetrical structure, and it is one of the lowest surface energy materials [19,20]. In order to explain the change of surface hydrophobicity from the chemical mechanism, the Fourier transform infrared spectroscopy (FT-IR) spectroscopy (Nicolet iS5 FT-IR) was used to confirm the reason for the change of the contact angle. FT-IR analysis was further carried out. As shown in Figure 12, the peak at 3345 cm^{-1} is assigned to the stretching vibration of O–H [26,27]. The peak at 2901 cm^{-1}, 1426 cm^{-1}, 1368 cm^{-1}, and 1315 cm^{-1} is assigned to the stretching vibration and the flexural vibration of C–H [26,27]. It is obvious that the shielding of nano-structure PTFE film leads to the reduction of hydroxyl, which is beneficial to reduce the interaction between hydroxyl and water. In addition, PTFE has the excellent hydrophobic and oleophylic properties [19,20]. The above two aspects increase the contact angle of the sputtered insulation pressboard surface.

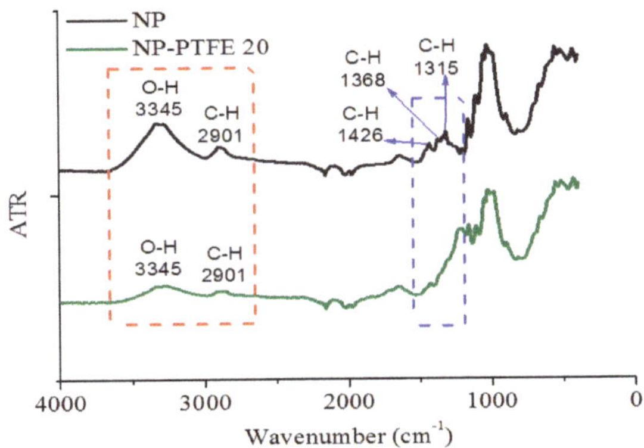

Figure 12. FT-IR of the new pressboard, new pressboard deposited PTFE for 20 min.

The hygroscopicity of the dried new pressboard and the pressboard coated for PTFE was also compared at here. According to the moisture equilibrium experiment that was done by our team [28], the dried pressboard samples were placed into a humidity chamber. The temperature of the humidity chamber was set to 60 °C and the relative humidity was set to 60%. The absorption time was set as 0 min, 20 min, 40 min, 60 min, and 90 min. The moisture content of the dried new pressboard (NP) and the pressboard coated for PTFE (NP-PTFE20) is shown in Figure 13. It can be seen from Figure 13 that the moisture absorption rate for NP-PTFE20 sample is slower than that of the NP sample.

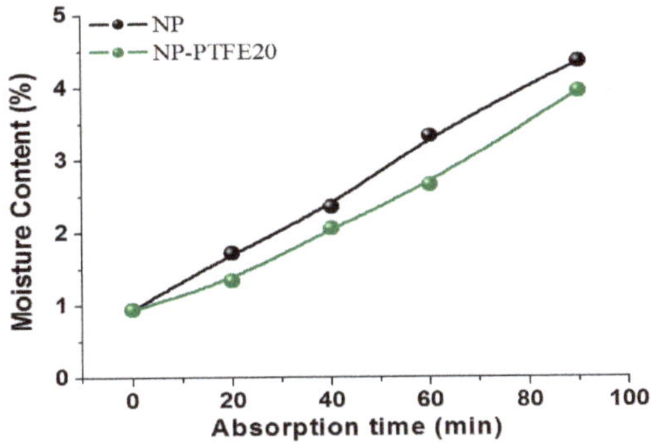

Figure 13. Moisture content of the NP and NP-PTFE20 samples under the absorption moisture condition.

For the moisture balance between paper and oil, the pressboard samples absorbed moisture for different times (0 min, 20 min, 40 min, 60 min, and 90 min) were placed into grinding bottles that were filled with new insulating oil. The grinding bottles were then sealed and placed in a constant temperature oven under controlled temperature 70 °C. The moisture concentration in oil and paper was constantly measured until the equilibrium state was considered to be achieved, when the measured moisture concentration remained constant. In this paper, the time that is required for reaching moisture equilibrium state was 13 days at 70 °C. The result for the moisture balance between paper and oil is shown in Figure 14. It can be seen that for the NP-PTFE20 sample, the moisture tends to stay in the oil.

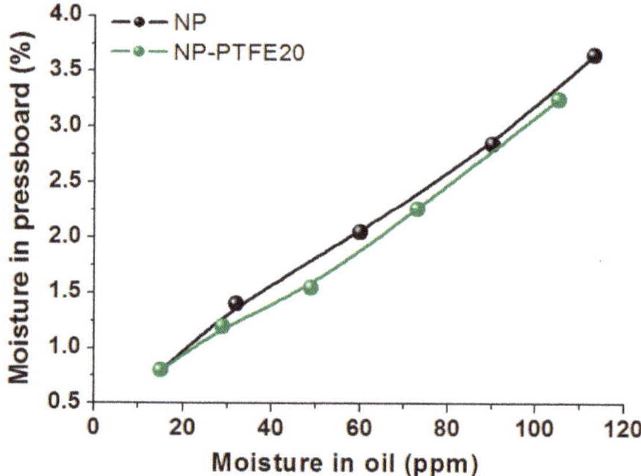

Figure 14. Moisture balance between paper and oil at 70 °C.

3.6. Oil Absorption and Impregnation

In order to investigate the PTFE functional film influence on the oil impregnation, we measured the contact angle between oil and paperboard, and compared the process of oil drop that is entering into the pressboard with and without PTFE functional film. As shown in Figure 15, the NP-PTFE20 sample has higher contact angle between oil and paperboard. The oil drop entering into the NP is very quickly, while the oil drop entering into the NP-PTFE20 sample is very slow. It takes about 3 h to fully enter the interior of the NP-PTFE20 sample. Therefore, the oil impregnation process is slow for the pressboard PTFE functional film.

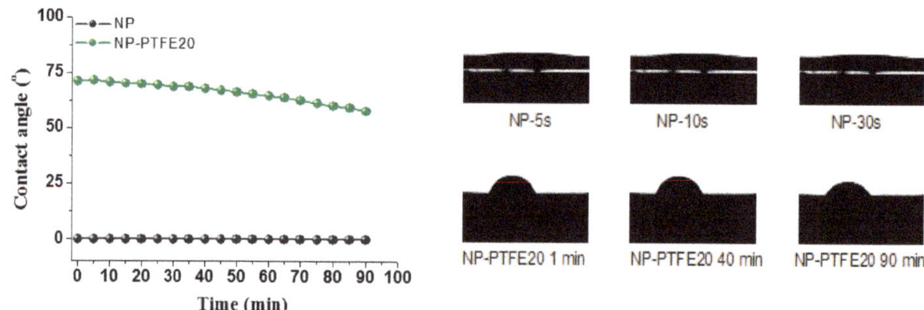

Figure 15. Contact angle between oil and paperboard.

In order to obtain the difference in the amount of oil that is impregnated into the NP and NP-PTFE 20 sample, the thermogravimetry (TG) and the derivative thermogravimetry (DTG) curves of the NP and NP-PTFE 20 sample impregnated with oil was measured, as shown in Figure 16. The heating rate is 7 °C/min. Each sample is 5.0 mg. The tested temperature is from 33 °C to 500 °C under a nitrogen flow of 50 mL/min. There are two peaks that can be seen for both NP and NP-PTFE20. This is because there is oil in the oil impregnated pressboard, and the thermal properties of oil and the pressboard are different. The first stage of weight loss in TG curve and the first peak in DTG curve is belonging to oil decomposition [29]. Subtracting the 0.95% moisture content, it can be deduced in Figure 16 that the NP sample contain 23.86% oil in the oil impregnated pressboard (mass ratio). The NP-PTFE20 sample contain 21.28% oil in the oil impregnated pressboard (mass ratio). The PTFE film fills many gaps on the surface, resulting in oil absorption being reduced.

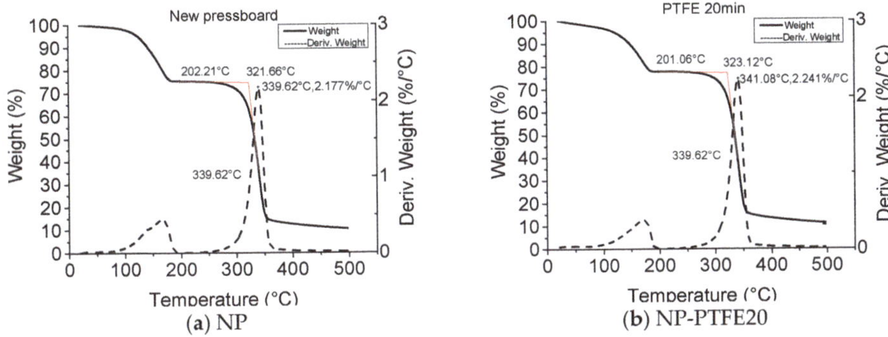

Figure 16. TG for NP and NP-PTFE20.

For the oil impregnation experiment, the impregnation test model is shown in Figure 17. The size of the pressboard is 100 mm × 30 mm × 0.5 mm. The impregnation experiment was conducted at 30 °C, 1 atm. The impregnation length versus the impregnation time is shown in Figure 17. The samples that were used was new pressboard (NP) and new pressboard coated PTFE for 20 min (NP-PTFE20). Because the PTFE was coated on the two side surface of the pressboard, the oil impregnated the inner part from the bottom and other side surface of the samples where there is no PTFE. Thus, the oil impregnation rate of NP-PTFE20 is slower than that of NP sample.

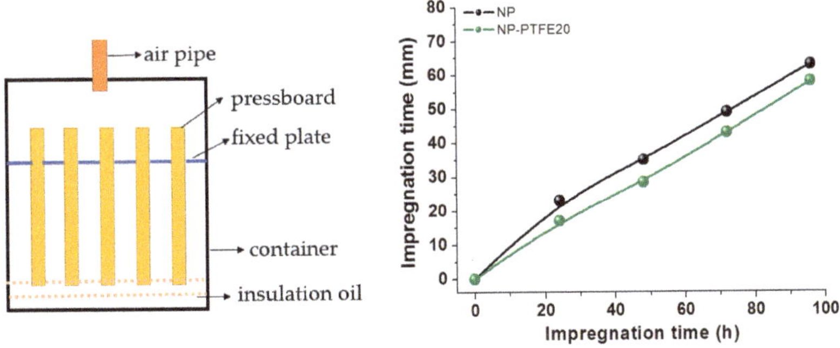

Figure 17. Oil impregnation experiment and result.

4. Conclusions

The present study confirmed the finding about improving the DC pre-pressure breakdown and the hydrophobicity properties of the cellulose insulation polymer by sputtering nano-structure PTFE functional film on the surface. The conclusions are as follows:

The nano-structure PTFE functional film was successfully fabricated on the cellulose insulation pressboard surface by RF magnetron sputtering. When compared with the fresh cellulose insulation pressboard, for the pressboard sputtered PTFE for 10 min and 20 min, the new peaks attributed to O–C–CF_3/CF_2, CF, CF_2, and CF_3 appear in their C 1s XPS spectroscopy, the new peak that is attributed to O–C–CF_3/CF_2 appears in the O 1s XPS spectroscopy, and for F 1s XPS spectroscopy, the new peaks that are attributed to F–C, F_2–C appear.

The SEM and XRD present that the nanoscale size PTFE particles were attached on the pressboard surface and exists in the amorphous form. The PTFE particles are about a few dozen nanometers in size for the surface sputtering 20 min. There are only three broad XRD peaks at $2\theta = 15°$, $2\theta = 22°$ and $2\theta = 34°$, which belongs to cellulose. AFM result shows that the sputtered pressboard surface is still rough.

The DC pre-pressure breakdown enhancement is 5% and 11% for NP-PTFE10 and NP-PTFE20, respectively. The contact angle of the new pressboard is 0°. However, the cellulose pressboard surface deposited PTFE for 10 min and 20 min is 118.2° and 116.6°, respectively. FTIR spectroscopy shows that the transmissivity of the peak at 3345 cm^{-1} for O–H and the peaks at 2901 cm^{-1}, 1426 cm^{-1}, 1368 cm^{-1}, and 1315 cm^{-1} for C–H decreases for the sputtered sample. The rough PTFE functional film and the reduction of the hydrophilic hydroxyl of the surface due to the shielding effect of PTFE improve the DC pre-pressure breakdown and hydrophobicity properties of the cellulose insulation pressboard obviously.

The moisture absorption rate for NP-PTFE20 sample is slower than that of the NP sample. According to the result of the moisture balance between paper and oil, it shows that the moisture tends to stay in the oil. The NP sample contains 23.86% oil in the oil impregnated pressboard (mass ratio), and the NP-PTFE20 sample contains 21.28%. The oil impregnation rate of NP-PTFE20 is slower than that of the NP sample.

Author Contributions: J.H. designed the experiments, performed the breakdown, contact angle, XPS, SEM and FITR measurement and writing; C.L. and Y.L. performed the RF magnetron sputtering experiment, AFM and XRD analysis; J.H. and R.L. analyzed the data; C.T. contributed literature search; Q.L. contributed discussion and paper modification.

Funding: This research was funded by National Natural Science Foundation of China (51707022), China Postdoctoral Science Foundation (2017M612910), Chongqing Special Funding Project for Post-Doctoral (Xm2017040) and Funds for Innovative Research Groups of China (51321063).

Conflicts of Interest: The authors declare no conflict of interest.

References

1. CIGRE Joint Working Group A2/B4.28. *HVDC Converter Transformers Guide Lines for Conducting Design Reviews for HVDC Converter Transformers*; CIGRE: Paris, France, 2010.
2. CIGRE Joint Working Group A2/B4.28. *HVDC Converter Transformers Design Review, Test Procedures, Ageing Evaluation and Reliability in Service*; CIGRE: Paris, France, 2010.
3. Smith, R.C.; Liang, C.; Landry, M.; Nelson, J.K.; Schadler, L.S. The mechanisms leading to the useful electrical properties of polymer nanodielectrics. *IEEE Trans. Dielectr. Electr. Insul.* **2008**, *15*, 187–196. [CrossRef]
4. Tanaka, T. Dielectric Breakdown in Polymer Nanocomposites. In *Polymer Nanocomposites*; Springer: Berlin, Germany, 2016.
5. Yan, S.; Liao, R.; Yang, L.; Zhao, X.; Yuan, Y.; He, L. Influence of nano-Al_2O_3 on electrical properties of insulation paper under thermal aging. In Proceedings of the IEEE International Conference on High Voltage Engineering and Application, Chengdu, China, 19–22 September 2016; pp. 1–4.
6. He, L.; Liao, R.; Lv, Y.; Yang, L.; Zhao, X.; Yan, S. Effect of nano-Al_2O_3 on the thermal aging physicochemical properties of insulating paper. In Proceedings of the IEEE International Conference on Condition Monitoring and Diagnosis, Xi'an, China, 25–28 September 2016; pp. 254–257.
7. Liao, R.; He, L.; Lü, Y.; Zhao, X.; Yuan, Y. Influence of nano-Al_2O_3 on properties of oil-paper insulation during thermal aging process. *Trans. China Electrotech. Soc.* **2017**, *32*, 207–215.

8. Liu, H.; Chi, M.; Chen, Q.; Gao, Z.; Zhu, X.; Wei, X. Analysis of dielectric characteristics of nano-Al$_2$O$_3$ modified insulation pressboard. *Zhongguo Dianji Gongcheng Xuebao/Proc. Chin. Soc. Electr. Eng.* **2017**, *37*, 4246–4253.

9. Liao, R.; Yuan, L.; Zhang, F.; Yang, L.; Wang, K.; Duan, L. Preparation of montmorillonite modified insulation paper and study on its electrical characteristics. *High Volt. Eng.* **2014**, *40*, 33–39.

10. Chi, M.; Tao, K.; Chen, Q.; Liu, H.; Gao, P.; Gao, Z. Dielectric properties of nano-montmorillonite modified insulation pressboard. *High Volt. Eng.* **2017**, *43*, 2842–2848.

11. Hao, J.; Chen, G.; Liao, R. Influence of moisture and temperature on space charge dynamics in multilayer oil-paper insulation. *IEEE Trans. Dielectr. Electr. Insul.* **2012**, *19*, 1456–1464. [CrossRef]

12. Oommen, T.V. Moisture equilibrium charts for transformer insulation drying practice. *IEEE Trans. Power App. Syst.* **1984**, *PAS-103*, 3063–3067. [CrossRef]

13. Emsley, A.M.; Stevens, G.C. Review of chemical indicators of degradation of cellulosic electrical paper insulation in oil-filled transformers. *IEEE Proc. Sci. Meas. Technol.* **1994**, *141*, 324–334. [CrossRef]

14. Emsley, A.M.; Xiao, X.; Heywood, R.J.; Ali, M. Degradation of cellulosic insulation in power transformers. Part 3: Effects of oxygen and water on ageing in oil. *IEEE Proc. Sci. Meas. Technol.* **2000**, *147*, 115–119. [CrossRef]

15. Lundgaard, L.E.; Hansen, W.; Linhjell, D.; Painter, T.J. Aging of oil-impregnated paper in power transformers. *IEEE Trans. Power Deliv.* **2004**, *19*, 230–239. [CrossRef]

16. Yong, J.; Fang, Y.; Chen, F.; Huo, J.; Yang, Q.; Bian, H.; Du, G.; Hou, X. Femtosecond laser ablated durable superhydrophobic PTFE films with penetrating microholes for oil/water separation: Separating oil from water and corrosive solutions. *Appl. Surf. Sci.* **2016**, *389*, 1148–1155. [CrossRef]

17. Toosi, S.F.; Moradi, S.; Kamal, S.; Hatzikiriakos, S.G. Superhydrophobic laser ablated PTFE substrates. *Appl. Surf. Sci.* **2015**, *349*, 715–723. [CrossRef]

18. Yang, L.J.; Liao, R.J.; Sun, C.X.; Zhu, M.Z. Influence of vegetable oil on the thermal aging of transformer paper and its mechanism. *IEEE Trans. Dielectr. Electr. Insul.* **2012**, *18*, 692–700. [CrossRef]

19. Hou, W.; Wang, Q. Stable polytetrafluoroethylene superhydrophobic surface with lotus-leaf structure. *J. Colloid Interface Sci.* **2009**, *333*, 400–403. [CrossRef] [PubMed]

20. Park, B.H.; Lee, M.H.; Kim, S.B.; Jo, Y.M. Evaluation of the surface properties of PTFE foam coating filter media using XPS and contact angle measurements. *Appl. Surf. Sci.* **2011**, *257*, 3709–3716. [CrossRef]

21. Liao, R.J.; Tang, C.; Yang, L.J.; Grzybowski, S. Thermal aging micro-scale analysis of power transformer pressboard. *IEEE Trans. Dielectr. Electr. Insul.* **2008**, *15*, 1281–1287. [CrossRef]

22. Liu, L.; Wang, X.; Xie, W.; Mingming, Y.U.; Fang, L.; Li, H.; Yang, M.; Xiao, Y.; Ren, M.; Sun, J. Properties of polysulfonamide fiber/polytetrafluoroethylene composites. *J. Shanghai Univ.* **2017**, *23*, 185–191.

23. Vandencasteele, N.; Reniers, F. *Plasma-modified polymer* surfaces: characterization using XPS. *J. Electron Spectrosc. Relat. Phenom.* **2010**, *178*, 394–408. [CrossRef]

24. Hare, E.F.; Shafrin, E.G.; Zisman, W.A. Properties of Films of Adsorbed Fluorinated Acids. *J. Phys. Chem.* **1954**, *58*, 236–239. [CrossRef]

25. De Toit, F.J.; Sanderson, R.D.; Engelbrecht, W.J.; Wagener, J.B. The effect of surface fluorination on the wettability of high density polyethylene. *J. Fluor. Chem.* **1995**, *74*, 43–48. [CrossRef]

26. Alia, M.; Emsley, A.M.; Herman, H.; Heywood, R.J. Spectroscopic studies of the ageing of cellulosic paper. *Polymer* **2001**, *42*, 2893–2900. [CrossRef]

27. Hinterstoisser, B.; Salmen, L. Application of dynamic 2D FTIR to cellulose. *Vib. Spectrosc.* **2000**, *22*, 111–118. [CrossRef]

28. Liao, R.; Lin, Y.; Guo, P.; Liu, H.B.; Xia, H.H. Thermal aging effects on the moisture equilibrium curves of mineral and mixed oil-paper insulation systems. *IEEE Trans. Dielectr. Electr. Insul.* **2015**, *22*, 842–850. [CrossRef]

29. Liao, R.; Hao, J.; Chen, G.; Ma, Z.Q.; Yang, L.J. A comparative study of physicochemical, dielectric and thermal properties of pressboard insulation impregnated with natural ester and mineral oil. *IEEE Trans. Dielectr. Electr. Insul.* **2011**, *18*, 1626–1637. [CrossRef]

Article

Silk as a Natural Reinforcement: Processing and Properties of Silk/Epoxy Composite Laminates

Youssef K. Hamidi [1,*], M. Akif Yalcinkaya [2], Gorkem E. Guloglu [2], Maya Pishvar [2], Mehrad Amirkhosravi [2] and M. Cengiz Altan [2]

[1] Mechanical Engineering Program, University of Houston–Clear Lake, Houston, TX 77058, USA
[2] School of Aerospace and Mechanical Engineering, University of Oklahoma, Norman, OK 73019, USA; akifyalcinkaya@ou.edu (M.A.Y.); gguloglu@ou.edu (G.E.G.); pishvar@ou.edu (M.P.); mehrad@ou.edu (M.A.); altan@ou.edu (M.C.A.)
* Correspondence: hamidi@ou.edu; Tel.: +1-281-283-3818

Received: 3 October 2018; Accepted: 26 October 2018; Published: 30 October 2018

Abstract: With growing environmental awareness, natural fibers have recently received significant interest as reinforcement in polymer composites. Among natural fibers, silk can potentially be a natural alternative to glass fibers, as it possesses comparable specific mechanical properties. In order to investigate the processability and properties of silk reinforced composites, vacuum assisted resin transfer molding (VARTM) was used to manufacture composite laminates reinforced with woven silk preforms. Specific mechanical properties of silk/epoxy laminates were found to be anisotropic and comparable to those of glass/epoxy. Silk composites even exhibited a 23% improvement of specific flexural strength along the principal weave direction over the glass/epoxy laminate. Applying 300 kPa external pressure after resin infusion was found to improve the silk/epoxy interface, leading to a discernible increase in breaking energy and interlaminar shear strength. Moreover, the effect of fabric moisture on the laminate properties was investigated. Unlike glass mats, silk fabric was found to be prone to moisture absorption from the environment. Moisture presence in silk fabric prior to laminate fabrication yielded slower fill times and reduced mechanical properties. On average, 10% fabric moisture induced a 25% and 20% reduction in specific flexural strength and modulus, respectively.

Keywords: epoxy; natural fiber composites; silk fibers

---

1. Introduction

During the last several decades, fiber-reinforced polymer composites have experienced remarkable growth in various sectors, ranging from packaging and sporting goods to automotive and aerospace industries. This increased usage is essentially due to their lightweight, higher mechanical properties, and superior corrosion resistance compared to conventional materials [1,2].

Recently, growing environmental awareness has led to stricter policies regarding sustainability and encouraged industry to pursue ecofriendly products [3–5]. In this context, natural fibers have attracted increased attention over the past several years as alternatives to traditional reinforcements, namely glass, carbon, and aramid fibers. Currently, glass fibers are the most commonly used reinforcement in composites [3], since they offer a stable supply chain and relatively low-cost products with high mechanical performance. However, these inorganic fibers introduce several drawbacks, including non-biodegradability, high abrasion of processing equipment, and potential dermal and respiratory irritations [6]. In contrast, natural fibers offer a lower density, less abrasiveness, as well as promising biodegradability and sustainability [4,5,7]. For instance, plant-based natural fibers such as sisal, flax, jute, and hemp have been widely investigated in the literature as potential low-cost, ecofriendly alternatives to synthetic fibers [5–7]. Nonetheless, composites reinforced with plant fibers exhibit lower mechanical performance compared to those reinforced with glass fibers, which limits their use in structural applications [4,8,9]. Furthermore, plant fibers tend to exhibit thermal instability at elevated temperatures, lower impact strength, and mechanical degradation during processing. Despite these drawbacks, the commercial use of plant fiber composites in non-load-bearing applications has significantly increased, predominantly in the automotive industry [3,10].

In contrast to fibers extracted from plants, silk is an animal-based fiber that offers attractive features such as low density, flame resistance, and high elongation even at low temperatures [4,8,11]. More importantly, silk exhibits higher mechanical performance than plant fibers, and, in some cases, comparable specific mechanical properties to glass fibers [4,8,9]. Silk denotes a group of protein-based fibers, called fibroin, produced by several arthropods like silkworms, spiders, and scorpions [3,12]. Fibroin generally has an irregular, almost-triangular cross-section, with a width in the range of 8 to 13 μm [13]. Owing to its biocompatibility and bioresorbable properties as well as high strength and toughness, silk fibers are used in a variety of clinical applications, such as braided suture threads for surgical procedures and scaffolds for cartilage and bone repair [13,14]. Aside from biomedical applications, silk from the cocoons of the domesticated mulberry silkworm, *Bombyx mori*, is of particular interest in textile industry due to its availability [9,14,15]. Silk cocoons are generally degummed, spun into rovings and yarns, then woven into textile fabrics [14,15]. These woven silk fabrics may be used as a woven reinforcement in composites for structural applications [14–16]. In addition, some researchers have explored using abundant silk waste from textile industry to reinforce polymer composites [17–19].

Despite these promising features, silk fibers have received only limited interest as a reinforcement for polymer composite products, and practically no commercial use exists beyond biomedical applications [4,9,11,13,15]. One plausible explanation for this limited use is the higher cost of silk compared to plant fibers in a very cost-competitive environment, especially for nonstructural composite parts. While silk might be more expensive than conventional reinforcements, waste silk fabric can be processed and utilized in composite laminates in a cost-effective manner [17–19]. Another possible limiting factor of silk fibers seems to be the incompatibility between the hydrophilic natural fibers and the hydrophobic polymer matrix that requires some form of surface treatment to improve the interfacial bonding [3,6]. In addition, silk is known to be prone to environmental factors, such as moisture and UV radiation, that significantly alter the mechanical performance of the fibers [20,21]. Therefore, silk fibers and fabrics might require special storage and transportation conditions for best performance. Nevertheless, the high specific properties of silk fibers make it a suitable replacement for glass fibers in composite applications where lightweight and energy-absorbance are important, such as automotive, aerospace, and wind turbine structures [15,16].

The limited available literature on silk-reinforced composites mainly investigated either discontinuous silk fiber reinforced thermoplastics, or continuous silk fiber reinforced thermosetting composites. Due to their recyclability, discontinuous natural fibers, also referred to as short fibers, were traditionally used to reinforce injection-molded thermoplastics, particularly polypropylene [22]. More recently, short fiber silk composites were used to reinforce biodegradable thermoplastics, such as polylactic acid (PLA) [22], poly vinyl alcohol (PVA) [23], and polybutylene succinate [24]. For instance, Ho et al. [22] manufactured PLA composite reinforced with 5 wt.% short silk fibers by injection molding, and reported a 27% and 2% improvement over PLA in tensile and flexural moduli, respectively. Considerably higher improvements were reported in several mechanical properties for silk reinforced gelatin composites over neat gelatin [25], including a 260% increase in tensile strength, a 4-fold rise in tensile modulus, a 320% improvement in bending strength, a 450% increase in bending modulus, and a 260% improvement in impact strength. Although these improvements achieved using silk fibers are significant for certain ecofriendly applications, the obtained mechanical performance remains inferior to glass reinforced composites.

On the other hand, structural composite laminates, intended for energy-absorbing structures, are often fabricated using woven textile fabrics and stiffer thermosetting polymers [3,8,9,15,26,27]. Epoxy resins are frequently used owing to their lower cost, higher processability, higher mechanical properties, good adhesive performance, and chemical resistance [11,15]. For instance, Oshkovr and coworkers attempted to use woven silk/epoxy composite square tubes as energy-absorbers and evaluated their crashworthiness [8,26]. However, catastrophic failures were reported under compression tests in both studies. Although impressive single fiber properties might be reported in the literature [3,8,21], the actual improvement over unreinforced epoxy may be limited by defects in the silk fabric, such as fiber misalignment and waviness inadvertently introduced during weaving. Furthermore, Yang et al. [15] investigated the tensile, flexural, interlaminar shear, impact, dynamic, and thermal properties of the silk/epoxy composites at 30%, 40%, 50%, 60%, and 70% fiber contents. A linear increase of most properties was observed with increasing fiber content between 30% and 70%. Optimal tensile properties were observed at 70% fiber content, with 145%, 130%, and 70% improvement over neat epoxy in tensile stiffness, ultimate stress, and ultimate strain, respectively. In addition, impact strength was observed to increase significantly only for fiber contents above 60%. The same research group studied silk/epoxy laminates for two natural silk varieties: *Bombyx mori* and *Antheraea pernyi* silk [27]. The authors reported that at 60% fiber content, both silk types showed a 2-fold increase in both specific tensile modulus and strength compared to the unreinforced epoxy resin. For the 60% *Antheraea pernyi* silk/epoxy laminates, the breaking energy was found to be 11.7 MJ/m^3, an order of magnitude higher than the 1.1 MJ/m^3 measured for neat epoxy. Moreover, a 3-fold increase in specific flexural strength was also reported, reaching 316 MPa/g·cm^{-3}.

More recently, Shah et al. [3,7,10] attempted to make a case for silk as a reinforcing agent in composite laminates by comparing their mechanical performance to flax- and glass-reinforced composites. The authors fabricated silk/epoxy laminates with nonwoven silk preform at 36% fiber content and woven silk fabric at 45% fiber content. The authors reported tensile and flexural specific strengths of ~90 MPa/g·cm^{-3} and ~170 MPa/g·cm^{-3}, respectively [3]. These values were comparable, although not necessarily superior, to those of glass/epoxy laminates. Other researchers incorporated silk into glass reinforced composites in the pursuit of hybrid composites with improved impact properties [16,17,28–30]. For example, Zhao et al. [16] investigated silk fabric/glass mat/polyester hybrid laminates at 14.5% and 2.4% fiber content of glass and silk, respectively. However, the authors reported practically no effect of the limited silk fabric presence on impact and flexural properties.

Surprisingly, two important aspects were not addressed in the available literature on silk reinforced composites. First, silk fabric is mostly used as received and surface treatment is seldom attempted to improve the silk/epoxy interface [31–33]. Generally, fiber sizing can be used to tune the bulk properties of composite laminates [34,35]. Surface treatment of natural fibers has been shown to significantly improve the properties of composite laminates [36–39]. Second, no attempt was made to investigate the effect of manufacturing processes and relevant process parameters on the produced silk composites. In fact, most of the reported investigations employed a rather simple, hand lay-up method to manufacture silk/epoxy laminates [16,17,30,40]. While simple fabrication methods such as hand lay-up can be attractive for their relative ease and low cost, they are operator-dependent, prone to process-induced defects, and often result in low-quality composite parts with higher void occurrence [40]. Presence of these defects, in turn, is known to significantly degrade the mechanical performance of composites [41]. A few other studies [9,15,27] used hand lay-up followed by hot pressing in order to increase the fiber content of the laminates, and thus improve the composite performance.

As described in the previous paragraph, remarkable improvements over neat epoxy were only achieved for silk/epoxy composites with fiber contents of 60% and higher [15,27]. Studies conducted at low or moderate fiber contents did not yield considerable improvement in mechanical performances. Consequently, investigating more appropriate manufacturing processes for silk/epoxy composite applications, such as variants of liquid composite molding (LCM), might be of interest. Only a couple of articles used vacuum-assisted resin transfer molding (VARTM) to manufacture silk/epoxy laminates [3,40]. In fact, Shah et al. [3] were able to achieve comparable mechanical performance to glass/epoxy laminates at a fiber content of only 45%. A lower occurrence of process-induced defects within the silk composites is believed to play a significant role in reaching this performance. More recently, our research group investigated fabrication challenges for silk/epoxy laminates [40], which showed that compared to hand lay-up, VARTM was more appropriate for silk/epoxy laminate fabrication as it allows uniform impregnation of the silk preform by the liquid resin, yielding higher part quality and reduced void formation.

In order to investigate the suitability of silk as an alternative reinforcement to glass fibers in polymer composites, the processability of silk reinforced composites was verified by fabricating silk/epoxy laminates using VARTM. In addition, the effects of manufacturing process and microstructural parameters such as post-fill external pressure and silk fabric anisotropy on the process-induced microstructure of silk epoxy laminates were studied. The mechanical performance of the fabricated laminates was also compared with those of neat epoxy and glass/epoxy laminates. Finally, the effect of storage conditions of silk fibers and moisture absorbed by silk on the manufacturing and performance of silk/epoxy composites was investigated.

2. Materials and Methods

2.1. Materials

INF 114 epoxy (PRO-SET) was used as the resin and polyamine INF 211 (PRO-SET) was chosen as the hardener. Before laminate fabrication, the resin and curing agent were mixed for 5 min at 350 rpm at a ratio of 100:27.4 by weight and degassed for 10 min under vacuum.

A woven silk fabric (Satin Ahimsa, Aurora Silk, Inc., Portland, OR, USA) was used in this study. The silk was produced from degummed cultivated *Bombyx mori* mulberry silk and had an areal density of 81.4 g/m^2. For each laminate, twelve layers of 152 mm-wide × 203 mm-long (6″ × 8″) silk fabric were prepared, stacked, and placed on the mold before infusion. Preparation of silk layers involved cutting the fabric to the desired size, ironing to remove wrinkles, and then drying in a vacuum oven at 50 °C. Since silk fabric exhibited an unbalanced weave pattern, laminates were fabricated with layers cut along both planar directions to investigate the effect of fabric anisotropy on the fill time and mechanical performance. Hence, two separate sets of laminates were investigated for each case: one set with fabric layers cut such that its length is parallel to the roll direction (longitudinal), and another set cut such that its length is perpendicular to the roll direction (transverse).

2.2. Manufacturing Procedure

An improved variant of vacuum-assisted resin transfer molding (VARTM) was used to fabricate silk/epoxy laminates. Figure 1 illustrates the experimental molding setup which can facilitate external air pressure on a typical VARTM mold to increase the limited compaction pressure in conventional VARTM (i.e., higher than 1 atm) [42]. Depending on the applied pressure, the fiber volume fraction in fabricated laminates can be significantly increased, which yields high-quality laminates with improved mechanical properties.

As seen in Figure 1, 12 layers of woven silk fabric (i.e., preform) were sealed with a vacuum bag and the epoxy/hardener mixture was infused into the mold from the inlet resin reservoir towards the vacuum source (i.e., exit). In order to reduce the effect of ambient temperature fluctuations, the mold temperature was kept constant at 30 °C. At this temperature, the viscosity of the resin was in the range of 180 to 200 mPa s. After the preform was completely wetted, infusion was continued for an additional 5 min (i.e., resin flushing) to mobilize and remove the process-induced voids with resin outflow from the exit gate [42]. Once the resin flushing was completed, the inlet port was closed to remove the excess resin from the exit of the mold (i.e., resin bleeding). In certain fabrication scenarios an external chamber pressure of 300 kPa (~44 psi) was applied during the post-filling stage to further compact the preform and improve the mechanical properties. All fabricated laminates were cured at 60°C for 8 hours after resin gelation occurred in 5 h at 30 °C.

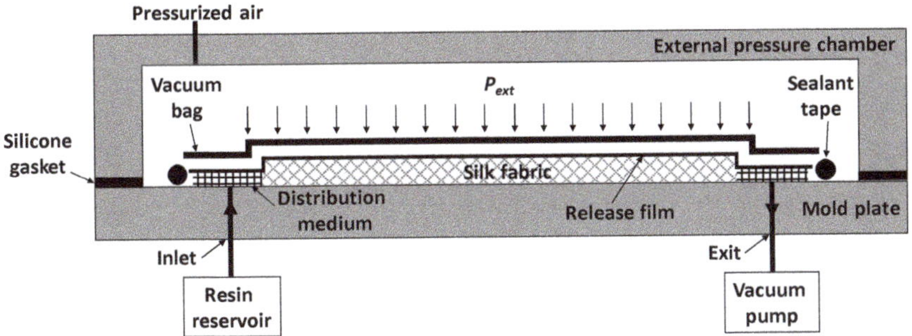

Figure 1. Experimental vacuum-assisted resin transfer molding (VARTM) setup with external pressure chamber to fabricate woven silk/epoxy composite laminates.

Table 1 presents the various fabrication scenarios used in this study and lists the laminate designations, reinforcement types, and fabric/lay-up details as well as the thicknesses of the fabricated laminates. For comparison purposes, neat epoxy samples (E) were manufactured by gravity casting and cured following the same cure schedule. In addition, glass/epoxy laminates (G), fabricated using the same resin in a recent study by our research group [43], were considered. Five different fabrication scenarios for silk composites (S) were performed. These scenarios were designed to assess the effects of fabric anisotropy, external pressure, and fabric moisture on the properties of the fabricated laminates. Furthermore, fabric anisotropy effects on wetting characteristics such as fill time, critical in defining the manufacturing cycle, were also investigated.

Table 1. Laminate designations, reinforcement types, and fabric/lay-up details. The external pressure is given in gauge pressure.

Designation	Reinforcement Type	Impregnation Direction	Fabric Condition	External Pressure (kPa)	Thickness (mm)
E	None	None	None	0	2.76 ± 0.07
G [43]	Glass Mat	Random	Dry	0	1.45 ± 0.02
S/L	Silk Woven	Longitudinal	Dry	0	2.39 ± 0.01
S/T	Silk Woven	Transverse	Dry	0	2.43 ± 0.01
S/L/P	Silk Woven	Longitudinal	Dry	300	2.22 ± 0.01
S/T/P	Silk Woven	Transverse	Dry	300	2.18 ± 0.01
S/T/M	Silk Woven	Transverse	Moist	0	2.51 ± 0.01

Figure 2a shows a representative image of the fabric and the disposition of longitudinal (y) and transverse (x) directions with respect to the fabric roll. Additionally, Figure 2b shows that the silk fabric was tightly woven with low porosity between the fibers with several silk threads in orthogonal directions. Abbreviations "L" and "T" were used to indicate whether the infusion was performed along the longitudinal or transverse directions of the fabric roll, respectively. In contrast, symbols x and y, depicted in Figure 2, were exclusively used in the designation to indicate testing directions of the composite samples. In addition, "P" indicates that external pressure was applied during the post-filling stage. "M" designates that the silk fabric was exposed to moisture prior to molding. For each fabrication scenario, two identical, 152 mm × 203 mm laminates were fabricated to ensure the repeatability of fabrication procedure.

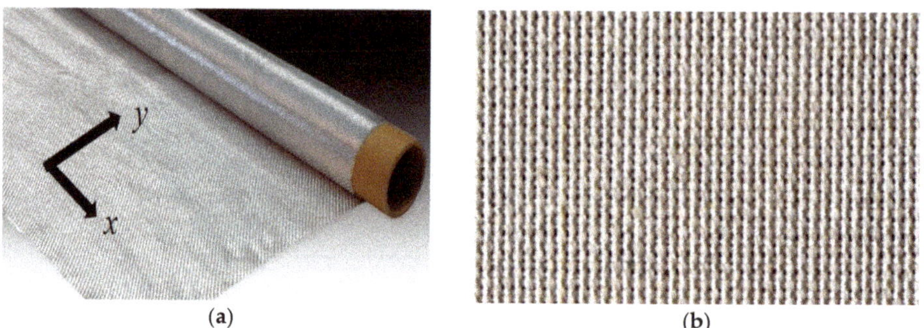

(a)   **(b)**

Figure 2. Representative images of the fabric, showing (**a**) longitudinal (abbreviated L for filling and y for testing) and transverse (abbreviated T for filling and x for testing) directions with respect to fabric roll and (**b**) a macroscopic image of the weave pattern.

2.3. Sample Preparation

Each molded laminate was sectioned using a diamond saw into several rectangular specimens. According to ASTM standards (D792, E1131, D790, D7028, and D2344), five samples for density measurement, five for thermogravimetric analysis, twelve for flexural testing (six samples along the flow direction (y), and six along (x) perpendicular to the flow), ten for SEM imaging (five along and five perpendicular to the flow direction) were cut in particular dimensions. When the sample thickness allowed, ten rectangular samples were cut for short beam tests (five along and five perpendicular to the flow direction).

2.4. Density Measurement and Volume Fraction Determinations

The density of five samples from each laminate was measured according to ASTM D792. Similarly, the density of the neat epoxy (i.e., the matrix) was measured as $\rho_{Epoxy} = 1.140 \pm 0.001$ g/cm^3. The density of silk fibers, on the other hand, was measured using a gas pycnometer (AccuPyc II 1340) as $\rho_{Silk} = 1.361 \pm 0.002$ g/cm^3. Using both densities of silk and epoxy, as well as the measured density of each fabricated composite laminate, $\rho_{Laminate}$, both fiber volume fraction, V_f, and resin volume fraction, V_r can be calculated as

$$V_f = \frac{\rho_{Laminate}}{\rho_{Silk}} \frac{M_{Silk}}{M_{Laminate}} \tag{1}$$

$$V_r = \frac{\rho_{Laminate}}{\rho_{Epoxy}} \times \frac{M_{Laminate} - M_{Silk}}{M_{Laminate}}, \tag{2}$$

where $M_{Laminate}$ is the measured mass of the laminate and M_{Silk} is the measured mass of the twelve silk fabric layers. Using both fiber and resin volume fractions, void volume fraction, V_v, can be calculated for each laminate as

$$V_v = 1 - (V_f + V_r) \tag{3}$$

2.5. Thermogravimetric Analysis

Thermogravimetric analysis (TGA) was used to verify the thermal stability and identify the maximum temperature at which the silk/epoxy composite laminates can be used. TGA thermograms were obtained by using a Thermogravimetry-Differential Scanning Calorimetry (TG-DSC) instrument (TA Instruments Q50, New Castle, DE, USA) at 10 °C/min heating rate in a nitrogen atmosphere.

2.6. Mechanical Testing

Flexural tests were performed according to the ASTM D790 standard to measure the flexural properties of each laminate. As mentioned earlier, specimens from each laminate were used to characterize the flexural properties along the longitudinal (y) and transverse (x) directions. Short beam tests were also executed on rectangular samples along the y- and x-directions cut from each silk/epoxy laminates. Interlaminar shear strength (ILSS) values were obtained in accordance with ASTM D2344. The mechanical testing was carried out at a rate of 2 mm/min.

2.7. Dynamic Mechanical Analysis

A dynamic mechanical analyzer (DMA) (TA Instruments DMA-Q800, New Castle, DE, USA) was used to measure the glass transition temperature, T_g, of selected configurations of the composites, namely neat epoxy (E), glass/epoxy (G), silk/epoxy impregnated along the transverse direction (S/T), and moist silk/epoxy impregnated along the transverse direction (S/T/M). For these particular measurements, 51 mm × 12.7 mm (2″ × 0.5″) specimens were prepared. The storage modulus, loss modulus, tanδ, and glass transition temperatures were determined for each sample.

2.8. Scanning Electron Microscopy (SEM) Imaging

SEM imaging was performed for the analysis of the microstructure of composite laminates. Specimen cut from the laminates were mounted in an acrylic resin to expose the through-the-thickness cross-section. Once polished, the specimens were sputter coated with 5 nm of gold/palladium to avoid charge build-up during SEM imaging. SEM images at different magnifications were obtained using a Zeiss Neon 40 EsB microscope (Carl Zeiss AG, Oberkochen, Germany). Additionally, fracture surfaces of the flexural test samples, as well as the silk preforms, were examined by SEM.

3. Results and Discussion

3.1. Morphology and Microstructure of Silk Fabric

Specialty material suppliers for composite reinforcement usually provide fabrics with balanced weave patterns, thus practically ensuring that the longitudinal and transverse properties would be similar. Having a fabric with a balanced planar weave fabric also ensures similar impregnation dynamics along both principal fabric directions. On the other hand, commercial fabrics for textile applications generally exhibit unbalanced weave patterns. For instance, the SEM micrograph of the silk fabric used by Yang et al. [15] showed a clearly unbalanced weave. Therefore, the morphology and microstructure of the silk fabrics that are often used for textile applications need to be investigated in detail to reveal possible anisotropy in their weave patterns.

SEM images presented in Figure 3 depict the microstructure of the silk fabric used in this study. Figure 3a shows the disposition of silk yarns along the longitudinal (y) and transverse (x) directions with respect to the fabric roll. The weave pattern was observed to be anisotropic, as clearly shown in Figure 3a,b. In fact, yarns along the transverse (x) direction have an average width of ~290 μm, which is approximately 60% higher than its counterpart along the longitudinal (y) direction (~180 μm). This unbalance in the weave pattern will likely induce disparate filling patterns and anisotropy in mechanical performance along these orthogonal directions.

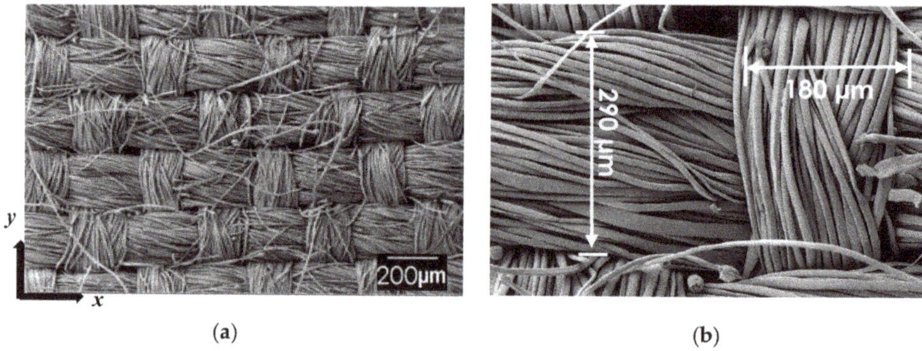

(a)   (b)

Figure 3. Sample SEM images of the microstructure of silk fabric, showing (**a**) an SEM image of multiple yarns (50×) and (**b**) higher magnification SEM image of two orthogonal yarns (200×).

3.2. Morphology and Microstructure of Silk/Epoxy Composite Laminates

Figure 4 depicts representative SEM images of the through-the-thickness microstructure of the fabricated silk/epoxy laminates. Figure 4a is a representative image along the longitudinal direction (y), obtained at 30×. Figure 4b, on the other hand, shows a representative image along the transverse direction (x) obtained at the same magnification. On these micrographs, fiber tows can be seen in both parallel and perpendicular directions to the cross-section. While a good wetting of the silk fabric was observed for all considered cases, voids are occasionally observed within the resin. Void occurrence was observed to be lower for laminates fabricated with external pressure. Interestingly, tows along the x and y directions showed contrasting undulation through-the-thickness of the composite. Figure 4a shows a pronounced sinusoidal waviness of silk tows continuously along the longitudinal direction, while tows along the transverse direction were less wavy (Figure 4b). Fiber waviness is known to have a substantial influence on the mechanical performance of the composites [41], thus mechanical properties were expected to be significantly lower along the longitudinal direction compared to the transverse direction.

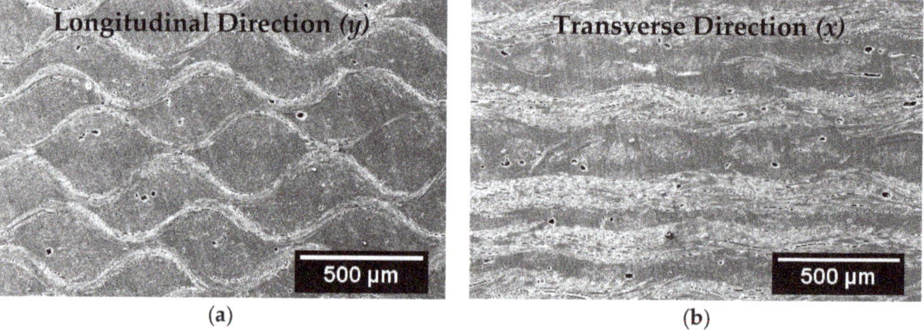

(a)   (b)

Figure 4. Representative SEM images of the through-the-thickness microstructure of fabricated silk/epoxy composite laminates: (**a**) along the longitudinal (y) direction and (**b**) along the transverse (x) direction. Both images are at 30×.

3.3. Thermal Stability of Silk/Epoxy Composite Laminates

Figure 5 shows TGA thermograms of neat epoxy, silk and glass fibers, in addition to the different composite laminates. The weight of neat epoxy was practically constant below 350 °C. Thermal decomposition of epoxy occurred mostly between 350 °C and 500 °C, after which no significant change in weight was observed. For silk fibers, no discernable change was observed at the processing

152

temperature of 30 °C, indicating that the properties of silk fibers would not be altered during manufacturing. Furthermore, an initial weight loss of ~5%–10% was observed below 100 °C, which is believed to correspond to desorption of moisture absorbed by the silk fabric from the environment. Thermal decomposition of the silk started at ~300 °C and continued at a decreasing rate until ~800 °C. In contrast, glass fibers showed a minor thermal decomposition between ~350 and ~450 °C, after which glass fiber weight did not change up to ~800 °C. This slight weight loss is possibly due to the removal of the organic sizing on the glass fibers. The divergence in thermal degradation behaviors of the three composite constituents would originate from the dissimilarities in their chemical compositions as well as their surface treatments.

Figure 5 also presents TGA spectra of both glass/epoxy and silk/epoxy laminates. For both composite types, the thermal degradation pattern showed three main stages. The first stage took place before thermal degradation occurred, where the weight was fairly stable. The second stage started with the decomposition of organic constituents of the composite. For G laminates, this stage started at ~350 °C, initiated by the degradation of the neat epoxy. Similarly, S laminates decomposition started at ~300 °C concurrently with the degradation of silk fibers. For both G and S composites, the thermal decomposition continued until ~450 to 500 °C due to the combined decomposition of the epoxy matrix and either the silk fibers or the sizing of glass fibers. In the final stage up to ~800 °C, the weight loss was observed to be minimal. These results showed that no thermal degradation was observed for silk/epoxy laminates up to ~300 °C, which was almost comparable to that of glass/epoxy laminates.

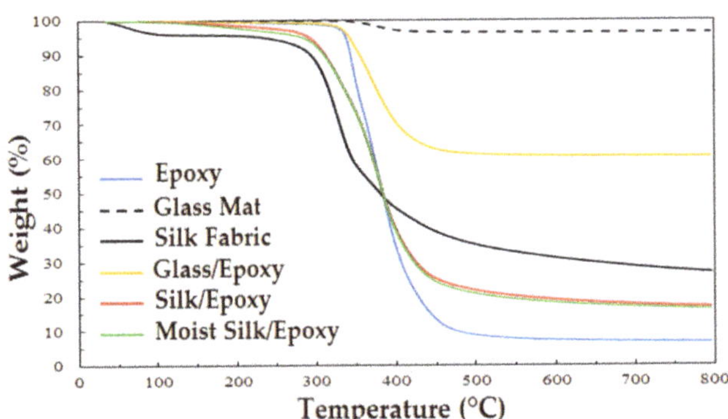

Figure 5. TGA thermographs of neat epoxy, glass mats, silk fabric, as well as glass/epoxy and silk/epoxy laminates, under a nitrogen atmosphere at a heating rate of 10 °C/min.

3.4. Fabric Anisotropy Effects on Fiber Wetting

The anisotropy of the silk fabric significantly influenced the impregnation characteristics during fabrication by VARTM. As presented in Table 2, impregnation along the transverse (x) direction was found to be much faster with an average filling time of 20 min. In contrast, when the impregnation was along the longitudinal (y) direction of the fabric roll, an average fill time of 68 min was observed. This ~140% increase in fill time was attributed to the anisotropic permeability of the silk fabric. As shown in Figure 3, fiber tows were ~60% thicker along the transverse direction than the longitudinal direction. This difference in yarn distribution would induce contrasting preform fabric permeabilities along the transverse and longitudinal directions, yielding this substantial increase in fill time.

The fabric anisotropy not only affected the mold filling time but also the final void content in the fabricated laminates as the impregnation velocity significantly affect the void formation [44]. As Table 2 documents, slow impregnation while fabricating the S/L and S/L/P laminates yielded lower void contents than those impregnated more rapidly (i.e., S/T and S/T/P, respectively). For instance, silk/epoxy laminates impregnated along the longitudinal direction (S/L) filled in 68 min and yielded a void content of 0.7%. In contrast, the laminates impregnated under identical conditions along the transverse direction (S/T) filled in 20 min, with a void content of 1.1%. A similar trend was observed for laminates manufactured with external pressure, as void contents of 0.5% and 1.2% observed for S/L/P and S/T/P, respectively. In LCM processes, voids are known to form mainly when air is trapped due to the unbalance between the velocities of the viscous macroflow occurring between fiber tows and the capillary flow inside fiber tows [44]. According to the modified capillary number analysis, any change in the equilibrium between these competing flows would affect the air entrapment void formation mechanism, thus affecting void presence in the final composite part [1]. The highest void content was observed for the fastest filling of 3 min corresponding to the glass/epoxy laminate case. Nonetheless, the comparison does not hold as the sized glass fibers possess an entirely different epoxy affinity and the glass fabric is in the form of a random mat.

Table 2. Density, fiber content, void content, and fill time of the manufactured composite laminates (n = 10, 95% confidence interval).

Designation	Density (g/cm^3)	Fiber Content (%)	Void Content (%)	Average Fill Time
E	1.140 ± 0.001	0	N/A	N/A
G	1.730 ± 0.010	45.7 ± 0.1	1.86 ± 0.72	3 min
S/L	1.227 ± 0.003	42.8 ± 0.1	0.66 ± 0.21	68 min
S/T	1.223 ± 0.001	43.5 ± 0.1	1.12 ± 0.06	20 min
S/L/P	1.237 ± 0.002	46.8 ± 0.2	0.53 ± 0.17	68 min
S/T/P	1.233 ± 0.002	48.4 ± 0.1	1.24 ± 0.17	20 min
S/T/M	1.226 ± 0.002	45.7 ± 0.4	1.30 ± 0.12	32 min

Furthermore, slower filling slightly increased the amount of resin intake during mold filling, resulting in a slight drop in fiber content (from 44% to 43% for S/T and S/L, respectively). Applying a post fill external pressure helped improve the fiber content in silk laminates from ~43% to ~47%. It is worth noting that the increase in fiber volume fraction induced a slight increase in the density of the composite laminate. Applying external pressure after the impregnation is completed, i.e., after formation of voids, can have opposing effect on the final void content of the part. On one hand, it helps in suppressing voids due to increased pressure and, on the other hand, it may cause void entrapment and prevent removal of mobile voids during flushing or bleeding of the resin. As a result, when impregnation was performed along the longitudinal direction, applying 300 kPa pressure reduced void content from 0.66% to 0.53%. However, in the case that impregnation was along the transverse direction, void content increased from 1.12% to 1.24%.

3.5. Fabric Anisotropy Effects on Mechanical Properties of Silk/Epoxy Laminates

Figure 6a shows both flexural strength and modulus of neat epoxy, random mat glass/epoxy laminates, and silk/epoxy laminates infused along both the longitudinal and transverse directions, each tested along orthogonal directions x and y. Moreover, Figure 6b depicts strain to failure and the breaking energy calculated as the area below flexural stress-strain curves of the tested specimen.

A strong anisotropy was observed in the flexural performance of silk/epoxy laminates. In fact, flexural strength, modulus, and breaking energy were observed to show disparate properties when tested along the longitudinal and transverse directions. As depicted in Figure 6a, flexural strength dropped by 46% from 229 MPa along the x direction to 123 MPa along the y direction for the silk/epoxy laminate infused along the longitudinal direction. A similar drop of 46% from 222 to 121 MPa was observed when infusing along the transverse direction. Flexural modulus showed an analogous tendency with 31 and 30% reductions between the x and y directions when infused along the longitudinal and transverse directions, respectively. Furthermore, Figure 6b depicts a similar trend for the breaking energy. For the silk epoxy laminate infused along the transverse direction (S/T), breaking energies of 13.8 and 3.6 MJ/m^3 were calculated respectively along the x and y directions, representing an almost 74% difference. Similarly, a 70% difference (11.8 to 3.6 MJ/m^3) was observed in the breaking energy for S/L along the longitudinal direction. Strain to failure is the flexural property least affected by the fabric microstructure with only 14% difference along the x and y directions in both S/T and S/L.

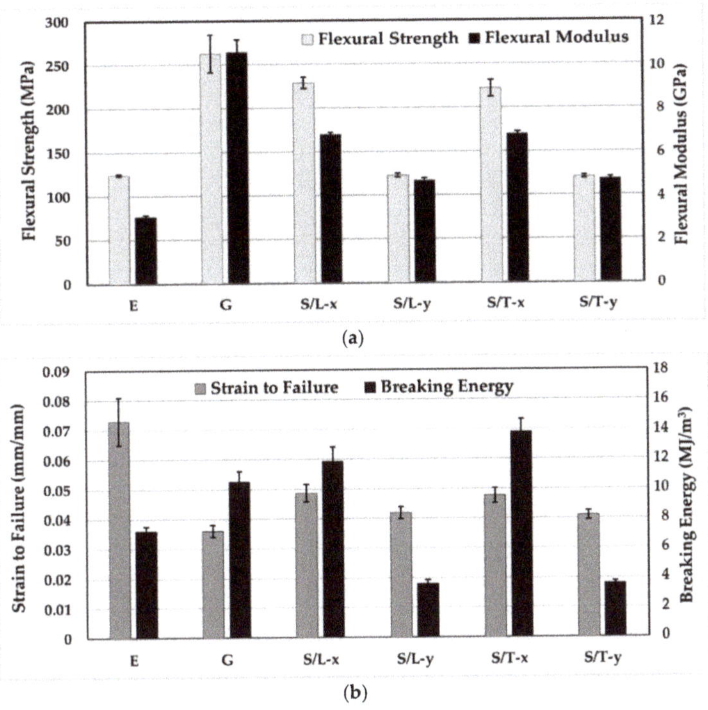

Figure 6. Flexural properties of silk/epoxy composite laminates in comparison with neat epoxy and glass/epoxy laminates [43]: (**a**) Flexural strength and modulus and (**b**) strain to failure and breaking energy.

As shown in Figure 3, yarns along the transverse (x) direction are ~60% thicker than yarns along the longitudinal (y) direction, which implies a larger number of silk fibers would be under loading along the x direction compared to those along the y direction. Furthermore and as shown in Figure 4a, higher waviness of the silk tows throughout the laminate was observed in the longitudinal (y) direction. Localized wrinkles are known to dramatically reduce the composite mechanical properties depending on the wrinkle severity, defined as the ratio of the amplitude to the wavelength of the undulation [41]. The wrinkles of the silk fabric along the y direction shown in Figure 4a exhibited a wrinkle severity in the range of 0.25 to 0.30. Single wrinkles with a severity in that range are reported to induce as much as 80% reduction in mechanical properties [41]. Continuous undulations are known to cause even further deterioration of composite performance [41]. In a sense, the undulated silk fabric along the longitudinal direction may be acting more as a defect rather than a reinforcement.

The results given in Figure 6 indicate that the mechanical performance of silk/epoxy laminates did not seem to depend on the infusion direction. In fact, regardless of the infusion direction, the laminates tend to exhibit statistically similar mechanical properties. The slight differences in void and fiber contents did not seem to affect the resulting flexural and breaking energy performance. For instance, the S/T laminate exhibited a ~44% fiber content, similar to the ~43% fiber content of S/L. A small difference in the void contents was observed in these laminates, with 1.1 and 0.7% void contents for S/T and S/L, respectively. Hence, the efficiency of manufacturing silk composites can be eventually improved in an industrial setting by an appropriate selection of infusion direction that reduces filling times and manufacturing cycles.

3.6. Silk/Epoxy Laminate Performance Compared to Neat Epoxy and Glass/Epoxy Laminates

In addition to silk/epoxy laminates, Figure 6 depicts the flexural performance of glass/epoxy laminates from a past paper [43] and neat epoxy. Silk was found to improve the flexural strength and modulus, as well as the breaking energy of epoxy along x direction. For instance, more than 78% and 121% improvements were obtained in flexural strength and modulus, respectively, compared to pure epoxy along x direction. In contrast, change in flexural strength was negligible along the y direction. Conversely, flexural modulus showed more than 53% increase over the neat epoxy along the y direction.

An expected drop in strain to failure was observed for silk/epoxy laminates compared to neat epoxy, as the addition of fibers limits the ability of the epoxy resin to elongate. However, the measured values of 0.041 to 0.049 were definitely higher than the strain to failure of 0.036 exhibited by glass/epoxy laminates. This result was also predictable as silk fibers enjoy a remarkably high failure strain of ~20% compared to the ~2.5% of glass fibers [3,8,9]. Lai and Goh reported an even higher strain to failure of ~34% [21]. These large failure strains would allow the silk reinforced laminates to experience higher failure strains compared to glass/epoxy composites, yielding improved breaking energy.

Aside from strain to failure, the glass/epoxy laminates performed slightly better than silk/epoxy at comparable fiber contents (~45%) in terms of flexural properties. Along the x direction, silk composites enjoyed flexural strengths merely 12 to 15% lower than glass laminates, as shown in Figure 6a. Similarly, silk flexural moduli were ~36% lower than glass/epoxy laminates. In contrast, and owing to the superior elongation of silk, the breaking energy of silk/epoxy were 35% higher than that of glass/epoxy laminates.

In high-end applications where both lightweight and high performance are pursued, specific properties are often used to assess the composite parts [3,15]. Figure 7 illustrates the specific flexural strength vs. specific flexural modulus for the silk/epoxy and glass/epoxy composites investigated in this study. For comparison purposes, results reported in the literature for silk/epoxy laminates [3,15] and flax/epoxy laminates [3] with comparable fiber contents are also included.

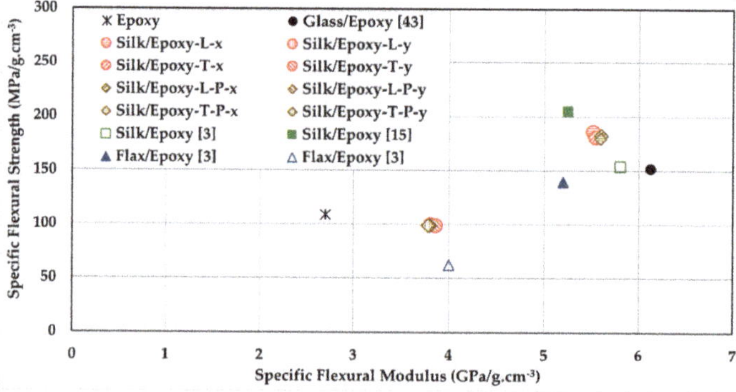

Figure 7. Specific flexural properties of silk/epoxy laminates compared to glass/epoxy laminates and neat epoxy samples, as well as reported properties for silk/epoxy and flax/epoxy from the literature.

First, as shown in Figure 7, the measured specific flexural strength and modulus compared well with the values reported in the literature for silk/epoxy laminates [3,15], and performed better than flax/epoxy laminates [3]. It is worth noting that both references reported only the mechanical properties of silk/epoxy laminates along strongest direction, although the unbalanced weave patterns were observed in the silk fabric used in this study.

The lowest value of specific flexural strength of silk/epoxy in this study was measured at ~100 MPa/g·cm^{-3} along the y direction, which is slightly lower than unreinforced epoxy. In contrast, the specific flexural strength along the x direction exhibited more than 66% improvement over epoxy.

Moreover, silk reinforcement induced ~41 and ~104% increases in specific flexural modulus along the x and y directions, respectively.

Compared to glass/epoxy, silk/epoxy laminates showed a 19 to 23% higher specific flexural strength along the x direction, and a concurrent ~10% lower specific modulus. This improvement in specific properties is significant, given the large difference between the properties of individual silk and glass fibers [4,8,9,15,16,21]. Along the y direction, the specific flexural strength and modulus were ~35 to 37% lower than glass/epoxy laminates. This was expected as the silk was not observed to effectively reinforce the epoxy along the y direction. Applying a post-fill external pressure did not seem to affect the specific flexural properties of the silk/epoxy laminates. The slight improvement in absolute flexural strength and modulus was offset by the concurrent increase in laminate density. These results showed that silk/epoxy may provide comparable, if not better, specific properties to glass/epoxy at least along the transverse direction. Once balanced weave patterns are produced without yarn undulations, silk fabrics can deliver the same reinforcement along both principal directions.

3.7. Silk/Epoxy Interface

A possible explanation of the lower absolute flexural properties of silk/epoxy laminates compared to glass/epoxy laminates may be an inferior bonding between silk fibers and the epoxy matrix. In fact, unlike glass fibers that have a commercial sizing suitable for epoxy matrix bonding, the silk fabric was used as received and did not undergo any surface treatment. Thus, a weak fiber/matrix adhesion is expected due to the incompatibility between the hydrophilicity of untreated silk fibers and the hydrophobicity of the epoxy [3,6]. The silk/epoxy interface was investigated by the careful review of the SEM images of silk/epoxy laminate cross-sections, as depicted in Figure 8.

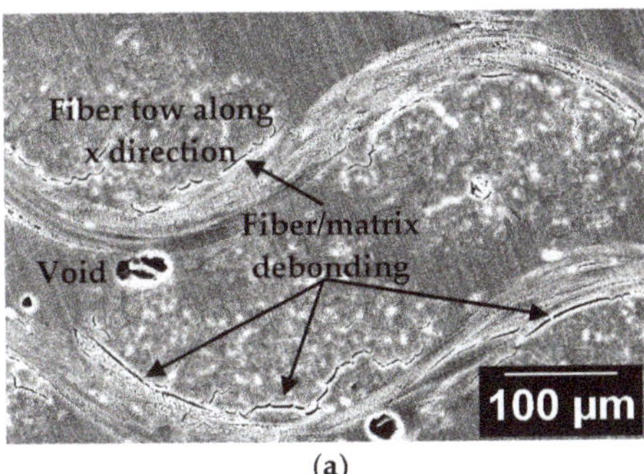

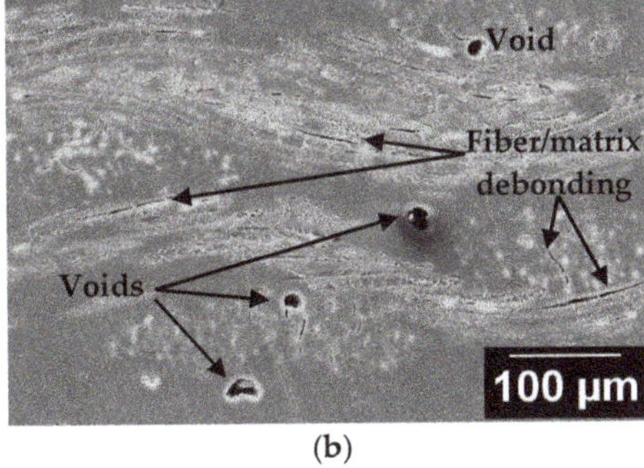

(a) (b)

Figure 8. Sample SEM images at 200× of the microstructure of silk/epoxy laminate cross-sections (**a**) along the y direction and (**b**) along the x direction.

Inspection of the silk/epoxy interface under SEM showed extensive fiber/matrix debonding in all fabricated laminates, indicating an overall weak fiber/matrix adhesion. In addition, fiber/matrix debonding was observed to have a higher occurrence along the y direction, as depicted in Figure 8a,b. A weak fiber/matrix adhesion is expected due to the incompatibility between the hydrophilicity of untreated silk fibers and the hydrophobicity of the epoxy [3,6]. In addition, the pronounced undulations of silk yarns along the longitudinal (y) direction might increase the fiber/matrix debonding as the curvatures create a more tortuous path for the impregnating epoxy resin. Furthermore, fiber/matrix debonding was visually observed to be much less frequent in laminates manufactured with post-fill external pressure, as the application of external pressure might have increased the resin pressure, thus causing a reduction on the porosity as well as slightly improving the fiber/matrix interface [42,45].

Another method to investigate the fiber/matrix adhesion is to analyze fracture surfaces of the silk/epoxy laminates. As illustrated in Figure 9a,b, fiber pullouts were observed in SEM images of fracture surfaces of tested silk/epoxy specimens, even in the laminates fabricated with external pressure. These findings corroborate the weak fiber/matrix adhesion and highlight the potential benefit of an appropriate surface treatment for silk fibers.

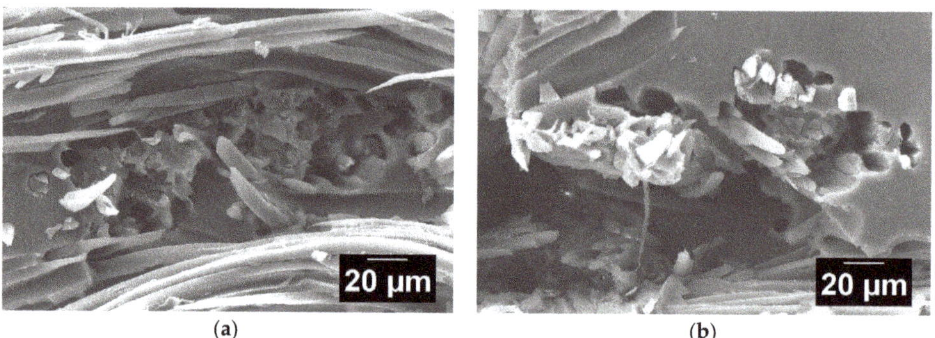

Figure 9. Sample SEM images at 400× of the fractured surfaces of silk/epoxy laminates fabricated (**a**) without external pressure and (**b**) with external pressure.

3.8. Effect of Post-Fill External Pressure on Silk/Epoxy Performance

The results presented earlier indicate that the external pressure did not affect the flexural properties of silk/epoxy laminates. Statistically indistinguishable values of strength, modulus, and strain to failure were measured with and without applying pressure for all considered cases. Specific flexural strength and modulus also showed no sensitivity to external pressure as the data points are observed to overlap in Figure 7. This insensitivity could be due to minor increases in fiber volume fraction at the external pressure level applied in this study, which could only yield a modest increase of ~4 to 5% in fiber content.

Breaking energy and interlaminar shear strength (ILSS), on the other hand, showed discernible improvement with the application of post-fill external pressure. Figure 10 presents the percent property improvement of both the breaking energy and ILSS due to the application of external pressure. For example, the energy required to break silk/epoxy laminates along the x direction increased by ~17% from ~12 MJ/m^3 to ~14 MJ/m^3 after applying the external pressure for the slow-filled silk/epoxy laminates infused along the longitudinal direction. Lower improvements (3%–8%) in breaking energy were observed for the remaining cases. In addition, Figure 10 shows a consistent improvement in ILSS (3%–9%) after the application of external pressure. Again, the highest improvement in ILSS (i.e., 9% to 44 MPa) was observed in the slow-filled silk/epoxy laminates along the x direction. These improvements in breaking energy and ILSS can be attributed to a lower occurrence of silk/epoxy debonding at higher pressures observed during SEM analysis.

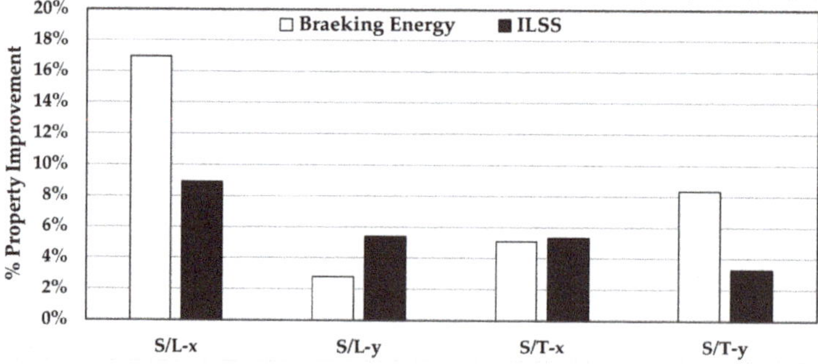

Figure 10. Effect of applying a 300 kPa post-fill external pressure to silk/epoxy laminates on the breaking energy and interlaminar shear strength (ILSS).

3.9. Dynamic Mechanical Analysis of Silk/Epoxy

Dynamic mechanical analysis (DMA) is usually used to investigate the mechanical properties and viscoelastic behavior of polymer-based materials as a function of temperature, frequency, and time [15]. In this study, DMA experiments were performed for two purposes: (a) to characterize the thermomechanical properties of the silk/epoxy composite laminates and (b) to compare glass transition temperature of silk/epoxy and glass/epoxy composites.

Figure 11 shows representative changes of the storage modulus E′, loss modulus E″, and tanδ (the ratio of loss modulus E″ to storage modulus E′) over a temperature range of 30 to 150 °C for the epoxy, glass/epoxy, and silk/epoxy laminates. Glass/epoxy showed a higher storage modulus compared to silk/epoxy and neat epoxy for all considered temperatures, which is expected since glass fibers are much stiffer than silk. Glass transition temperature was calculated using the storage modulus, loss modulus, and tanδ for all tested samples, and the obtained average values are presented in Table 3.

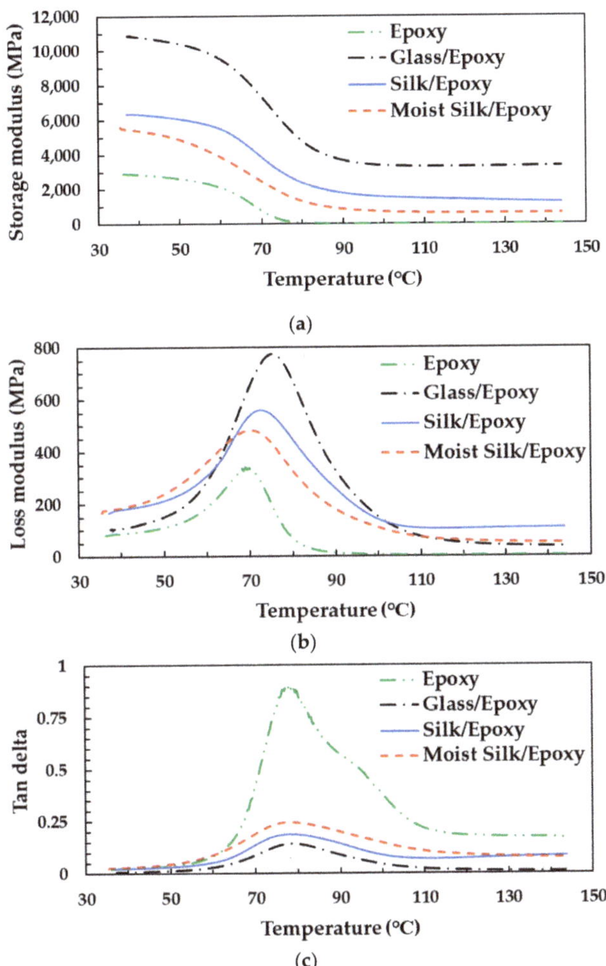

Figure 11. Representative changes of the (**a**) storage modulus E′, (**b**) loss modulus E″, and (**c**) tanδ for epoxy, glass/epoxy, and silk/epoxy laminates over a temperature range of 30 to 150 °C.

Table 3. Glass transition temperature, T_g, of silk/epoxy compared to glass/epoxy and neat epoxy.

Designation	T_g by Storage Modulus (°C)	T_g by Loss Modulus (°C)	T_g by tanδ (°C)
E	59.9	69.6	78.4
G	59.9	75.2	78.2
S/T	60.9	74.0	79.5
S/T/M	57.4	71.9	79.3

Silk/epoxy laminates were found to exhibit a glass transition temperature comparable, if not higher, than glass/epoxy. When measured using tanδ for instance, silk/epoxy exhibited a T_g of 79.5 °C compared to 78.2 °C for glass/epoxy. Comparable T_g values are also obtained when using both the storage and loss moduli. In addition, using silk as reinforcement was observed to increase the glass transition temperature of the epoxy resin. These findings showed that silk/epoxy laminates offer comparable glass transition temperatures to glass/epoxy, which could make silk a useful alternative to glass fibers in T_g sensitive applications.

3.10. Silk Moisture Effects on Silk/Epoxy Laminate Performance

Unlike glass and other inorganic fibers that do not absorb moisture [46], organic fibers readily absorb ambient moisture, which would alter the mechanical performance of the produced composites. In order to investigate the effect of storage conditions of the silk fabrics on the mechanical performance, silk/epoxy laminates were manufactured with silk fabric containing ~10% moisture by weight. As revealed in Figure 5, this moisture content approximately corresponds to the initial weight drop observed in silk fabrics during TGA tests. Therefore, after the usual drying cycle in a vacuum oven at 50 °C, the silk fabric layers were exposed to a humid environment at 35 °C, and their weight gains were monitored. Figure 12 presents a representative moisture absorption curve of silk fabric. A moisture content of about 10% is usually reached within 48 h.

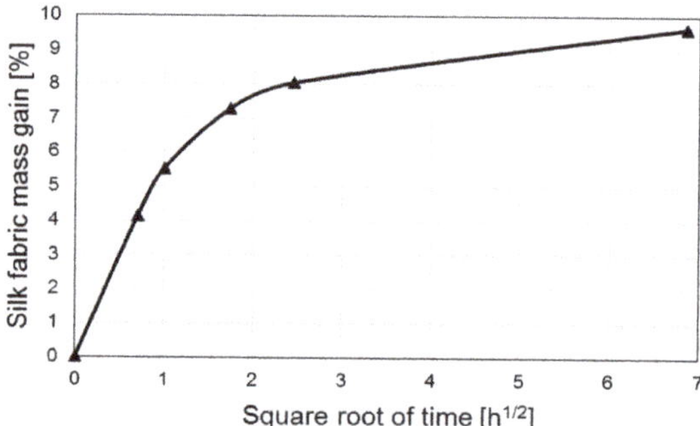

Figure 12. Moisture absorption curve of silk fabric exposed to a humid environment at 35 °C.

Immediately after the moisture content reached 10%, a vacuum bag lay-up was prepared and the moist fabric was infused following the same VARTM procedure used in this study. Since the fabric is exposed to vacuum within the mold for about 5 min prior to impregnation, a mock molding was performed. During this mock molding, a similar lay-up was prepared, thus exposing the moist silk fabrics to vacuum for 5 min. The fabrics were taken out of the lay-up and weighed after the 5 min vacuum period. The difference in the fabric weights was negligible, thus indicating no appreciable moisture loss during this period. The faster infusion case (S/T) was chosen as a baseline for investigating the silk moisture effects. Hence, two laminates were fabricated by infusing resin along the transverse (x) direction of the moist silk fabric.

Interestingly, presence of moisture in the silk fabric was observed to affect the impregnation dynamics. As presented in Table 2, the average fill time increased from 20 to 32 min due to the presence of 10% moisture in the silk fibers. Conceivable causes of this ~60% increase in fill time include accelerated cross-linking and viscosity change of the epoxy resin due to moisture. Decreased preform permeability due to slight swelling of the fabric can also slow the impregnation. Moisture-induced preform swelling combined with a slower filling also resulted in a slight increase in part thickness from 2.43 to 2.51 mm, as shown in Table 1. Moisture contact with the advancing epoxy front during impregnation and evaporation of the water during curing might also have caused the slight increase in void content from 1.1% to 1.3% (Table 2).

As presented earlier in Figure 5, the TGA behavior of laminates fabricated with dry and moist silk fabric overlap, indicating a similar thermal degradation behavior. Furthermore, dynamic mechanical analysis was performed on dry and moist silk reinforced composites. As presented in Figure 11, moisture presence in silk fabric is found to decrease the thermomechanical properties of the silk/epoxy, including storage modulus, loss modulus, and glass transition temperature. For instance, the peak loss modulus decreased from ~561 MPa for dry silk/epoxy to ~485 MPa for moist silk/epoxy. A concurrent drop from ~6500 MPa to ~5500 MPa was observed in storage modulus at 30 °C. Glass transition temperature was also lower for moist silk/epoxy regardless of the calculation method as shown in Table 3. For instance, when calculated using the loss modulus, moisture presence in the silk fabric caused the glass transition to drop from ~74 to ~72 °C. A similar drop from ~61 to ~57 °C was observed using the storage modulus.

Additionally, silk fabric moisture was found to detrimentally affect the mechanical performance of silk/epoxy laminates, as depicted in Figure 13. The presence of 10% moisture in the silk fabric prior to fabrication was found to induce 23 and 20% reductions in specific flexural strength and modulus of the laminates along the transverse (x) direction, respectively. Similarly, in absolute values, moisture presence caused the flexural strength and modulus to drop respectively by 23 and 20% along the x direction, as depicted in Figure 13a. As depicted in Figure 13b, silk moisture also caused approximately 29 and 33% reductions in the breaking energy along both x and y directions, while the strain to failure dropped by 2 to 8%. Silk moisture also critically affected the interlaminar shear strength as shown in Figure 13c. A 20% reduction in ILSS was registered between dry and moist silk/epoxy laminates along the x direction. Hence, the storage conditions of the silk fabrics that are to be used in composite laminates should be carefully monitored. Drying silk fabrics to remove the absorbed moisture can significantly alleviate these adverse effects on the composite performance.

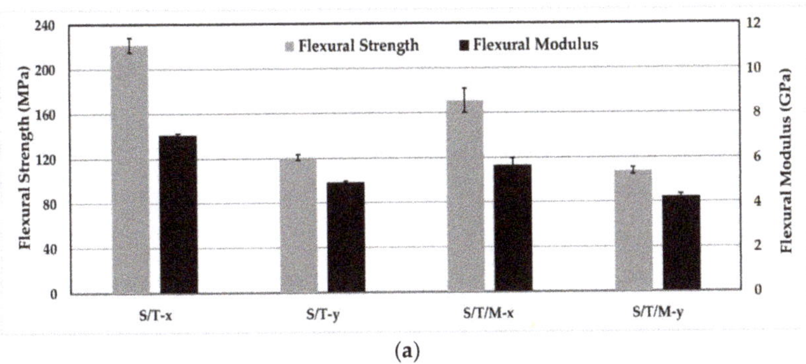

(a)

Figure 13. *Cont.*

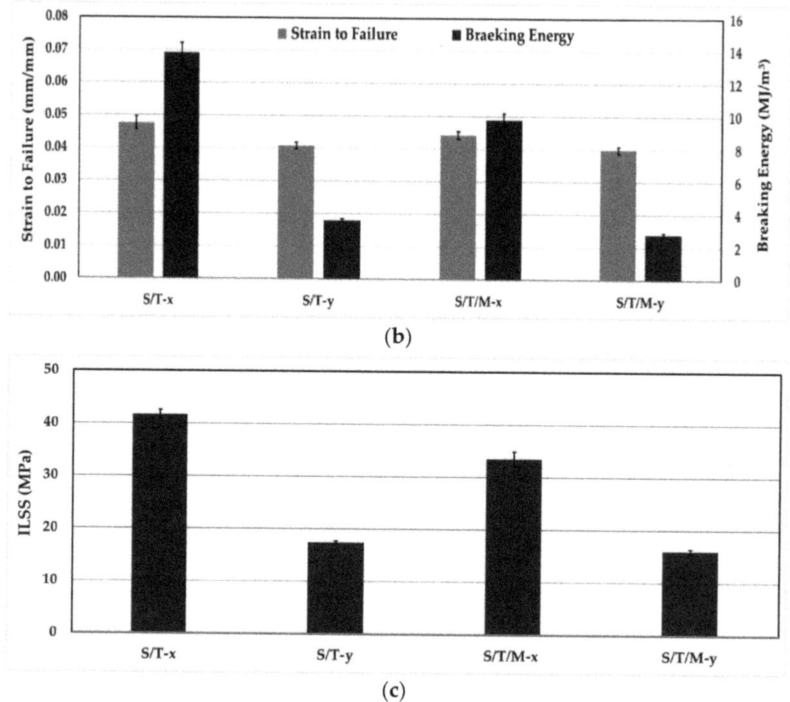

(b)

(c)

Figure 13. Effect of the presence of 10% moisture in the silk fabric on the mechanical performance of silk/epoxy laminates: (**a**) Flexural strength and modulus; (**b**) strain to failure and breaking energy; and (**c**) ILSS.

4. Conclusions

The processing and properties of silk/epoxy composite laminates were investigated to assess the viability of silk as a natural alternative to glass as a reinforcement in polymer composites. Vacuum-assisted resin transfer molding (VARTM) was used to manufacture composite laminates reinforced with plain weave silk.

Silk/epoxy laminates exhibited similar thermal stability, glass transition temperature, and thermomechanical properties to glass/epoxy composites. In addition, specific mechanical properties of silk/epoxy laminates were comparable to those of glass/epoxy. Silk composites even exhibited a 23% improvement over glass in specific flexural strength. In addition, applying a 300 kPa post-fill external pressure did not show a considerable effect on mechanical properties, except the breaking energy and interlaminar shear strength (ILSS). However, silk/epoxy laminates showed anisotropic mechanical properties due to the unbalanced weave pattern of the silk fabric. Therefore, developing silk fabrics with a balanced planar weave may prove beneficial when used as an ecofriendly alternative to glass reinforcement in structural composites. SEM analysis of the silk/epoxy composites revealed extensive fiber/matrix debonding. Treating the silk surface seems to be necessary for higher silk/epoxy interfacial adhesion, and for further improvement of mechanical properties.

Finally, silk fabric was found to be prone to moisture absorption from the environment, which considerably degraded the mechanical properties of the fabricated laminates. The moisture presence in silk fabric prior to laminate fabrication yielded slower fill times and caused an overall reduction in thermomechanical properties. A 10% fabric moisture content induced 23 and 20% reductions in specific flexural strength and modulus, respectively. These results stress the elevated sensitivity of silk composites to the storage conditions of the silk preforms.

Materials **2018**, *11*, 2135

Author Contributions: Conceptualization, Y.K.H., M.A.Y., G.E.G., M.P., M.A., and M.C.A.; Methodology, Y.K.H., M.A.Y., G.E.G., M.P., M.A., and M.C.A.; Validation, Y.K.H., M.A.Y., G.E.G., M.P., M.A., and M.C.A.; Formal Analysis, Y.K.H., M.A.Y., G.E.G., M.P., M.A., and M.C.A.; Investigation, M.A.Y., G.E.G., M.P., and M.A.; Resources, M.C.A.; Writing—Original Draft Preparation, Y.K.H.; Writing—Review & Editing, Y.K.H., M.A.Y.,G.E.G., M.P., M.A., and M.C.A.; Visualization, Y.K.H., M.A.Y., G.E.G., M.P., and M.A.; Supervision, Y.K.H. and M.C.A.; Project Administration, Y.K.H. and M.C.A.; Funding Acquisition, M.C.A.

Funding: This research received no external funding.

Conflicts of Interest: The authors declare no conflicts of interest.

References

1. Hamidi, Y.K.; Altan, M.C. Process-induced defects in resin transfer molded composites. In *Comprehensive Composite Materials II, Vol. 2*; Beaumont, P.W.R., Zweben, C.H., Eds.; Elsevier: Amsterdam, The Netherlands, 2018; pp. 95–106. [CrossRef]

2. Hamidi, Y.K.; Altan, M.C.; Grady, B.P. Polymer composites. In *Encyclopedia of Chemical Processing*; Lee, S., Ed.; Decker Publisher: New York, NY, USA, 2005; pp. 2313–2322. ISBN 978-0-8247-5563-8.

3. Shah, D.U.; Porter, D.; Vollrath, F. Can silk become an effective reinforcing fibre? A property comparison with flax and glass reinforced composites. *Compos. Sci. Technol.* **2014**, *101*, 173–183. [CrossRef]

4. Pickering, K.L.; Aruan Efendy, M.G.; Le, T.M. A review of recent developments in natural fibre composites and their mechanical performance. *Compos. Part A Appl. Sci. Manuf.* **2016**, *83*, 98–112. [CrossRef]

5. Ho, M.; Wang, H.; Lee, J.-H.; Ho, C.; Lau, K.; Leng, J.; Hui, D. Critical factors on manufacturing processes of natural fibre composites. *Compos. Part B Eng.* **2012**, *43*, 3549–3562. [CrossRef]

6. Reddy, B.M.; Reddy, Y.V.M.; Reddy, B.C.M. Mechanical properties of burmese silk orchid fiber reinforced epoxy composites. *Mater. Today Proc.* **2017**, *4*, 3116–3121. [CrossRef]

7. Shah, D.U. Natural fibre composites: Comprehensive Ashby-type materials selection charts. *Mater. Des.* **2014**, *62*, 21–31. [CrossRef]

8. Ataollahi, S.; Taher, S.T.; Eshkoor, R.A.; Ariffin, A.K.; Azhari, C.H. Energy absorption and failure response of silk/epoxy composite square tubes: Experimental. *Compos. Part B Eng.* **2012**, *43*, 542–548. [CrossRef]

9. Ude, A.U.; Ariffin, A.K.; Azhari, C.H. Impact damage characteristics in reinforced woven natural silk/epoxy composite face-sheet and sandwich foam, coremat and honeycomb materials. *Int. J. Impact Eng.* **2013**, *58*, 31–38. [CrossRef]

10. Shah, D.U.; Porter, D.; Vollrath, F. Opportunities for silk textiles in reinforced biocomposites: Studying through-thickness compaction behavior. *Compos. Part A Appl. Sci. Manuf.* **2014**, *62*, 1–10. [CrossRef]

11. Saba, N.; Jawaid, M.; Alothman, O.Y.; Paridah, M.T.; Hassan, A. Recent advances in epoxy resin, natural fiber-reinforced epoxy composites and their applications. *J. Reinf. Plast. Compos.* **2016**, *35*, 447–470. [CrossRef]

12. Hsia, Y.; Gnesa, E.; Jeffery, F.; Tang, S.; Craig Vierra, C. Spider Silk Composites and Applications. In *Metal, Ceramic and Polymeric Composites for Various Uses*; Cuppoletti, J., Ed.; InTech: Rijeka, Croatia, 2011; pp. 303–324. ISBN 978-953-307-353-8.

13. Hardy, J.G.; Scheibel, T.R. Composite materials based on silk proteins. *Prog. Polym. Sci.* **2010**, *35*, 1093–1115. [CrossRef]

14. Pereira, R.F.P.; Silva, M.M.; de Zea Bermudez, V. Bombyx mori Silk Fibers: An Outstanding Family of Materials. *Macromol. Mater. Eng.* **2015**, *300*, 1171–1198. [CrossRef]

15. Yang, K.; Ritchie, R.O.; Gu, Y.; Wu, S.J.; Guan, J. High volume-fraction silk fabric reinforcements can improve the key mechanical properties of epoxy resin composites. *Mater. Des.* **2016**, *108*, 470–478. [CrossRef]

16. Zhao, D.; Dong, Y.; Xu, J.; Yang, Y.; Fujiwara, K.; Suzuki, E.; Furukawa, T.; Takai, Y.; Hamada, H. Flexural and hydrothermal aging behavior of silk fabric/glass mat reinforced hybrid composites. *Fibers Polym.* **2016**, *17*, 2131–2142. [CrossRef]

17. Priya, S.P.; Ramakrishna, H.V.; Rai, S.K.; Rajulu, A.V. Tensile, flexural, and chemical resistance properties of waste silk fabric-reinforced epoxy laminates. *J. Reinf. Plast. Compos.* **2005**, *24*, 643–648. [CrossRef]

18. Rama, S.R.; Rai, S.K. Performance analysis of waste silk fabric-reinforced vinyl ester resin laminates. *J. Compos. Mater.* **2011**, *45*, 2475–2480. [CrossRef]

19. Chen, S.; Cheng, L.; Huang, H.; Zou, F.; Zhao, H. Fabrication and properties of poly (butylene succinate) biocomposites reinforced by waste silkworm silk fabric. *Compos. Part A Appl. Sci. Manuf.* **2017**, *95*, 125–131. [CrossRef]

20. Agarwal, N.; Hoagland, D.A.; Farris, R.J. Effect of moisture absorption on the thermal properties of Bombyx mori silk fibroin films. *J. Appl. Poly. Sci.* **1997**, *63*, 401–410. [CrossRef]

21. Lai, W.L.; Goh, K.L. Consequences of ultra-violet irradiation on the mechanical properties of spider silk. *J. Funct. Biomater.* **2015**, *6*, 901–916. [CrossRef] [PubMed]

22. Ho, M.; Lau, K.; Wang, H.; Bhattacharyya, D. Characteristics of a silk fiber reinforced biodegradable plastic. *Compos. Part B Eng.* **2011**, *42*, 117–122. [CrossRef]

23. Sheik, S.; Nagaraja, G.K.; Naik, J.; Bhajanthri, R.F. Development and characterization study of silk fiber reinforced poly (vinyl alcohol) composites. *Int. J. Plast. Technol.* **2017**, *21*, 108–122. [CrossRef]

24. Kim, B.K.; Kwon, O.H.; Park, W.H.; Cho, D. Thermal, mechanical, impact, and water absorption properties of novel silk fibroin fiber reinforced poly (butylene succinate) biocomposites. *Macromol. Res.* **2016**, *24*, 734–740. [CrossRef]

25. Shubhra, Q.T.H.; Alam, A.K.M.M.; Khan, M.A.; Saha, M.; Saha, D.; Khan, J.A.; Quaiyyum, M.A. The preparation and characterization of silk/gelatin biocomposites. *Polym. Plast. Technol. Eng.* **2010**, *49*, 983–990. [CrossRef]

26. Oshkovr, S.A.; Eshkoor, R.A.; Taher, S.T.; Ariffin, A.K.; Azhari, C.H. Crashworthiness characteristics investigation of silk/epoxy composite square tubes. *Compos. Struct.* **2012**, *94*, 2337–2342. [CrossRef]

27. Yang, K.; Wu, S.; Guan, J.; Shao, Z.; Ritchie, R.O. Enhancing the mechanical toughness of epoxy-resin composites using natural silk reinforcements. *Sci. Rep.* **2017**, *7*, 11939. [CrossRef] [PubMed]

28. Darshan, S.M.; Suresha, B.; Divya, G.S. Waste Silk Fiber Reinforced Polymer Matrix Composites: A Review. *Indian J. Adv. Chem. Sci.* **2016**, *1*, 183–189.

29. Priya, S.P.; Rai, S.K. Impact, compression, density, void content, and weight reduction studies on waste silk fabric/epoxy composites. *J. Reinf. Plast. Compos.* **2005**, *24*, 1605–1610. [CrossRef]

30. Priya, S.P.; Rai, S.K. Mechanical performance of biofiber/glass-reinforced epoxy hybrid composites. *J. Ind. Text.* **2006**, *35*, 217–226. [CrossRef]

31. Zulkifli, R.; Azhari, C.H. Effect of silane concentrations on mode I interlaminar fracture of woven silk/epoxy composites. *Adv. Mater. Res.* **2011**, *150*, 1171–1175. [CrossRef]

32. Loh, K.; Tan, W. Natural silkworm-epoxy resin composite for high performance application. In *Metal, Ceramic and Polymeric Composites for Various Uses*; Cuppoletti, J., Ed.; InTech: Rijeka, Croatia, 2011; pp. 325–340. ISBN 978-953-307-353-8.

33. Izaki, T.; Nakatani, H.; Nam, T.H.; Song, D.Y.; Yoshii, K.; Ogihara, S. Evaluation of mechanical properties of silk fiber reinforced biodegradable plastic composites (part 1, effect of fiber surface treatment on mechanical properties of unidirectional composites). *Trans. Jpn. Soc. Mech. Eng. A* **2011**, *77*, 2107–2117. [CrossRef]

34. Barraza, H.J.; Hamidi, Y.K.; Aktas, L.; Altan, M.C. Performance of glass woven fabric composites with admicellar-coated thin elastomeric interphase. *Compos. Interfaces* **2017**, *24*, 125–148. [CrossRef]

35. Barraza, H.J.; Olivero, K.A.; Hamidi, Y.; O'Rear, E.A.; Altan, M.C. Elastomeric sizing for glass fibers and their role in fiber wetting and adhesion in resin transfer molded composites. *Compos. Interfaces* **2002**, *9*, 477–508. [CrossRef]

36. Li, X.; Tabil, L.G.; Panigrahi, S. Chemical treatments of natural fiber for use in natural fiber-reinforced composites: A review. *J. Polym. Environ.* **2007**, *15*, 25–33. [CrossRef]

37. Valadez-Gonzalez, A.; Cervantes-Uc, J.M.; Olayo, R.; Herrera-Franco, P.J. Effect of fiber surface treatment on the fiber–matrix bond strength of natural fiber reinforced composites. *Compos. Part B Eng.* **1999**, *30*, 309–320. [CrossRef]

38. Bledzki, A.K.; Reihmane, S.; Gassan, J. The effect of alkaline treatment on mechanical properties of kenaf fibers and their epoxy composites. *J. Appl. Polym. Sci.* **1996**, *59*, 1329–1336. [CrossRef]

39. Sathishkumar, T.P.; Navaneethakrishnan, P.; Shankar, S.; Rajasekar, R.; Rajini, N. Characterization of natural fiber surfaces and natural fiber composites. *J. Reinf. Plast. Compos.* **2013**, *32*, 1457–1476. [CrossRef]

40. Hamidi, Y.K.; Yalcinkaya, M.A.; Guloglu, G.E.; Pishvar, M.; Amirkhosravi, M.; Altan, M.C. Manufacturing Silk/Epoxy Composite Laminates: Challenges and Opportunities. In Proceedings of the 34th International Conference of the Polymer Processing Society, Taipei, Taiwan, 21–25 May 2018.

41. Hamidi, Y.K.; Altan, M.C. Process induced defects in liquid molding processes of composites. *Int. Polym. Proc.* **2017**, *32*, 527–544. [CrossRef]

42. Yalcinkaya, M.A.; Sozer, E.M.; Altan, M.C. Fabrication of high quality composite laminates by pressurized and heated-VARTM. *Compos. Part A Appl. Sci. Manuf.* **2017**, *102*, 336–346. [CrossRef]

43. Amirkhosravi, M.; Pishvar, M.; Altan, M.C. Void Reduction in VARTM Composites by Compaction of Dry Fiber Preforms with Stationary and Moving Magnets. *J. Compos. Mater.* **2018**. [CrossRef]

44. Hamidi, Y.K.; Dharmavaram, S.; Aktas, L.; Altan, M.C. Effect of fiber content on void morphology in resin transfer molded composites. *J. Eng. Mater. Technol.* **2009**, *131*, 021014. [CrossRef]

45. Hamidi, Y.K.; Aktas, L.; Altan, M.C. Effect of packing on void morphology in resin transfer molded composites. *Polym. Compos.* **2005**, *26*, 614–627. [CrossRef]

46. Guloglu, G.E.; Hamidi, Y.K.; Altan, M.C. Fast recovery of non-Fickian moisture absorption parameters for polymers and polymer composites. *Polym. Eng. Sci.* **2016**, *57*, 921–931. [CrossRef]

Article

"Skin-Core-Skin" Structure of Polymer Crystallization Investigated by Multiscale Simulation

Chunlei Ruan

School of Mathematics and Statistics, Henan University of Science and Technology, Luoyang 471023, China; ruanchunlei@haust.edu.cn

Received: 16 March 2018; Accepted: 13 April 2018; Published: 16 April 2018

Abstract: "Skin-core-skin" structure is a typical crystal morphology in injection products. Previous numerical works have rarely focused on crystal evolution; rather, they have mostly been based on the prediction of temperature distribution or crystallization kinetics. The aim of this work was to achieve the "skin-core-skin" structure and investigate the role of external flow and temperature fields on crystal morphology. Therefore, the multiscale algorithm was extended to the simulation of polymer crystallization in a pipe flow. The multiscale algorithm contains two parts: a collocated finite volume method at the macroscopic level and a morphological Monte Carlo method at the microscopic level. The SIMPLE (semi-implicit method for pressure linked equations) algorithm was used to calculate the polymeric model at the macroscopic level, while the Monte Carlo method with stochastic birth-growth process of spherulites and shish-kebabs was used at the microscopic level. Results show that our algorithm is valid to predict "skin-core-skin" structure, and the initial melt temperature and the maximum velocity of melt at the inlet mainly affects the morphology of shish-kebabs.

Keywords: "skin-core-skin" structure; flow-induced crystallization; multiscale simulation; crystal morphology

---

1. Introduction

Currently, semicrystalline polymers are widely used in industry [1]. Usually, such polymers should be processed to become useful products. Common processing techniques include injection molding, extrusion molding, blow molding, and so others. Among these, injection molding is the most widely used. It involves a high-speed flow field and complex temperature field. These processing conditions are key factors in determining the microstructure of the products (crystal types, crystal orientation, etc.). On the other hand, the mechanical properties of the products (strength, modulus, etc.) are strongly dependent on these microstructures. Therefore, it is of great significance to control the microstructures by precisely applying external flow and temperature fields to improve the performance of the products.

Experimental results show that the crystalline structure in the final injection product takes on a typical "skin-core-skin" structure: the shish-kebab structure with high orientation appears in the skin layer, and the spherulitical structure with essentially no preferred orientation appears in the core layer [2]. Figure 1 shows the cross-section of the shish-kebab structure and spherulitical structure of an injection product [2]. It has been reported that in the skin layer, because of the high shear stress and shear strain, the extended polymer chains lead to extended chain crystals and, ultimately, shish-kebab structures. In the core layer, because of the absence of shear, the random polymer chains lead to lamellar, chain-folded crystals and, finally, spherulites. Hence, the shish-kebab structure is related to flow-induced crystallization, and the spherulitical structure is related to quiescent crystallization [2–4]. Compared with the spherulitical structure, the shish-kebab structure improves the performance of products in tensile strength, tensile elastic modulus in stress direction, but reduces the performance of products in impact strength in the direction perpendicular to the stress [5]. Therefore, it is important to predict the spherulitical structure and shish-kebab structure precisely.

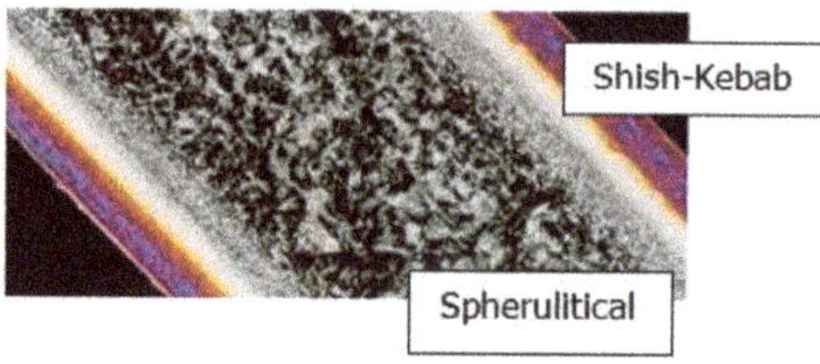

Figure 1. "Skin-core-skin" structure in an injection polymer product [2].

To date, several numerical works have reported capturing the evolution of the spherulitical structure in quiescent crystallization. For example, Raabe [6] and Spina et al. [7] used the cell automaton method to simulate spherulite growth in polymer crystallization; Ketdee [8] presented Monte Carlo simulations to predict the kinetics and morphology of isothermal polymer crystallization; Ruan et al. [9,10] applied the pixel coloring method to capture the spherulite evolution in isothermal and non-isothermal polymer crystallization; and Liu et al. [11,12] constructed a level set method to capture the spherulite development during the polymer cooling stage.

Compared with the spherulitical structure in quiescent crystallization, models and methods for the determination of shish-kebab structure in flow-induced crystallization are rare. Eder [13] considered shish-kebabs as growing cylinders and obtained a series of differential equations by using the Schneider rate equation [14]. Zuidema et al. [15] thought that recoverable strain was the driving force for the nucleation of shish-kebabs and modified the Eder model. Zinet et al. [4,16] and Mu et al. [17] used a modified Schneider rate equation to describe the growth of thermally and flow-induced nuclei. Guo et al. [18,19] introduced a molecular deformation factor, which distinguished spherulites and shish-kebabs by comparing the molecular deformation factor with the critical one. Wang et al. [20] presented a phase field method to simulate the shish-kebab structure in simple shear and temperature fields. Although the above works are based on crystal morphology, crystallization kinetics models are also needed. Crystallization kinetics models, such as the Avrami model, are often reported with lower accuracy at the later stage of polymer crystallization. Furthermore, these works did not construct a method to reveal the details of shish-kebabs. Recently, Ruan et al. [21–23] presented a Monte Carlo method to simulate spherulites and shish-kebabs in a parametrical study, with simple shear flow and Couette flow. They obtained detailed crystal morphology and reliable crystallization kinetics without using a kinetics model.

In this study, we extended the multiscale method to simulate the "skin-core-skin" structure of the polymer crystallization in a pipe flow, which is treated as a simplification of the injection processing. Unlike our previous work of Couette flow [22], the conservation at the macroscopic level in pipe flow is more complicated. Therefore, the SIMPLE (semi-implicit method for pressure linked equations) algorithm on collocated coarse grid was used to calculate the flow and temperature fields at the macroscopic level. Rhie–Chow-type interpolation was introduced to overcome the pressure-velocity decoupling. The morphological Monte Carlo method on fine grid was used to capture the crystal growth fronts and compute the relative crystallinity. Effects of external flow and temperature fields, (e.g., temperature cooling rate of the mold, initial melt temperature, maximum velocity of the melt in inlet) on the crystal morphology were investigated and are herein discussed.

2. Mathematical Model

In injection molding, polymer melts are injected into a mold to form different products. A changing flow domain is more suitable. Some software, such as C-mold and Moldflow, can address all stages of injection. In this work, we assumed the mold is a pipe, which is shown in Figure 2. Actually, the mold is supposed to have a thin-wall thickness in the y direction, which means the length in the x direction is substantially larger than the length in the y direction ($L \gg W$). Our aim was to simulate the crystallization during and after shear treatment, which refer to the injection and cooling stages, respectively. We assumed the melt experiences shear effects with a parabolic velocity in the inlet that lasts for shear time t_s (injection stage). We also assumed that after the shearing flow, the mold has experienced a large temperature change (cooling stage). Therefore, the mathematical model at the macroscopic level should be divided into two cases.

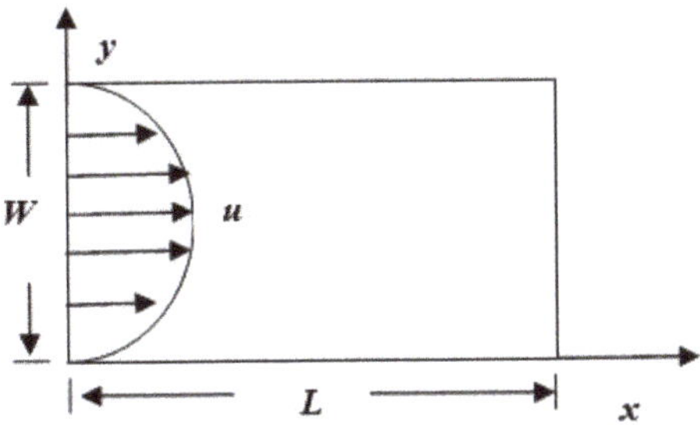

Figure 2. The pipe model for simulation.

2.1. Conservation Equations at the Macroscopic Level

We assumed the polymer melt is a non-isothermal, non-compressible, and non-Newtonian flow. Therefore, three conservation equations were needed. Furthermore, the melt is non-Newtonian, and a constitutive equation was needed.

(1) Conservation equations at the macroscopic level during shearing flow (injection stage)

The mass conservation equation is

$$\nabla \cdot \mathbf{u} = 0, \tag{1}$$

the momentum conservation equation is

$$\frac{\partial}{\partial t}(\rho \mathbf{u}) + \nabla \cdot (\rho \mathbf{u}\mathbf{u}) = -\nabla p + \nabla \cdot \boldsymbol{\tau}_c, \tag{2}$$

and the energy conservation equation is

$$\frac{\partial}{\partial t}(\rho c_p T) + \nabla \cdot (\rho c_p \mathbf{u} T) = \nabla \cdot (k_p \nabla T) + \rho \Delta H \frac{\partial \alpha}{\partial t} + (-p\mathbf{I} + \boldsymbol{\tau}_c) : \nabla \mathbf{u}, \tag{3}$$

where ρ is the density; $\mathbf{u}$ is the velocity; p is the pressure; c_p is the heat capacity; k_p is the thermal conductivity; T is the temperature; α is the relative crystallinity; ΔH is the crystallization enthalpy; $\mathbf{I}$ is the identity tensor; and $\boldsymbol{\tau}_c = \boldsymbol{\tau}_a + \boldsymbol{\tau}_{sc}$ is the composite tensor, with $\boldsymbol{\tau}_a$ being the stress of the amorphous phase and $\boldsymbol{\tau}_{sc}$ the stress of the semicrystalline phase.

We used the conception of Zheng et al. [24] to describe the non-Newtonian system. The amorphous phase is described by FENE-P (finite extensible nonlinear elastic model with a Peterlin closure approximation) dumbbells, while the semicrystalline phase is modeled as rigid dumbbells. The stress caused by FENE-P dumbbells is [24]

$$\boldsymbol{\tau}_a = nkT \left(\frac{\mathbf{C}}{1 - tr\mathbf{C}/b} - \mathbf{I} \right), \tag{4}$$

and the evolution of the conformation tensor is [24]

$$\lambda_\alpha(T)\overset{\triangledown}{\mathbf{C}} + \left[\frac{1}{1 - tr\mathbf{C}/b}\mathbf{C} - \mathbf{I}\right] = 0, \tag{5}$$

where n is the number density of the molecules, k is the Boltzmann constant, b is the dimensionless parameter of the nonlinear spring, tr is the trace of the matrix, $\mathbf{C}$ is the configuration tensor, and the upper-convected derivative of $\mathbf{C}$ is defined as $\overset{\triangledown}{\mathbf{C}} = D\mathbf{C}/Dt - (\nabla\mathbf{u})^T\cdot\mathbf{C} - \mathbf{C}\cdot(\nabla\mathbf{u})$. λ_α is the relaxation time of the fluid, which obeys the Arrhenius equation, namely [24],

$$\lambda_\alpha(T) = \exp\left[\frac{E_a}{R_g}\left(\frac{1}{T} - \frac{1}{T_r}\right)\right]\lambda_{a,r}, \tag{6}$$

where T_r is a reference temperature, and E_a/R_g is a constant that can be determined from the experimental data. The stress caused by rigid dumbbells is [24]

$$\boldsymbol{\tau}_{sc} = \frac{\eta_{sc}}{\lambda_{sc}}[\langle\mathbf{RR}\rangle + \lambda_{sc}\dot{\boldsymbol{\gamma}} : \langle\mathbf{RRRR}\rangle], \tag{7}$$

where λ_{sc} is the relaxation time of the rigid dumbbells, η_{sc} is the viscosity of the semicrystalline system, $\dot{\boldsymbol{\gamma}}$ is the deformation tensor, $\langle\mathbf{RR}\rangle$ is the second-order orientation tensor, and $\langle\mathbf{RRRR}\rangle$ is the fourth-order orientation tensor. The evolution of the orientation tensor $\langle\mathbf{RR}\rangle$ is defined as [24]

$$\overset{\triangledown}{\langle\mathbf{RR}\rangle} = -\frac{1}{\lambda_{sc}}\left(\langle\mathbf{RR}\rangle - \frac{\mathbf{I}}{2} - \dot{\boldsymbol{\gamma}} : \langle\mathbf{RRRR}\rangle\right) \tag{8}$$

The relationship between the viscosity of the semicrystalline system and the amorphous system is dependent on the following empirical equation [24]

$$\frac{\eta_{sc}(x, T)}{\eta_a(T)} = \frac{(\alpha/A)^{\beta_1}}{(1 - \alpha/A)^{\beta}} \qquad \alpha < A, \tag{9}$$

and the relaxation times of the rigid dumbbells and FENE-P dumbbells are [24]

$$\frac{\lambda_{sc}(x, T)}{\lambda_\alpha(T)} = \frac{(\alpha/A)^{\beta_1}}{(1 - \alpha/A)^{\beta}} \qquad \alpha < A, \tag{10}$$

where β_1, β_1 and A are the empirical constants. Equation (9) clearly shows that when $\alpha \to A$ (A being the "critical value" of the degree of crystallinity), the viscosity of the semicrystalline system approaches infinity.

To calculate the second-order orientation tensor $\langle\mathbf{RR}\rangle$ from Equations (7) and (8), one shall use a closure approximation—such as the hybrid [25], EBOF (eigenvalue-based orthotropic fitting) [26,27], or IBOF (invariant-based orthotropic fitting) [28] method—to gain an expression of $\langle\mathbf{RRRR}\rangle$ in terms of $\langle\mathbf{RR}\rangle$. Here, the hybrid expression was used, namely,

$$\langle\mathbf{RRRR}\rangle_{ijkl} = \langle\mathbf{RR}\rangle_{ij}\langle\mathbf{RR}\rangle_{kl}, \tag{11}$$

where $\langle\mathbf{RR}\rangle_{ij}$ and $\langle\mathbf{RR}\rangle_{kl}$ are the components of $\langle\mathbf{RR}\rangle$, and $\langle\mathbf{RRRR}\rangle_{ijkl}$ is the component of $\langle\mathbf{RRRR}\rangle$.

(2)   Conservation equations at the macroscopic level after shearing flow (cooling stage)

In the second case, we assumed there is no fluid flow and the melt is stationary. Therefore, the conservation equation was the energy equation, which can be written as follows,

$$\rho c_p \frac{\partial T}{\partial t} = \nabla\cdot(k_p\nabla T) + \rho\Delta H\frac{\partial\alpha}{\partial t}. \tag{12}$$

Actually, for high accuracy, the material parameters can be calculated with the mixture rule. For example, $\rho = \alpha \rho_{sc} + (1-\alpha)\rho_\alpha$, with ρ_{sc} as the density of the semicrystalline phase and ρ_α as the density of the amorphous phase.

2.2. Crystal Evolution Model at the Microscopic Level

In injection processing, both the spherulitical structure and shish-kebab structure appear in polymer products. The former one is caused by temperature and is known as quiescent crystallization; the latter one is caused by shear or stress and is known as flow-induced crystallization.

In the morphological simulation, crystals follow the steps of nucleation-growth-impingement. Therefore, it is important to model the nucleation and growth of spherulites and shish-kebabs. In our previous study [21,23], we deduced the evolution equations of spherulites and shish-kebabs based on the Eder model [13] and Schneider rate model [14]. Here, we used the equations obtained in our previous work [23].

We assumed the relationship between the nucleation of spherulites (N_s) and temperature is [29]

$$N_s(T) = N_0 \exp[a_n \Delta T + b_n],\tag{13}$$

where a_n and b_n are constants, and $\Delta T = T_m^0 - T$ is the degree of supercooling, with T_m^0 being the equilibrium melting temperature. We shall mention that different nucleation relations of N_s have been reported, and the reviews of Pantani et al. [5] are helpful. Usually, these relations may be quite restricting depending on the conditions or materials.

The growth rate of spherulites (G_s) is often adopted by the Hoffman–Lauritzen expression [30], which is

$$G_s(T) = G_0 \exp\left[-\frac{U^*}{R_g(T-T_\infty)}\right] \exp\left[-\frac{K_g}{T \Delta T f}\right],\tag{14}$$

where G_0 and K_g are constants, U^* is the activation energy of motion, R_g is the gas constant, $T_\infty = T_g - 30$ (where T_g is the glass transition temperature), and $f = 2T/(T_m^0 + T)$.

We assumed the driving force of the nucleation of shish-kebabs (N_{s-k}) is the first normal stress difference, which can be written as [29]

$$\dot{N}_{s-k} = \mathbf{C} N_1,\tag{15}$$

where $\mathbf{C}$ is a constant, and N_1 is the first normal stress difference that can be calculated by Equations (4) and (7). Notice that the driving force for flow-induced nucleation is not well understood, and different approaches have been proposed. Examples of the driving forces include the shear rate [13], shear strain [31], recoverable strain [15], dumbbell free energy [24]. Equation (15) is the simplest but is also widely used in simulations [29,32].

Shish-kebabs are assumed to grow as a cylinder, which means they can grow in two directions, namely, along the length and radial directions. According to Eder [13], the length growth rate ($G_{s-k,l}$) obeys the following equation

$$G_{s-k,l} = \dot{\gamma}^2 \cdot g_1/\dot{\gamma}_l^2,\tag{16}$$

where $g_1/\dot{\gamma}_l^2$ is a constant, and $\dot{\gamma}$ is the shear rate. The radial growth rate of the shish-kebabs ($G_{s-k,r}$) is always assumed to be equal to the growth rate of the spherulites, which is

$$G_{s-k,r} = G_s.\tag{17}$$

3. Multiscale Method

The conception of a multiscale method here is similar to the method we built up in the Couette flow case [22]. We used different methods at different scales and then coupled them together. The finite volume method at the macroscopic level is constructed to calculate the velocity, temperature, stress, etc. The finite volume method is a conservation method. It has advantages of small storage and cheap computational cost, as well as easy handling of the couplings of velocity-pressure, velocity-stress, etc. [33]. Therefore, the finite volume method is widely used in CFD (computational fluid dynamics). The Monte Carlo method at the microscopic level is constructed to capture the development of crystals. The Monte Carlo method is also known as a stochastic simulation and can address the stochastic birth-growth process of spherulites and shish-kebabs. The advantages of the Monte Carlo method are that it can avoid the use of a crystallization kinetics model and it can also predict the detailed morphology evolution. The finite volume method and Monte Carlo method were implemented on different grids—namely, the finite volume method was used on a coarse grid to solve macroscopic Equations (1)–(3), (5), (8), and (12) to obtain the velocity, pressure, stress, and temperature, and the Monte Carlo method was used on a fine grid to capture the evolution of crystals by using Equations (13)–(17) to obtain the relative crystallinity. We refer to our previous work for the arrangement of the coarse grid and fine grid [22,34].

In the modeling part, we put fully coupled mass, momentum, and energy conservation equations together with the constitutive equations of amorphous and semicrystalline phases during the shear treatment (injection stage). However, in the algorithm of the finite volume method, we present a decoupled one. We solved the non-isothermal Newtonian flow to achieve the velocity and temperature. Then, with the results of velocity and temperature, we solved the constitutive equations of amorphous and semicrystalline phases. In other words, the velocity, pressure, and temperature were coupled, while the stress was decoupled. This is because the stresses caused by the amorphous and semicrystalline phases are seriously dependent on the relative crystallinity: a slight increase in relative crystallinity causes a dramatic increase in viscosity, which leads to a large increase in stress. If we put this stress into the momentum equation, we cannot obtain a convergent result because of the large stress source term. Therefore, in our simulation, Equations (1)–(3), (5), and (8) were not solved simultaneously. Actually, when we solve Equations (1)–(3), the SIMPLE method is used [35] . We assumed the flow is a non-isothermal Newtonian flow, which is incompressible. The collocated finite volume method was used. Compared with the finite volume method on a staggered grid, the collocated finite volume method has the advantage of easy implementation on the same grid to overcome the decoupling of velocity-pressure and velocity-stress [33]. It is also noted that in our previous work on the Couette flow model, a continuity equation could not be calculated that avoided the decoupling of velocity-pressure. Detailed implementation of the collocated finite volume method were as follows.

Equations (1)–(3), (5), and (8) can be written as a general transport equation

$$\frac{\partial(\delta\varphi)}{\partial t} + \nabla\cdot(m\mathbf{u}\varphi) = \nabla\cdot(\Gamma\nabla\varphi) + S_\varphi, \tag{18}$$

where δ, m, and Γ are constant, and φ and S_φ are the functions that are defined in Table 1. The terms in Equation (18) represent the transient, convective, diffusive and source contributions.

Equation (18) is integrated over the finite volume cell shown in Figure 3 in space, and the use of the divergence theorem yields

$$\int_V \frac{\partial(\delta\varphi)}{\partial t}dV + \int_s^n [(m\mathbf{u}\varphi - \Gamma\nabla\varphi)_e - (m\mathbf{u}\varphi - \Gamma\nabla\varphi)_w]dy \\ + \int_w^e [(m\mathbf{u}\varphi - \Gamma\nabla\varphi)_n - (m\mathbf{u}\varphi - \Gamma\nabla\varphi)_s]dx = \int_s^n \int_w^e S_\varphi dx dy. \tag{19}$$

The transient term in Equation (19) is integrated in time and then divided by Δt. A linear approximation is used, which leads to

$$\frac{1}{\Delta t}\int_t \delta\frac{\partial\varphi}{\partial t}dVdt \approx \frac{\delta V\left(\varphi_P - \varphi_P^0\right)}{\Delta t}, \tag{20}$$

where the superscript "0" indicates the value at the previous time step. The upwind scheme and central differences are used to approximate the convective and diffusive fluxes across each face, respectively. This gives rise to the following discretization

$$A_P \phi_P = A_E \phi_E + A_W \phi_W + A_N \phi_N + A_S \phi_S + Q_P, \tag{21}$$

where A_P, A_E, A_W, A_N, and A_S are the coefficients of ϕ_P, ϕ_E, ϕ_W, ϕ_N, and ϕ_S, respectively, and Q_P is the source term. The Gauss–Seidel iteration method can be used to solve the above linear equations. Note that a Rhie–Chow-type [36] interpolation should be used to overcome the pressure-velocity decoupling. Details can be found in the work of Oliveira et al. [37] and Ruan et al. [38].

Table 1. Definition of constants and functions in the general equation.

Equation	δ	m	ϕ	Γ	S_ϕ
Continuity	0	1	1	0	0
Momentum	ρ	ρ	$\mathbf{u}$	η	$-\nabla p$
Energy	ρc_p	ρc_p	T	k	$\rho \Delta H \frac{\partial \alpha}{\partial t}$
FENE-P model	1	1	$\mathbf{C}$	0	$-\frac{1}{\lambda_\alpha(T)}\left[\frac{\mathbf{C}}{1-\frac{tr\mathbf{C}}{b}} - \mathbf{I}\right] + (\nabla\mathbf{u})^T{\cdot}\mathbf{C} + \mathbf{C}{\cdot}\nabla\mathbf{u}$
Rigid dumbbell model	1	1	$\langle\mathbf{RR}\rangle$	0	$-\frac{1}{\lambda_{sc}(T)}\left[\langle\mathbf{RR}\rangle - \frac{\mathbf{I}}{2}\right] - \dot{\boldsymbol{\gamma}} : \langle\mathbf{RRRR}\rangle + (\nabla\mathbf{u})^T\langle\mathbf{RR}\rangle + \langle\mathbf{RR}\rangle{\cdot}\nabla\mathbf{u}$

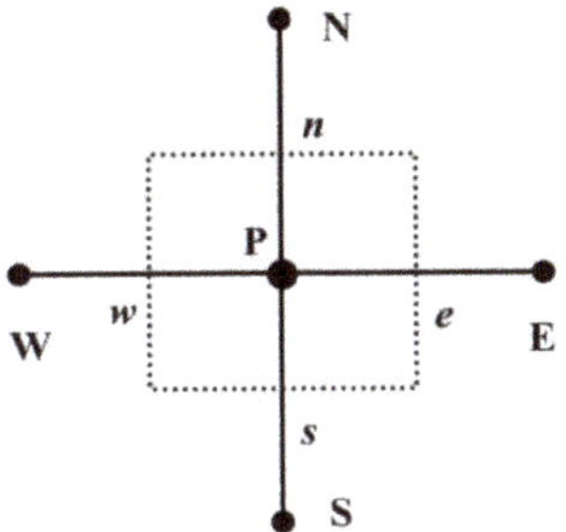

Figure 3. A general control volume.

The Monte Carlo method was employed on the fine grid to track the development of spherulites and shish-kebabs. With the crystal evolution model of Equations (13)–(17), the detailed morphology can be obtained by the Monte Carlo method. This is the main advantage that the morphological simulation has. Furthermore, by using the Monte Carlo method, the relative crystallinity can also be obtained from the volume fraction of crystals under the assumption that the semicrystalline phase is spatially uniform. Thus, the relative crystallinity was calculated by the following equation

$$\alpha = \text{number of cells that are occupied by crystals/total number of cells.} \tag{22}$$

Here, we shall not show the detailed implementation of the Monte Carlo method, but refer to our previous work [21,23] for more details.

Figure 4 shows the flowchart of the implementation of the multiscale method.

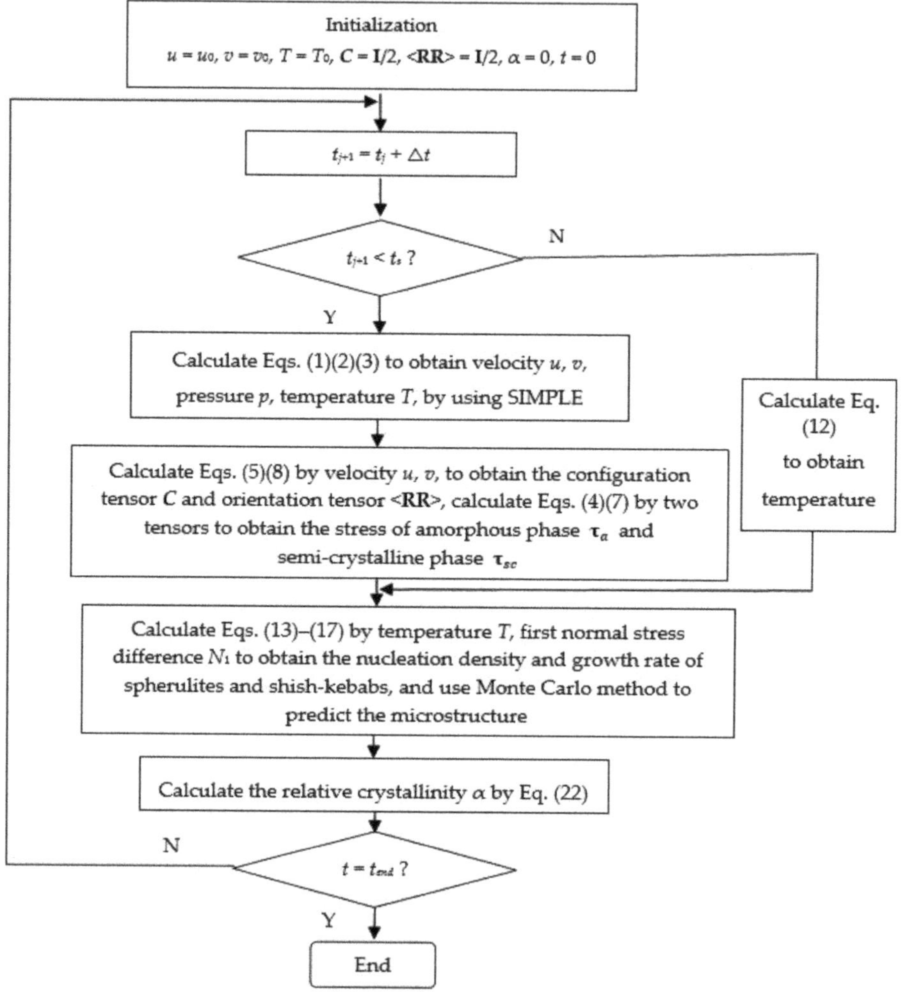

Figure 4. Flowchart for multiscale method in the simulation.

4. Results and Discussion

4.1. Problem Definition

Considering the injection mold shown in Figure 2, the length is $L = 16$ mm, and the thickness is $W = 8$ mm. We assumed the walls, with $y = 0$ mm and $y = 8$ mm, experience a constant cooling rate operation and set the boundary conditions as $T = T_0 - c \times t$, with T_0 as the initial temperature and c as the cooling rate. The other boundary conditions were assumed as $\partial T / \partial \mathbf{n} = 0$, with $\mathbf{n}$ as the unit normal vector. Note that the mold is a thin wall with thickness in the y direction. Because of the complexity of the multiscale algorithm, here, we set the length in the x direction as twice the length as in the y direction. The adiabatic boundary conditions of $x = 0$ mm and $x = 16$ mm were used to obtain a condition similar to the thin-wall thickness of mold in industry. Moreover, here, we set a shear flow to account for the flow and flow history of the injection processing. We assumed the velocity at the inlet is $u = Uy(W - y)$, where U is a constant, and last with the shear time t_s s; once the time reaches the "shear time" (t_s), the flow field is vanished. We fixed the shear time to $t_s = 10$ s and will not discuss this effect later.

The material here was the isotactic polypropylene homopolymer with tacticity 0.96. The parameters used were [5,15,24,29]: $a_n = 0.156$ K/m^3, $b_n = 15.1$ /m^3, $G_0 = 2.83 \times 10^2$ m/s, $U^*/R_g = 755$ K, $K_g = 5.5 \times 10^5$ K^2, $T_m^0 = 483$ K, $T_g = 269$ K, $g_l/\dot{\gamma_l}^2 = 2.69 \times 10^{-7}$, $C = 10^6$ /Pa/s^2/m, $\lambda_{\alpha,r} = 4.00 \times 10^{-2}$ s, $T_r = 476.15$ K, $E_\alpha/R_g = 5.602 \times 10^3$ K, b = 5, n = 1.26×10^{26} /m^3, k = 1.38×10^{-23}, $\beta = 9.2$, $\beta_1 = 0.05$, and A = 0.44. The other parameters we chose were $\rho = 900$ kg/m^3, $c_p = 2.14 \times 10^3$ J/kg/K, $k_p = 0.193$ W/m/K, $\Delta H = 107 \times 10^3$ J/kg, $T_0 = 490$ K, $c = 2$ K/min, and $U = 625$. In the implementation of the multiscale algorithm, the coarse grid was chosen as 16×18, and the fine grid was chosen as 500×500.

To show the validity of the model and the Monte Carlo method used at the microscopic level, isothermal crystallization was considered. Figure 5 shows the simulated data with the experimental data [29]. It can be seen that the numerical relationship between the shear rate and the half crystallization time is in good agreement with the experimental data. Therefore, our model and Monte Carlo method is valid.

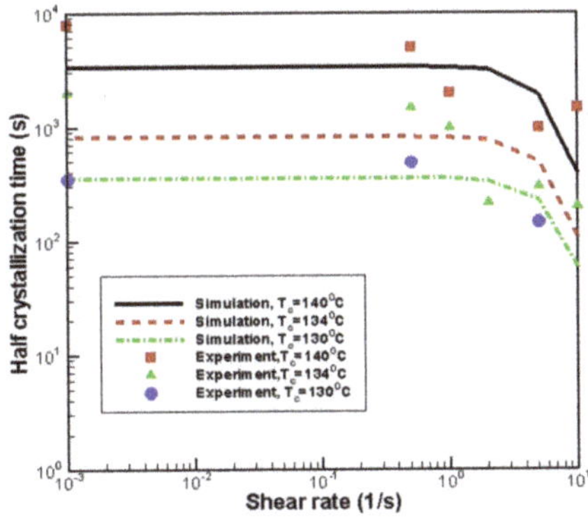

Figure 5. Comparison of simulation result with the experimental result [29].

4.2. Temperature, Relative Crystallinity Distribution, and "Skin-Core-Skin" Structure

Figure 6 shows the temperature and relative crystallinity evolution at the profile of $x = 8$ mm. Results obtained for our algorithm are compared with the Avrami model. Here, the Avrami model is $\alpha = 1 - \exp\left(-\alpha_f\right)$, with $\alpha_f = V_{sp} + V_{sh}$, where V_{sp} is the undisturbed total volume of spherulites and V_{sh} is the undisturbed total volume of shish-kebabs. The Schneider rate equation [14,15] was used to compute V_{sp} and the Eder model [13,15] was used to calculate V_{sh}. The "Avrami model" in the temperature curves means that the temperature is calculated with the α obtained by the Avrami model. As can be seen in Figure 6, the simulation data show a good agreement with the Avrami model. In addition, as shown in the temperature curves, there is a "platform" in the core layer near 2800–3600 s. According to the evolution of relative crystallinity, the relative crystallinity value of the core layer in this period increases rapidly and finally reaches 1. Therefore, this period is the time that crystallization happens. Because of the latent heat released by crystallization, the temperature "platform" forms in the core layer. Furthermore, the crystallization rate in the skin layer is faster than that in the core layer. This result is consistent with our previous study on quiescent crystallization [39].

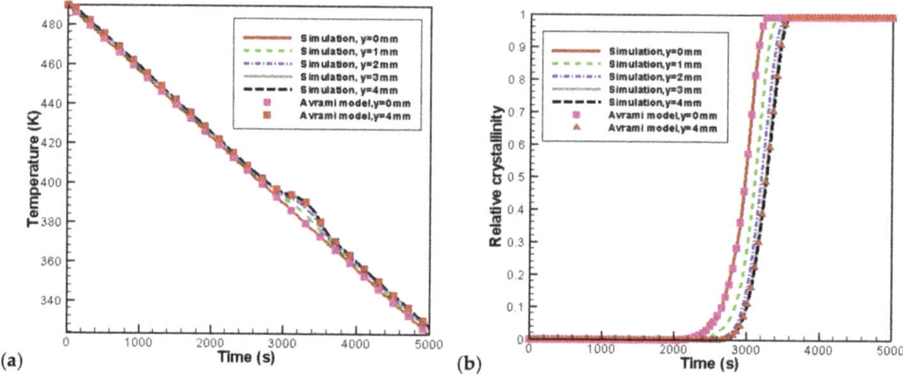

Figure 6. Evolution of temperature and relative crystallinity with time at the profile of x = 8 mm: (**a**) temperature, (**b**) relative crystallinity.

Figure 7 shows the evolution of temperature and relative crystallinity at different thicknesses in the profile at x = 8 mm. It is evident that crystallization occurs near 400–380 K. It is also clear that crystallization finishes earlier in the skin layer because of the lower temperature at the walls.

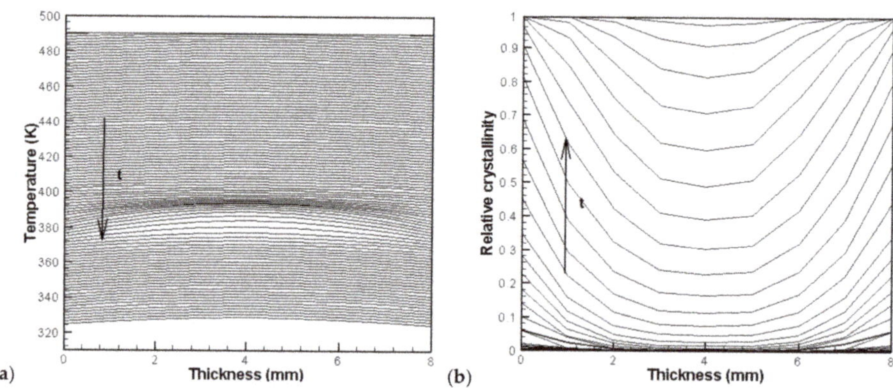

Figure 7. Distribution of temperature and relative crystallinity at different thicknesses: (**a**) temperature, (**b**) relative crystallinity.

Figure 8 shows the development of crystals in the control volume of the skin layer (8 mm, 0 mm) and in the control volume of the core layer (8 mm, 4 mm). It is clear that in the skin layer, the shish-kebab structure is dominant, while in the core layer, only the spherulitic structure appears. Crystals follow the steps of nucleation, growth, impingement, and, finally, fully filling the whole space. In fact, the shear rate near the skin layer is large, which is of benefit to the nucleation and growth of shish-kebabs. However, the shear rate is often absent or small in the core layer, which is not suitable to the development of shish-kebabs. However, a lower temperature is favorable for the nucleation and growth of spherulites. This development of crystal morphology is in agreement with the experimental finding of Koscher et al. [29].

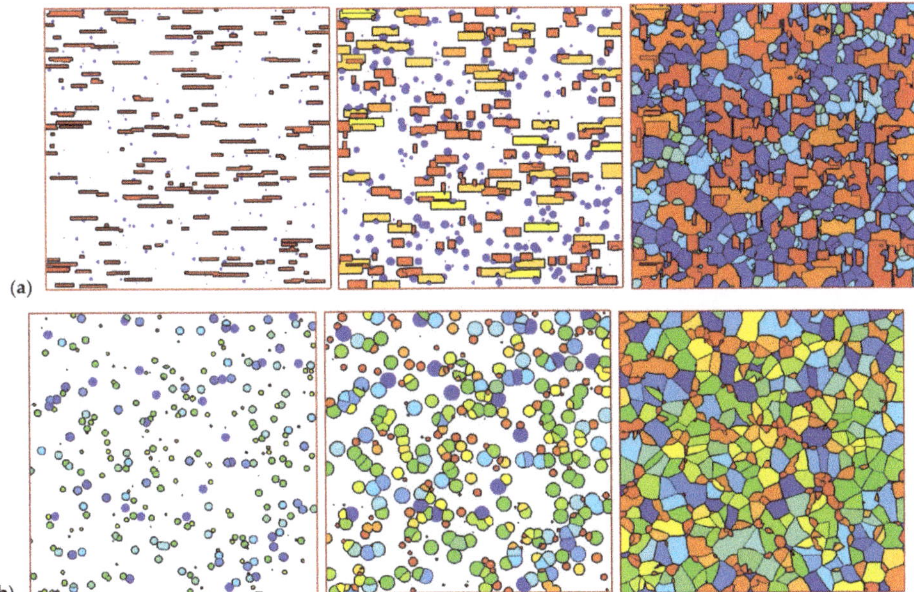

Figure 8. Morphology evolution in the polymer control volume: (**a**) skin volume and (**b**) core volume.

Figure 9 shows the final crystal morphology in the computational domain. The structure takes on a typical "skin-core-skin" structure: in the skin layer, the crystal structure is mainly composed of the anisotropic shish-kebab; in the core layer, the crystal structure is the isotropic spherulite. This structure is consistent with experimental findings [5,40–42]. Our approach is valid in revealing the crystal microstructures. It should be mentioned that in our simulation, we do not consider a "frozen layer" [5,40–42]. Here, we consider a moderate temperature cooling boundary condition; therefore, a "frozen layer" cannot be captured in this case.

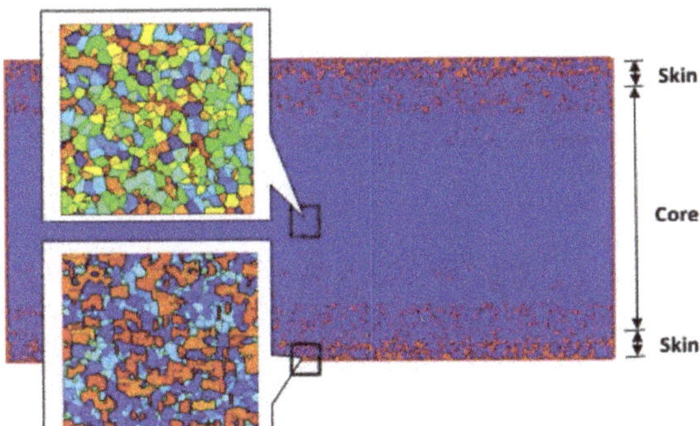

Figure 9. "Skin-core-skin" structure.

The Monte Carlo simulation also allowed us to show the details of spherulites and shish-kebabs. Figure 10 shows the number of spherulites and shish-kebabs at different thicknesses at the profile of $x = 8$ mm. It is evident that the number of shish-kebabs decreases from the skin to core, while the number of spherulites increases from the skin to core. The trend in number of shish-kebabs is caused by the change in shear rate. It is worth noting that the trend in the number of spherulites is in contrast to the quiescent case [40]. In the quiescent case, because of the highest cooling rate being in the skin layer, the number of spherulites is largest, which leads to the smallest size of spherulites. Although in this case, the temperature performance is similar to the quiescent case, the shish-kebab structure appears, which restricts the number of spherulites.

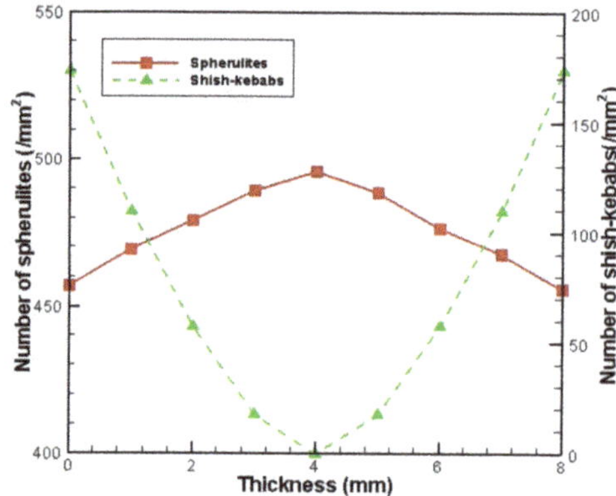

Figure 10. Number of spherulites and shish-kebabs at the different thicknesses at the profile of $x = 8$ mm.

4.3. Effects of Temperature Cooling Rate of the Mold Wall

Three cases of temperature cooling rate of the mold wall were examined, namely, $c = 1$ K/min, $c = 2$ K/min, and $c = 5$ K/min. It is noted that the high cooling rate is related to the low wall temperature in real injection processing.

Figure 11 shows the temperature evolution and the rates of crystallization at the profile of $x = 8$ mm. To show the difference between the skin and core layers, we chose the skin point (8 mm, 0 mm) and core point (8 mm, 4 mm) as examples. As seen in Figure 11, the case with a high cooling rate leads to the fast decrease in temperature and high crystallization rate.

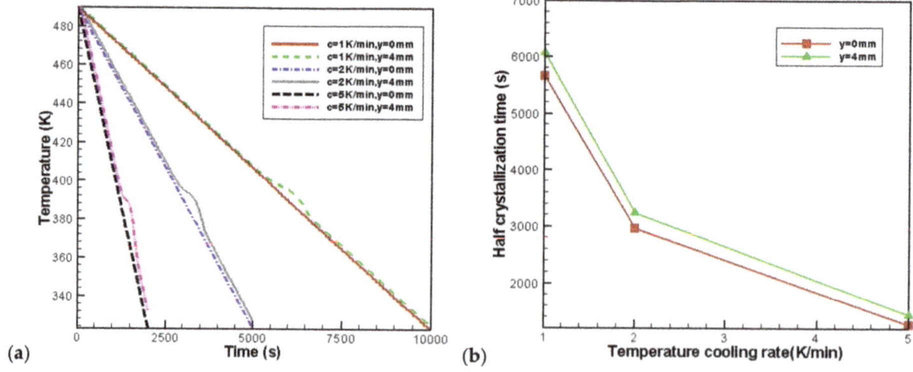

Figure 11. Effect of temperature cooling rate of the mold wall on the temperature and the rates of crystallization at the profile of $x = 8$ mm: (**a**) temperature, (**b**) half crystallization time.

Table 2 shows the parameters related to the microstructure. The average diameter of spherulites, number of shish-kebabs and the relative crystallinity caused by spherulites and shish-kebabs are displayed. It is evident that in the case of the higher cooling rate, the average size of spherulites decreases. However, the number of shish-kebabs does not change with the cooling rate. Therefore, it can be concluded that the temperature cooling rate of the wall mainly affects the nucleation and growth of spherulites. The predicted average diameter of spherulites also shows agreement with the experimental and numerical works of Pantanin et al. [5].

Table 2. Effect of the temperature cooling rate of the mold wall on the spherulites and shish-kebabs at the profile of $x = 8$ mm.

Thickness, Cooling Rate	Average Diameter of Spherulites (μm)	Relative Crystallinity Contributed by Spherulites	Number of Shish-Kebabs (/mm^2)	Relative Crystallinity Contributed by Shish-Kebabs
$y = 0$ mm, $c = 1$ K/min	48.63	52.93%	174	47.07%
$y = 0$ mm, $c = 2$ K/min	40.58	57.82%	174	42.18%
$y = 0$ mm, $c = 5$ K/min	29.39	65.53%	176	34.47%
$y = 4$ mm, $c = 1$ K/min	63.06	100%	0	0%
$y = 4$ mm, $c = 2$ K/min	50.66	100%	0	0%
$y = 4$ mm, $c = 5$ K/min	37.63	100%	0	0%

4.4. Effects of Initial Melt Temperature

Three cases of initial melt temperature were examined, namely, $T_0 = 490$ K, $T_0 = 500$ K, $T_0 = 510$ K. Figure 12 presents the evolution of the temperature and the rates of crystallization at the skin point (8 mm, 0 mm) and core point (8 mm, 4 mm) for different initial melt temperatures. It can be seen from Figure 12 that the higher the initial melt temperature, the later the occurrence of the temperature "platform" and crystallization. However, the curves only shift to the right when the initial melt temperature is increased.

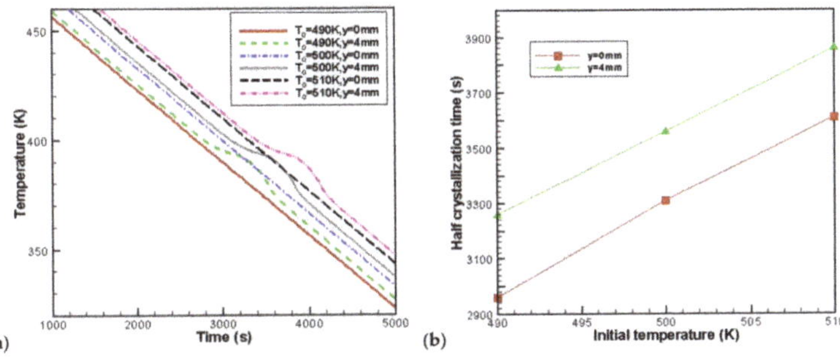

Figure 12. Effect of the initial melt temperature on the temperature and the rates of crystallization at the profile $x = 8$ mm: (**a**) temperature, (**b**) half crystallization time.

Table 3 shows the effects of the initial melt temperature on the microstructures. These effects are reflected by the average diameter of spherulites, number of shish-kebabs, and the relative crystallinity contributed by the spherulites and shish-kebabs at the skin and core volume. It is clear that when the initial melt temperature increases, both the contribution and the number of shish-kebabs are reduced. This is because the higher melt temperature reduces the first normal stress difference. According to Equation (15), the number of shish-kebabs reduces, which leads to a reduction in the contribution of shish-kebabs to the relative crystallinity. It is also evident that the initial melt temperature has minor effects on spherulites.

Table 3. Effect of initial melt temperature on spherulites and shish-kebabs at the profile $x = 8$ mm.

Thickness, Initial Melt Temperature	Average Diameter of Spherulites (μm)	Relative Crystallinity Contributed by Spherulites	NUMBER of Shish-Kebabs ($/mm^2$)	Relative Crystallinity Contributed by Shish-Kebabs
$y = 0$ mm, $T_0 = 490$ K	40.58	57.82%	174	42.18%
$y = 0$ mm, $T_0 = 500$ K	39.83	60.77%	140	39.23%
$y = 0$ mm, $T_0 = 510$ K	40.32	68.15%	112	31.85%
$y = 4$ mm, $T_0 = 490$ K	50.66	100%	0	0%
$y = 4$ mm, $T_0 = 500$ K	51.77	100%	0	0%
$y = 4$ mm, $T_0 = 510$ K	50.53	100%	0	0%

4.5. Effects of the Maximum Velocity of Melt at the Inlet

The effects of maximum velocity at the inlet were also investigated. We changed U from 125 to 1250 to obtain the velocity at the inlet. The velocity also affects the maximum shear rate. The shear rate was calculated with the velocity as $\dot{\gamma} = |\partial u / \partial y|$. Figure 13 shows the final crystal morphology of the control volume at $x = 8$ mm. When the maximum velocity is small (small shear rate), the shish-kebab structure in the skin layer is not apparent. With the increase in velocity (or shear rate), the shish-kebab structure becomes clear, and the thickness of the skin layer becomes wide. Hence, the velocity at the inlet has significant effects on the crystal morphology.

Figure 13. Final crystal morphology with different maximum velocities at the inlet. (a) $\dot{\gamma}_{max} = 1\,s^{-1}$, (b) $\dot{\gamma}_{max} = 3\,s^{-1}$, (c) $\dot{\gamma}_{max} = 5\,s^{-1}$, (d) $\dot{\gamma}_{max} = 7\,s^{-1}$, (e) $\dot{\gamma}_{max} = 10\,s^{-1}$.

We now restrict our attention to the control volume of skin point (8 mm, 0 mm) and core point (8 mm, 4 mm) with different velocities (shear rate). Figures 14 and 15 show the morphologies of the skin and core volumes. As we can see in Figure 14, the morphology of the skin volume changes clearly when the velocity is increased. The number and anisotropy of shish-kebabs become higher in the larger shear rate case of the skin volume. The crystal morphology of the core volume in Figure 15 shows that shear rate has a minor effect on the number and size of spherulites.

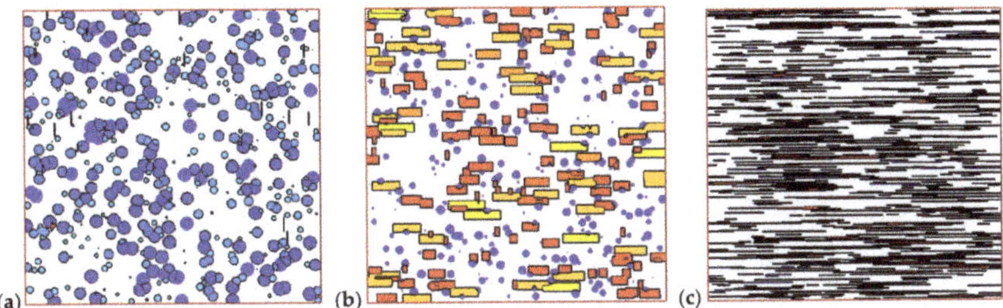

Figure 14. Morphology of the polymer skin volume: (a) $\dot{\gamma}_{max} = 1\ \mathrm{s}^{-1}$, (b) $\dot{\gamma}_{max} = 5\ \mathrm{s}^{-1}$, (c) $\dot{\gamma}_{max} = 10\ \mathrm{s}^{-1}$.

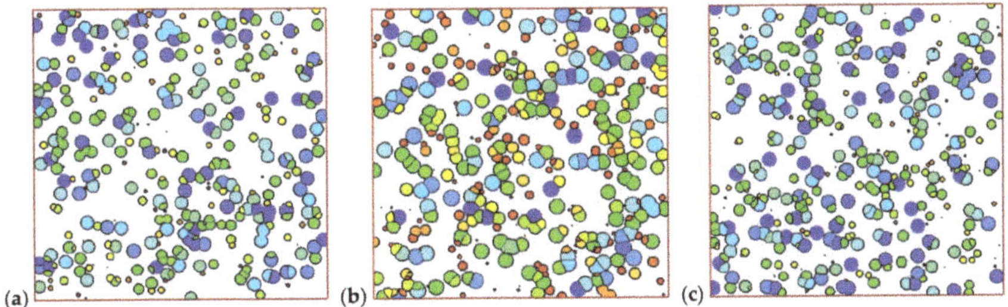

Figure 15. Morphology at the polymer core volume: (a) $\dot{\gamma}_{max} = 1\ \mathrm{s}^{-1}$, (b) $\dot{\gamma}_{max} = 5\ \mathrm{s}^{-1}$, (c) $\dot{\gamma}_{max} = 10\ \mathrm{s}^{-1}$.

Table 4 shows the average diameter of spherulites, number of shish-kebabs, and the relative crystallinity caused by spherulites and shish-kebabs at the skin and core volume. The number of shish-kebabs and the contribution of shish-kebabs to relative crystallinity decrease from the skin to core. Furthermore, the number of shish-kebabs and the contribution of shish-kebabs to the relative crystallinity increase with the maximum velocity, and the impact is significant.

Table 4. Effect of the maximum velocity at the inlet on spherulites and shish-kebabs at the profile of $x = 8$ mm.

Thickness, Shear Rate	Average Diameter of Spherulites (μm)	Relative Crystallinity Contributed by Spherulites	Number of Shish-Kebabs (/mm²)	Relative Crystallinity Contributed by Shish-Kebabs
$y = 0$ mm, $\dot{\gamma} = 1\ \mathrm{s}^{-1}$	45.05	98.64%	16	1.36%
$y = 0$ mm, $\dot{\gamma} = 5\ \mathrm{s}^{-1}$	40.58	57.82%	174	42.18%
$y = 0$ mm, $\dot{\gamma} = 10\ \mathrm{s}^{-1}$	16.63	3.09%	722	96.91%
$y = 4$ mm, $\dot{\gamma} = 1\ \mathrm{s}^{-1}$	49.98	100%	0	0%
$y = 4$ mm, $\dot{\gamma} = 5\ \mathrm{s}^{-1}$	50.66	100%	0	0%
$y = 4$ mm, $\dot{\gamma} = 10\ \mathrm{s}^{-1}$	49.66	100%	0	0%

5. Conclusions

We have extended the multiscale simulation for polymer crystallization in a pipe flow related to simplified injection processing. The "skin-core-skin" crystal structure was obtained. Both the spherulitical structure and shish-kebab structure can be captured by our algorithm. We have also shown the effects of the mold temperature cooling rate, initial melt temperature, and the maximum velocity of the melt at the inlet on microstructures. The results indicate that the temperature cooling rate of the mold (or mold temperature) especially affects the morphology of spherulites, whereas the initial melt temperature and maximum velocity of the melt at the inlet mainly affect the morphology of the shish-kebabs. We hope the results presented here can provide more insight into the microstructural details of crystallization and thus be more helpful to industrial applications.

In this work, we did not consider the changing flow domain in the injection stage, and used the viscoelastic flow without a free surface as the flow field for simplicity. To model the crystallization more accurately, a melt with the free surface should be taken into account. Moreover, the temperature boundary condition used here reflects a moderate cooling rate. In real injection processing, a low mold temperature that generates a higher temperature gradient should be applied. Our future work will be concentrated on improving our multiscale method and combining it with other software for simulating real injection processing.

Acknowledgments: The financial supports provided by the Natural Sciences Foundation of China (Nos. 11402078, 51375148, U1304521) and the Scientific and Technological Research Project of Henan Province (No. 14B110020) are fully acknowledged.

Conflicts of Interest: The authors declare no conflict of interest.

References

1. Kennedy, P.K.; Zheng, R. *Flow Analysis of Injection Molds*; Hanser Publishers: Munich, Germany, 2013.
2. Zuidema, H. *Flow Induced Crystallization of Polymers*; Eindhoven Technische University: Eindhoven, The Netherlands, 2001.
3. Boutaous, M.; Bourgin, P.; Zinet, M. Thermally and flow induced crystallization of polymers at low shear rate. *J. Non-Newton. Fluid Mech.* **2010**, *165*, 227–237. [CrossRef]
4. Zinet, M.; EI Otmani, R.; Boutaous, M.; Chantrenne, P. Numerical modeling of nonisothermal polymer crystallization kinetics: flow and thermal effects. *Polym. Eng. Sci.* **2010**, *50*, 2044–2059. [CrossRef]
5. Pantanin, R.; Coccorullo, I.; Speranza, V.; Titomanlio, G. Modeling of morphology evolution in the injection molding process of thermoplastic polymers. *Prog. Polym. Sci.* **2005**, *30*, 1185–1222. [CrossRef]
6. Raabe, D.; Godara, A. Mesoscale simulation of the kinetics and topology of spherulite growth during crystallization of isotactic polypropylene (iPP) by using a cellular automaton. *Model. Simul. Mater. Sci. Eng.* **2005**, *13*, 733–751. [CrossRef]
7. Spina, R.; Spekowius, M.; Hopmann, C. Multiphysics simulation of thermoplatic polymer crystallization. *Mater. Des.* **2016**, *95*, 455–469. [CrossRef]
8. Ketdee, S.; Anantawaraskul, S. Simulation of crystallization kinetics and morphological development during isothermal crystallization of polymers: Effect of number of nuclei and growth rate. *Chem. Eng. Commun.* **2008**, *195*, 1315–1327. [CrossRef]
9. Ruan, C.; Guo, L.; Liang, K.; Li, W. Computer modeling and simulation for 3D crystallization of polymers. II. Non-isothermal case. *Polym.-Plast. Technol. Eng.* **2012**, *51*, 816–822. [CrossRef]
10. Ruan, C.; Guo, L.; Liang, K.; Li, W. Computer modeling and simulation for 3D crystallization of polymers. I. Isothermal case. *Polym.-Plast. Technol. Eng.* **2012**, *51*, 810–815. [CrossRef]
11. Liu, Z.J.; Ouyang, J.; Ruan, C.L.; Liu, Q. Simulation of polymer crystallization under isothermal and temperature gradient conditions using praticle level set method. *Crystals* **2016**, *6*, 90. [CrossRef]
12. Liu, Z.J.; Ouyang, J.; Zhou, W.; Wang, X.D. Numerical simulation of the polymer crystallization during cooling stage by using level set method. *Comput. Mater. Sci.* **2015**, *97*, 245–253. [CrossRef]
13. Eder, G.; Janeschitz-Kriegl, H. *Materials Science and Technology*; Wiley-VCH: Weinheim, Germany, 1997.

14. Schneider, W.; Koppl, A.; Berger, J. Non-isothermal crystallization of polymers: System of rate equations. *Int. Polym. Process.* **1988**, *2*, 151–154. [CrossRef]

15. Zuidema, H.; Peters, G.W.M.; Meijer, H.E.H. Development and validation of a recoverable strain-based model for flow induced crystalization of polymers. *Macromol. Theory Simul.* **2001**, *10*, 447–460. [CrossRef]

16. Boutaous, K.; Carror, C.; Guillet, J. Polypropylene during crystallization from the melt as a model for the rheology of molten-filled polymers. *J. Appl. Polym. Sci.* **1996**, *60*, 103–117.

17. Mu, Y.; Zhao, G.; Chen, A.; Wu, X. Numerical investigation of the thermally and flow induced crystallization behavior of semi-crystalline polymers by using finite element-finite difference method. *Comput. Chem. Eng.* **2012**, *46*, 190–204. [CrossRef]

18. Guo, X.; Isayev, A.I.; Demiray, M. Crystallinity and microstructure in injection modelings of Isotactic Polypropylenes. Part II: Simulation and experiment. *Polym. Eng. Sci.* **1999**, *39*, 2132–2149. [CrossRef]

19. Guo, X.; Isayev, A.I.; Guo, L. Crystallinity and microstructure in injection modelings of Isotactic Polypropylenes. Part I: A new approach to modeling and model parameters. *Polym. Eng. Sci.* **1999**, *39*, 2096–2114. [CrossRef]

20. Wang, X.D.; Ouyang, J.; Zhou, W.; Liu, Z.J. A phase field technique for modeling and predicting flow induced crystallization morphology of semi-crystalline polymers. *Polymers* **2016**, *8*, 230. [CrossRef]

21. Ruan, C.; Liu, C.; Zheng, G. Monte carlo simulation for the morphology and kinetics of spherulites and shish-kebabs in isothermal polymer crystallization. *Math. Probl. Eng.* **2015**, 506204. [CrossRef]

22. Ruan, C.; Liang, K.F.; Liu, E.L. Macro-micro simulation for polymer crystallization in Couette flow. *Polymers* **2017**, *9*, 699. [CrossRef]

23. Ruan, C. Kinetics and morphology of flow induced polymer crystallization in 3D shear flow investigated by Monte Carlo simulation. *Crystals* **2017**, *7*, 51. [CrossRef]

24. Zheng, R.; Kennedy, P.K. A model for post-flow induced crystallization: General equations and predictions. *J. Rheol.* **2004**, *48*, 823–842. [CrossRef]

25. Advani, S.G.; Tucker, C.L., III. The use of tensors to describe and predict fiber orientation in short fiber composites. *J. Rheol.* **1987**, *31*, 751–784. [CrossRef]

26. Chung, D.H.; Kwon, T.H. Improved model of orthotropic closure approximation for flow induced fiber orientation. *Polym. Compos.* **2001**, *22*, 636–649. [CrossRef]

27. Cintra, J.S.; Tucker, C.L., III. Orthotropic closure approximations for flow-induced fiber orientation. *J. Rheol.* **1995**, *39*, 1095–1122. [CrossRef]

28. Chung, D.H.; Kwon, T.H. Invariant-based optimal fitting closure approximation for the numerical prediction of flow-induced fiber orientation. *J. Rheol.* **2002**, *46*, 169–194. [CrossRef]

29. Koscher, E.; Fulchiron, R. Influence of shear on polypropylene crystallization: Morphology development and kinetics. *Polymer* **2002**, *43*, 6931–6942. [CrossRef]

30. Hoffman, J.D.; Miller, R.L. Kinetics of crystallization from the melt and chain folding in polyethylene fractions revisited: Theory and experiment. *Polymer* **1997**, *38*, 3151–3212. [CrossRef]

31. Tanner, R.I. A suspension model for low shear rate polymer solidification. *J. Non-Newton. Fluid Mech.* **2002**, *102*, 397–408. [CrossRef]

32. Rong, Y.; He, H.; Cao, W.; Shen, C.; Chen, J. Multi-scale molding and numerical simulation of the flow-induced crystallization. *Comput. Mater. Sci.* **2013**, *67*, 35–39. [CrossRef]

33. Owens, R.G.; Phillips, T.N. *Computational Rheology*; Imperial College Press: London, UK, 2002.

34. Ruan, C.; Ouyang, J.; Liu, S. Multi-scale modeling and simulation of crystallization during cooling in short fiber reinforced composites. *Int. J. Heat Mass Transf.* **2012**, *55*, 1911–1921. [CrossRef]

35. Tao, W.Q. *Numerical Heat Transfer*; Xi'an Jiao Tong University Press: Xi'an, China, 2001.

36. Rhie, C.M.; Chow, W.L. Numerical study of the turbulent flow past an airfoil with trailing edge separation. *AIAA J.* **1983**, *21*, 1525–1532. [CrossRef]

37. Oliveira, P.J.; Miranda, A.I.P. A numerical study of steady and unsteady viscoelastic flow past bounded cylinders. *J. Non-Newton. Fluid Mech.* **2005**, *127*, 51–66. [CrossRef]

38. Ruan, C.; Ouyang, J. Microstructures of polymer solutions of flow past a confined cylinder. *Polym.-Plast. Technol. Eng.* **2010**, *49*, 510–518. [CrossRef]

39. Ruan, C. Multiscale numerical study of 3D polymer crystallization during cooling stage. *Math. Probl. Eng.* **2012**, *2012*, 802420. [CrossRef]

40. Chang, B.B.; Wang, B.; Zheng, G.Q.; Dai, K.; Liu, C.T.; Shen, C.Y. Tailoring microstructure and mechanical properties of injection molded isotactic-polypropylene via high temperature preshear. *Polym. Eng. Sci.* **2015**, *55*, 2714–2721. [CrossRef]
41. Jan-Willem, H.; Markus, G.; Gerrit, W.M.P. Structure-property relations in molded, nucleated isotactic polypropylene. *Polymer* **2009**, *50*, 2304–2319.
42. Zheng, G.Q.; Yang, W.; Yang, M.B.; Chen, J.B.; Li, Q.; Shen, C.Y. Gas-assisted injection molded polypropylene: the skin-core structure. *Polym. Eng. Sci.* **2008**, *45*, 976–986. [CrossRef]

Review

Additive Manufacturing of Metallic and Ceramic Components by the Material Extrusion of Highly-Filled Polymers: A Review and Future Perspectives

Joamin Gonzalez-Gutierrez [1,*], Santiago Cano [1], Stephan Schuschnigg [1], Christian Kukla [2], Janak Sapkota [1,*] and Clemens Holzer [1]

[1] Polymer Processing, Department of Polymer Engineering and Science, Montanuniversitaet Leoben, Otto Gloeckel-Strasse 2, 8700 Leoben, Austria; santiago.cano-cano@unileoben.ac.at (S.C.); stephan.schuschnigg@unileoben.ac.at (S.S.); clemens.holzer@unileoben.ac.at (C.H.)

[2] Industrial Liaison Department, Montanuniversitaet Leoben, Peter Tunner Strasse 27, 8700 Leoben, Austria; christian.kukla@unileoben.ac.at

* Correspondence: joamin.gonzalez-gutierrez@unileoben.ac.at (J.G.-G.); janak.sapkota@unileoben.ac.at (J.S.); Tel.: +43-384-2402-3541 (J.G.-G.)

Received: 26 April 2018; Accepted: 16 May 2018; Published: 18 May 2018

Abstract: Additive manufacturing (AM) is the fabrication of real three-dimensional objects from metals, ceramics, or plastics by adding material, usually as layers. There are several variants of AM; among them material extrusion (ME) is one of the most versatile and widely used. In MEAM, molten or viscous materials are pushed through an orifice and are selectively deposited as strands to form stacked layers and subsequently a three-dimensional object. The commonly used materials for MEAM are thermoplastic polymers and particulate composites; however, recently innovative formulations of highly-filled polymers (HP) with metals or ceramics have also been made available. MEAM with HP is an indirect process, which uses sacrificial polymeric binders to shape metallic and ceramic components. After removing the binder, the powder particles are fused together in a conventional sintering step. In this review the different types of MEAM techniques and relevant industrial approaches for the fabrication of metallic and ceramic components are described. The composition of certain HP binder systems and powders are presented; the methods of compounding and filament making HP are explained; the stages of shaping, debinding, and sintering are discussed; and finally a comparison of the parts produced via MEAM-HP with those produced via other manufacturing techniques is presented.

Keywords: additive manufacturing; fused filament fabrication; material extrusion; 3D-printing; highly-filled polymers; metals and ceramics

1. Introduction

Additive manufacturing (AM) is a technology for fabricating real three-dimensional (3D) objects, using metals, ceramics, or plastics, which may be used in various applications [1]. AM is defined by International Organization for Standardization (ISO) and American Society for Testing and Materials (ASTM) as the "process of joining materials to make parts from 3D model data, usually layer upon layer, as opposed to subtractive and formative manufacturing methodologies". The processes encompassed in AM are the 3D analog of the very common 2D digital printers; therefore, AM is also commonly referred as 3D printing. However, in the last 30 years, AM has also been referred to as direct digital manufacturing, additive layer manufacturing, additive fabrication, additive techniques, additive processes, free-formed fabrication, solid free-formed fabrication, rapid manufacturing, and rapid prototyping [2]. The term additive manufacturing has been accepted by the ASTM F42 Technical Committee and the ISO Technical Committee TC261 and this has contributed to the international adoption of this term [3]. There are several variants of AM processes available today, but the whole process can be summarized as follows [1,3]:

1. Design concepts are generated from scratch or using 3D scanners, computed tomography (CT) scans, or magnetic resonance imaging (MRI) in the case of medical implants.
2. A 3D computer aided design (CAD) model is prepared.
3. The CAD model is analyzed and optimized with the aid of computer optimization techniques such as finite element analysis (FEA).
4. The CAD model is commonly transformed into a Standard Triangulation or Tessellation Language (STL) file and imported into an AM setup. Nevertheless, the STL format lacks many features such as color or materials in the parts. For these reasons a new format was implemented by ASTM ISO, the Additive Manufacturing File Format (AMF) [4]. Besides AMF, more than 30 other alternatives to the STL file exist, three important examples of which are OBJ, PLY, and 3MF [5]. This last one (3MF) is supported among others by Microsoft, Autodesk, Dessault Systemes, 3D Systems, Materialise, Ultimaker, Mcor, PTC, FIT, GE, EOS, HP, Siemens PLM, nTopology, SLM solutions, Stratasys, and Shapeways [6]. Only time will tell which file format becomes the standard file in the future.
5. The geometric shape in the STL or other format files is sliced into thin layers and the movement of the depositing or fusing unit ("printing head"), and substrate ("printing platform"), as well as other parameters are programmed by specialized software that prepares the G-code, which is a numerical control programming language.
6. The AM machine builds the tridimensional object layer by layer with the specified parameters.
7. The built part is removed from the building platform and the removal of support structures used to build complex geometries is conducted. The excess unbound building material needs to be removed in a cleaning step depending on the type of AM technique used.
8. After the object is removed and cleaned, it might require further post-processing such as polishing, coating, or thermal treatment to obtain a functional part.

The main advantage of AM over conventional manufacturing processes is dealing with geometric and material complexities that cannot be created, technically and/or economically, using subtractive and formative manufacturing processes [7]. AM has the possibility to create structures that can be very light, stable, and at the same time contain features with a high degree of functionality [2]. The cost of producing a part using AM techniques is almost independent of the number of parts that are needed to be produced since there are no tooling costs associated with the process [3]; thus, AM is ideal for unique parts that are manufactured in low production volumes. For this reason, AM has strong usage in medical and dental applications [8–10]. AM allows the simulation of implant designs prior to their manufacturing and allows for customization for each individual patient. Thus, AM helps to reduce the costs and time required to manufacture fitting implants [1]. However, the medical field is not the only one that benefits from the use of AM; AM has many significant applications in the automotive, aerospace, and energy fields [7]. This is reflected in a drastic increase of AM fabricators, the number of parts produced, and materials used in AM between 2010 and 2015, with an annual growth rate of approximately 30% [3]. The constant evolution of production and design techniques using AM will make the technology even more cost-effective and efficient in the future. As such, the use of AM with industrial metals and ceramics will continue to grow.

Over the last three decades, many AM technologies have been developed. The standard EN ISO/ASTM 52921:2017 [11] defines the different AM technologies as shown in Table 1. This table also gives alternative names, the materials that are processable, and the strengths and weaknesses of each technique. Alternative names for the different types of additive manufacturing include: 3D printing (3DP) [1], selective laser sintering (SLS) [12], laser engineered net shaping (LENS) [12], selective laser melting (SLM) [13], direct laser metal fabrication (DLMF) [14], electron beam melting (EBM) [15], stereolithography (SLA) [16], high speed sintering (HSS) [17,18], laminated object manufacturing (LOM) [19], and fused deposition modeling (FDM) [20], also known as fused filament fabrication (FFF) [21]. Details of each of these processes have been described in the cited references [1–3,7–16,19–23]. As it can be seen, the different techniques can be used for different applications and with different materials. Thus, it can be said that one technique complements another.

The focus of this review is on material extrusion additive manufacturing with highly-filled polymers (MEAM-HP) with a particular emphasis on its application for the fabrication of metallic and ceramic components. MEAM-HP, in this case, is a multi-step/indirect process, which makes use of a sacrificial polymeric binder material to shape metallic and ceramic powder particles. The polymeric binder is usually removed in subsequent (catalytic, solvent, and/or thermal debinding) treatments and the powder particles are bonded together in a conventional sintering step.

This review paper is organized in six subsequent sections following this introduction. Section 2 explains the different types of MEAM currently available. Section 3 introduces the process of MEAM-HP and the materials used for the production of metal, ceramics, or metal-ceramic components. Section 4 describes the procedure of building parts with MEAM. Section 5 describes the post-shaping operations needed to obtain metal, ceramics, or metal-ceramic components. Section 6 offers a comparison between MEAM-HP and other processing technologies used to produce similar parts. Finally, Section 7 offers a summary of the review paper and perspectives for further improving MEAM-HP.

Table 1. Different additive manufacturing (AM) technologies and descriptions according to EN ISO/ASTM 52921:2017 [11].

AM Process Category	Technologies	Description	Typical Materials	Strengths/Weaknesses
Material extrusion	Fused filament fabrication (FFF) Fused deposition modeling (FDM™) Robocasting	Process in which material is selectively extruded through a nozzle or orifice.	Pellets or filaments of thermoplastic polymers, composites, and highly-filled polymers with metals or ceramics. Highly-filled inks containing a ceramic or metallic powder.	Inexpensive equipment. Great variety of materials. Easy to use.Small to large building spaces. Multi-material parts are possible. / Rougher surface, limited by nozzle radius. Accuracy and speed can be low. Anisotropy of properties. Support structures are needed.
Vat photo-polymerization	Stereolithographic apparatus (SLA™) Digital light processing (DLP™) Scan, spin, and selectively photocure (3SP™) Continuous liquid interface production (CLIP™)	Process which uses photopolymerization. Liquid photopolymer is selectively cured by light (ultraviolet)-activated polymerization.	UV-curable photopolymer resins (with various fillers).	High level of complexity and accuracy. Smooth surface finish. Accommodates large build areas. / Photo-resins only. Liquid monomers can be harmful. Material creeping can occur after curing. Lengthy post-processing needed. Support structures might be needed.Expensive equipment.
Material jetting	Polyjet ™ Smooth curvatures printing (SCP™) Multi-jet modeling (MJM™)	Process in which droplets of build material are selectively deposited.	Photopolymers, thermoplastic polymers, waxes, composites.	High level of accuracy. Allows for full color parts. Multi-material parts are possible. / Support structures are required. Limited number of materials.
Binder jetting	3D printing (3DP™) ExOne VoxelJet	Process in which a liquid adhesive/bonding agent is selectively deposited to join powder materials.	Powdered plastics, metals, ceramics, glass, and sand.	Allows for full color parts. High productivity. Wide range of materials available. / Properties are dependent on the binder used. Post-processing is needed. Powders can be harmful.
Sheet lamination	Laminated object manufacture (LOM) Selective deposition lamination (SDL) Ultrasonic additive manufacturing (UAM)	Process in which sheets of material are bonded to form an object.	Paper, plastic sheets, metal foils/tapes	High volumetric build rates. Relatively low cost (non-metals). Allows for combination of metal foils, including embedding components. / Finishing depends on material used. Post-processing is required. Limited materials. Properties are dependent on the adhesive used.

Table 1. *Cont.*

AM Process Category	Technologies	Description	Typical Materials	Strengths/Weaknesses
Powder bed fusion	Selective laser sintering (SLS[TM]) Direct metal laser sintering (SLM[TM]) Electron beam melting (EBM[TM]) Selective heat sintering (SHS[TM]) Multi-jet fusion (MJF[TM]) HP Jet fusion[TM] High speed sintering	Process in which thermal energy selectively fuses regions of a powder bed.	Plastics, metals, ceramics powders, and sand.	High level of complexity. Powder acts as support material. Wide range of materials available. / Equipment is more expensive. Special powders are required that are more expensive and can be harmful. Powders can age/oxidase fast. Post-processing is generally needed.
Directed energy deposition	Laser metal depositionLaser engineered net shaping (LENS[TM]) Direct metal deposition (DMD[TM])	Process in which focused thermal energy is used to fuse materials by melting as the material is being deposited.	Metal wire and powders, and powder ceramics	Not limited by direction or axis. Effective for adding features and repairs. Multiple materials can be deposited in a single part. Highest single-point deposition rate. / Powders can be harmful. Finishes depend on material. Post-processing is needed. Limited materials.

2. Material Extrusion Additive Manufacturing (MEAM)

Material extrusion additive manufacturing (MEAM) consists of softening a material and pushing it through an orifice in order to deposit that material in layers to build a 3D structure [23]. Extrusion-based additive manufacturing processes are among the most widely used AM processes, particularly when working with polymers and thermoplastic composites [24]. Compared to other AM processes, the equipment used for MEAM can be inexpensive and very easy to operate [2,25,26]. Therefore, the main advantage of MEAM is the rapid or cheap reproduction of standard components or prototypes with a variety of polymeric materials, even with low melting temperature metallic alloys [2,27].

Unlike other AM techniques, extrusion-based additive manufacturing techniques are well suited for multi-material deposition and can be used for a wide range of thermoplastic materials [2,3,22,23,26]. In general, most of the MEAM machines are equipped with a single extrusion head, but there is the possibility of adding two or more extrusion units to allow for multi-material fabrication [2,24]. Meanwhile, the growing interest in additive manufacturing is focusing currently to create high value of the technology by developing and validating new materials and novel applications of fabricated parts.

2.1. Types of Material Extrusion Additive Manufacturing

The basic principle of material extrusion additive technology involves the loading and liquefaction of the material, moving the material through a nozzle or orifice by applying force or pressure, plotting liquefied material according to a pre-defined path in a controlled manner, and layer-by-layer bonding of the material to itself or a secondary build material to form a coherent solid structure [2]. After a layer is completed, the build platform moves down or the extrusion head moves up, and a new layer of material is deposited and adhered onto the previous layer. Whenever necessary, support structures are included in the process to enable the fabrication of complex geometrical features. This basic principle enables the production of complex parts without a shaping tool other than a die with a simple geometry, generally round. Depending on the type of extruder used, one can classify material extrusion additive manufacturing into different types [28], which will be described in the following section and schematically shown in Figure 1.

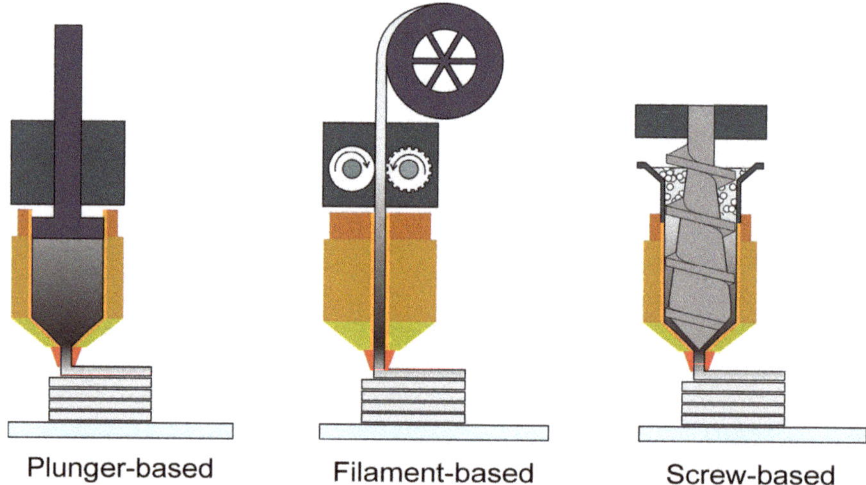

Figure 1. Different types and approaches for extrusion-based additive manufacturing.

2.1.1. Material Extrusion with Plungers

Two companies based in the USA, Desktop Metal Inc. and Markforged Inc. [29,30], currently offer MEAM machines that use special profiles (rods) made up of metal or ceramic powder with a thermoplastic binder system. Desktop Metal calls their process bound metal deposition[TM] and Markforged calls theirs atomic diffusion additive manufacturing (ADAM). The profiles are fit into cartridges and are then fed into a plasticizing unit where the highly-filled thermoplastic composite is soft enough for extrusion. The soft material accumulates in a reservoir and finally a mechanical drive system (e.g., plunger) pushes the soft material and deposits it onto the building platform in a layer-by-layer manner [29–31]. It can be seen that these machines are very similar to the machines used in robocasting [32,33], with the exception that the building materials have a thermoplastic material as a binder, while in robocasting water is used as a binder. Another particular difference is that the machine patented by Desktop Metal Inc. has an ultrasonic vibrator with sufficient energy to ultrasonically bond an extruded building material onto the previously deposited layers [29]. Alternatively, the machine by Markforged Inc. has a laser scanning displacement sensor on the printed head that acts as an in-process inspection tool to ensure that the correct dimensions are being printed [30,31].

In general, ram extrusion machines with cartridges are meant to be used for shaping parts that eventually will be made out of only metal or ceramic, thus the rods have a large amount of powder and the printed parts are sintered to obtain a dense part. The rods use similar materials as used in the well-established process of powder injection molding (PIM) [34]. On their website, Markforged Inc. offers their proprietary binder system with powder of stainless steel (316L and 17-4PH), and advertises in-development face feedstocks with Inconel (625), titanium alloy (Ti-6Al-4V), tool steel (A-2 and D-2), and aluminum (6061 and 7075) [30]. The Markforged Inc. binder is thermally debound before sintering [35]. Desktop Metal Inc. advertises the development of feedstock materials with powders of stainless steel, high-performance steel, copper, tool steel, carbide, aluminum, heavy alloys, titanium, magnetics, low expansion metals, and superalloys [36]. The binder used by Desktop Metal Inc. is solvent debound before thermal debinding and sintering is done [36].

2.1.2. Material Extrusion with Filaments

Material extrusion of filaments was first patented by the company Stratasys [2,37] and commercialized as fused deposition modeling or FDM[TM]. However, such a name could be applied to other AM techniques that melt materials and deposit them onto a platform or onto previously deposited layers of material, such as pneumatic extrusion, microinjection molding of droplets (e.g., Freeformer [38]), screw extrusion of pellets, and ram extrusion with rods. Therefore, an alternative terminology was introduced as fused filament fabrication or FFF [39]. Fused filament fabrication (FFF) is the most widely used MEAM technique. The main reasons for its popularity are its safe and simple fabrication process (i.e., no powders, lasers, solvents, nor volatile compounds are needed), the low cost of the equipment, and the availability of a great variety of filaments for printing. In the FFF process, the filament is extruded through a nozzle and deposited on a building platform one layer at a time, where it solidifies. When a heated chamber and/or heated building platform are available, the printing chamber and platform are kept at temperatures below the material's melting point, but higher than room temperature to promote adhesion to the printed bed and to reduce thermally induced stresses [2,40,41]. Please note that even if a heated building platform and/or chamber are not available, it is still possible to perform MEAM with certain materials at room temperature [42].

FFF machines are ram extruders, with the filament being the ram that pushes the softened material out of the printing head. In conventional FFF machines the filament is first pulled by the driving wheels and then it is pushed by the same wheels into a liquefier and later into a nozzle. Therefore, sufficient mechanical strength is required for the filament to retain its shape after being forced through the drive wheels [43] to transfer the force into the liquefier. This transfer of force can be altered by a number of factors. First, the motors must generate sufficient torque. Next, the wheels must have enough friction with the filament to transfer the force from the wheels to the filament. At the same time, the filament must be strong enough to avoid shearing due to the pinching from the wheels. Finally, the filament must not buckle between the drive wheels and the entrance to the liquefier. That is, the force transferred from the drive wheels to the filament should be efficiently transferred into the center of the liquefier in the direction of the melt flow, with minimal loss due to filament buckling and compression [2,22,43]. In addition to these requirements, the filament should also be flexible enough to be spooled, so that the filament can be easily stored in a compact place and fed in a continuous manner into the liquefier [15,43]. As it can be expected, not all materials can fulfill all of these conditions, yet numerous thermoplastics-based materials are available as filaments for FFF.

The most common non-filled thermoplastic materials used in FFF are acrylonitrile butadiene styrene (ABS) and polylactic acid (PLA). However, other examples of non-filled thermoplastics filaments commercially available include: acrylonitrile styrene acrylate (ASA), polyamide (PA), polycarbonate (PC), polyphenylsulfone (PPSF, PPS, or PPSU), polyetherimide (PEI), thermoplastic polyurethane (TPU), polyethylene terephthalate (PET), thermoplastic elastomer (TPE), high impact polystyrene (HIPS), polyvinyl alcohol (PVA), polyether ether ketone (PEEK), polyvinylidene fluoride (PVDF), polyoxymethylene (POM), polyhydroxyalkanoate (PHA) blended with PLA, and some other blends of the previously mentioned polymers [1,24,26,44]. Examples of composite materials commercially available for FFF include: ABS reinforced with carbon fibers; PLA with carbon fiber, graphite, stainless steel, bronze, brass, copper, bamboo fibers, wood fibers, and iron particles; and PET with carbon fibers. The filler content of these composites is between 5 and 40 vol % [1,24,26,44]. Highly-filled polymeric materials for FFF will be discussed in Section 3 of this review.

The process of ram extrusion of filaments was pioneered by Stratasys and in 1991 they introduced the first AM system of this kind. Their FDM system had two extrusion heads and used two spools of material; one material was used to build the part and the second was used for the support material. Based on the FDM system, a novel system for the manufacturing of multi-material parts was presented by the Rutgers research group, the fused deposition of multiple ceramics (FDMC) [45]. Four extrusion nozzles were included in the system, i.e., four materials could be deposited at the same layer. Different demonstrators, such as piezoelectric components with layers of soft and hard piezoelectric ceramics, were produced. Expiration of the Stratasys patents on the FDM process and growing demand for customized products has driven other companies, such as Beijing Tiertime Technology Co., Ltd., to become emerging competitors in this market [24]. In addition, personal fabrication markets are being encouraged with open source RepRap projects and several small and medium companies are producing FFF machines, such as German RepRap, Aleph Objects, MakerBot Systems, 3D System Inc., Delta Micro Factory Corp., Hage Sondermaschinenbau GmbH & Co KG, EVO-tech, BigRep GmbH, Printbot, Indmatec GmbH, Rokit Inc., Ultimaker, Sharebot srl, MarkForged Inc., 3D Platform, Titan Robotics Ltd., Vixel8, Xery 3D, Prusa Printers, Robox, Zortrax, and Felix printers [24].

2.1.3. Material Extrusion with Screws

The production of rods or filaments represents an additional task that requires special extrusion lines and know-how to obtain filaments or rods with constant cross-sectional area and minimum ovality, which are prerequisites to deposit the adequate amount of material and therefore for a reliable process in fused filament fabrication (FFF) machines. However, not all materials can be made into filaments that can be spooled, but at the same time are rigid enough that they can be pushed by the feeding mechanisms of FFF machines (see Section 2.1.2). Therefore, several research groups and companies are looking into screw-extrusion AM machines that can utilize pellets.

A screw extruder is divided into several zones. In the solid conveying zone pellets are transported to the melting zone, where pellets are softened under heat and friction, and the metering zone in which the molten material is submitted to high pressure before its eviction through the nozzle. The rotating screw has a pumping effect and thus it moves the material from the feeding zone to the nozzle [28].

Controlling the flow of the extruder to deposit the material in a precise manner could be a more challenging task and requires other tools as compared to ram extrusion. Also, the size of the pellets should be controlled in order to obtain a uniform flow of the extruded material [46]. Nevertheless, solutions have been found and below some examples of screw extruder AM machines are described.

Bellini et al. [46,47] developed a system called mini extruder deposition (MED), which consists of a mini screw extruder mounted on three high precision linear motor tables. The three tables were connected to three digital servo drives to monitor the torque, velocity, and rotational speed. The servo drives were also equipped with digital notch filters to eliminate mechanical resonance. The driver's position, speed, and acceleration of the three axes can simultaneously be controlled. A separate controller was used to regulate the heaters and the motor of the extruding screw. Material temperature was checked at the entrance of the liquefier and closer to the nozzle. Even though the developed preliminary configuration shows opportunities for the use of a wider range of materials, it can only be considered as a starting point for further development, due to the limited information provided by the researchers and the lack of follow-up publications.

Cruz et al. [48] developed their own screw-based extrusion system. The equipment consisted of a vertical single screw extruder with a screw length of 90 mm, a screw diameter of 15 mm, and a die with a diameter of 2 mm. Two band heaters were placed around the barrel to ensure a constant temperature (up to 250 °C) during the plasticization process. The building platform was capable of moving in XYZ directions, controlled by step motors to control the trajectory and the material deposition. The printing process was controlled by a logical controller and a computer was used as an interface to enter the processing conditions (barrel temperature, screw rotational speed, and material rate of deposition) and monitor the process. The designed extruder was capable of processing a feedstock with 59 vol % of carbonyl iron; however, no further details in terms of printability and printed parts were shown.

Two companies have developed screw-based MEAM setups for making small parts and both are currently commercializing their machines. One of them is AIM3D GmbH (Rostock, Germany) [49,50]. The AM machine from AIM3D has two extruders that can take commercially available pellets from thermoplastics or metal injection molding (MIM) feedstock to build a three-dimensional object. The building volume is a cube measuring 255 mm on all sides. As indicated on their website, the only material that is beyond the beta phase of development is a MIM feedstock with stainless steel particles. The second company is Pollen AM Inc. (Paris, France) [51,52]. The Pollen AM MEAM is capable of printing with up to four different materials, and it is also capable of mixing two materials during the printing process. Materials available include unfilled thermoplastics and filled thermoplastic pellets with natural fibers, carbon fibers, minerals, and metal particles [52].

Cincinnati Inc. (Cincinnati, OH, USA) and Oak Ridge National Laboratories (Oak Ridge, TN, USA) have developed a screw extrusion machine for large size additive manufacturing. The setup is called big area additive manufacturing or BAAM. It consists of a single screw extruder mounted vertically on a machine frame, similar to the frames used for laser-based AM machines. The extruder has a feed-rate of 36 kg/h and a unique automatic taping mechanism, which is used to flat the deposited material to increase the contact between deposited layers. The setup is available in two sizes: $7.8 \times 3.7 \times 3.3$ m^3 and $10.8 \times 3.9 \times 4.4$ m^3. The motion system is driven by linear motors and the absolute position accuracy is ± 0.127 mm. Using BAAM, the manufacturers have been able to print sections of car bodies and sections of buildings. The materials that have been tested include pellets of acrylonitrile-butadiene-styrene (ABS), polyphenylenesulfide (PPS), polyetherketoneketone (PEKK), and polyetherimide (PEI), as well as composites materials containing carbon, glass fibers, and NdFeB particles [53,54].

3. Material Extrusion Additive Manufacturing of Highly-Filled Polymers

Highly-filled polymers are compounds of polymers with added particles at concentrations well above 20 vol % in which the interactions between fillers cannot be neglected [55]. In this review we will be talking about compounds with filler contents between 45 and 65 vol %, which can be used for the fabrication of metal or ceramic components. The use of highly-filled polymers for the production of metal or ceramic parts with complex geometries has a long history. Ceramic injection molding (CIM) was the first to be introduced in the 1930s, simultaneously in USA and Germany, for the production of spark plug bodies, but for the next three decades it was of minor interest to the ceramic industry. In the 1960s, CIM was also utilized for the production of ceramic tableware. It was only in the 1970s and 1980s that CIM provided a cost-effective manufacturing method for the mass production of ceramic parts for the automotive industry [56,57]. Metal injection molding (MIM) reached production in the 1970s. In 1979, MIM drew attention when two parts won awards [56]. One part was a screw seal used on a Boeing jetliner. The second part was a niobium alloy thrust chamber and injector for a liquid propellant rocket engine. By the middle 1980s, the MIM sector already had multiple actors [56]. When compared to other manufacturing technologies such as casting and forging, powder injection molding (PIM) is a relatively young technology with great potential.

The idea of using highly-filled polymers for the additive manufacturing of metal and ceramic parts was first introduced in the 1990s; it was named fused deposition of metals (FDMet) [58,59] and fused deposition of ceramics (FDC) [60,61], respectively. It was based on the Stratasys FDM technology, in which highly-filled polymers with metal or ceramic particles are initially extruded as filaments, and then these filaments are selectively extruded at a temperature higher than the melting point of the binder polymers. Later, as in the case of PIM, the shaping step is followed by the removal of the polymer from the samples using solvents, catalyzers, and/or by thermally decomposition; finally, fully densified metallic or ceramic components are obtained after sintering the parts [58,61,62]. The process is sometimes referred to as shaping, debinding, and sintering (SDS) and a schematic representation of the overall process is shown in Figure 2.

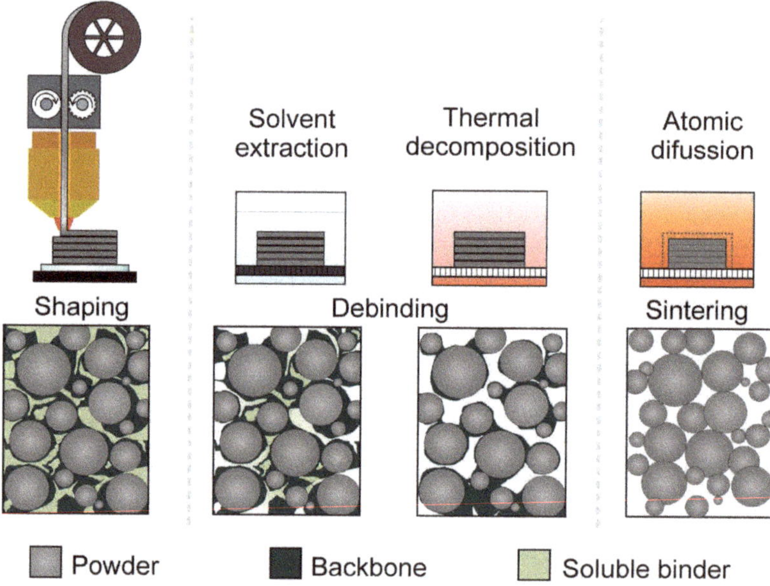

Figure 2. Schematic representation of the shaping, debinding, and sintering (SDS) process and respective morphology of the parts for the fabrication of metal, ceramic, or metal-ceramic components.

It is important to mention that the SDS process could use other additive manufacturing techniques such as indirect powder bed fusion, binder jetting, vat polymerization, and material jetting. For indirect powder bed fusion, sinterable particles can be coated with a thermoplastic and these coated particles are then fused together with low power lasers, since only the polymer needs to be sintered or melted [63,64]. For binder jetting, sinterable powders can be bound together with resins or adhesives [65–67]. As for vat polymerization and material jetting, slurries containing sinterable particles and photopolymerizing resins or thermoplastics can be used to shape parts [68–72]. More details about these processes are given in References [63–72]. The emphasis of this review is shaping with MEAM, as shown in Figure 2.

MEAM-HP has shown great promise as a cost-effective alternative for the fabrication of metal, ceramic, and metal-ceramic parts [73], particularly for companies currently working with PIM, which already have the equipment and know-how to carry out the subsequent steps of binder removal and sintering to obtain solid parts with complex geometries [73,74]. The production of small metal or ceramic parts by PIM and MEAM-HP are not mutually exclusive; on the contrary, they are complementary. PIM is a technology that becomes economically feasible when large quantities of parts are to be produced (>1000 parts per year), due to the costs associated with the design and manufacture of the mold used in the injection molding machine [57,75]. MEAM-HP is meant to be used for the production of small quantities (i.e., prototypes or custom-made parts) or parts with geometries that cannot be achieved by filling the cavity of a mold. Therefore, we believe that an industrial niche for MEAM-HP will emerge in the near future and thus it is worth investigating and improving the feedstock materials and equipment used in this AM process.

As mentioned before, PIM and MEAM-HP rely on the use of highly-filled polymers; the similarities and differences between their feedstocks will be described in the following section.

3.1. PIM and MEAM-HP Feedstocks

Feedstocks for MEAM-HP and PIM are multicomponent systems consisting of a polymeric blend, sinterable powder, and additives. All of these components are needed to fulfil the requirements at different steps of the overall SDS process. Details about these different feedstock components will be described in the following sections.

3.1.1. Binder Systems

The polymeric component of the feedstock is referred to as the binder system. The binder system greatly influences the production process and the quality of the sintered parts, even though it is completely removed during the debinding step. Usually a binder system consists of different types of polymers, waxes, and additives [55]. Generally, three main groups in the binder system can be identified [34,56]:

i.   The main binder component is the component present in the largest amount and it is removed first during the debinding step. The main binder component represents between 50 and 90 vol % of the total binder system

ii.   The backbone is the component used to hold together the shape of the part while the main binder component is removed during the first debinding stage. The backbone is thermally decomposed prior to sintering. The backbone represents 0 to 50 vol% of the total binder system.

iii.   Additives like dispersant agents, compatibilizers, and stabilizers help to disperse the filler particles in the polymeric binder, preventing agglomeration and phase separation. Additives represent between 0 to 10 vol % of the binder system.

In fact, it is possible to use a binder system with only one polymeric material, but then the debinding step is limited only to thermal degradation, which is a very slow process compared to solvent or catalytic debinding [34,76,77]. For this reason, most binders have at least two components. Tables 2–4 show some examples of binder systems components reported in the literature for PIM, MEAM-HP with filaments, and MEAM-HP with pellets or powders, respectively.

As can be seen in Table 2, the main components in the binder system for PIM are polymeric materials with low viscosity (e.g., waxes), materials that dissolve in water (e.g., polyethylene glycol, agar, etc.), or those that undergo catalytic degradation (e.g., polyoxymethylene) [78–80]. The majority of backbones are polyolefins (e.g., polyethylene, polypropylene). This is because polyolefins are resistant to many solvents used during debinding, add strength to the debound part, and degrade into hydrocarbons only before sintering. Finally, the most commonly used additive is stearic acid, which facilitates the dispersion of the filler particles. For PIM, low viscosity of the feedstock is required to fill the cavity at lower pressures in the injection molding machine and this is reflected in the binder composition [34].

Table 2. Examples of binder system compositions used in powder injection molding (PIM).

Main Component (50–90 vol %)	Backbone (10–50 vol %)	Additives (1–10 vol %)	Ref.
Carnauba wax	Polypropylene (PP)	Stearic acid	[81]
Paraffin wax	Ethylvinylacetate (EVA)	Stearic acid	[78]
Paraffin wax	High density polyethylene (HDPE)	Stearic acid	[78,81–83]
Paraffin wax	Polyethylene (PE), PP	Stearic acid	[84–86]
Paraffin wax	HDPE, PP, Polystyrene (PS)	Stearic acid	[87]
Paraffin wax	PE	Stearic acid, oleic acid	[88]
Polyethylene glycol (PEG)	Polymethyl methacrylate (PMMA)	Stearic acid	[89–91]
PEG	Polyvinylbutyral (PVB)	Stearic Acid	[92]
PEG	Polyethylene wax	Stearic acid	[93,94]
PEG	Polyimide diisocyanate	2, 6-di-tert-butyl-4-hydroxytoluene	[95]
Polyoxymethylene (POM)	Low density polyethylene (LDPE)	Stearic acid	[77]
POM	Polyolefins	Poly-1,3-dioxepane or poly-1,3-dioxolane or mixtures thereof	[96]
POM	PE	Butanediol formal	[97]
Agar (gel forming polysaccharide)	Glucose	Deionized water, calcium borate, methyl-p-hydroxybenzoate and propyl-p-hydroxybenzoate as biocides	[98]
PEG or polypropylene glycol or polyvinyl alcohol	PS and/or PE	Methylene chloride	[80]
Partially hydrolyzed cold water soluble polyvinyl alcohol	PE or PP	Glycerin, INT-33PA, steric acid, water	[99]

On the other hand, binder systems for MEAM-HP, which are used as filaments, have components that lead to flexible feedstocks that can be spooled. One way to make feedstocks flexible is to add elastomers [60–62,100] or amorphous polyolefins [45]. Another way is to add a stiffer polymer-like polyamide or a polyolefin and add other components that plasticize these polymers to increase their flexibility and at the same time can be dissolved to speed up debinding [101,102]. Other components in the MEAM-HP binders include tackifiers, waxes, and plasticizers. One example of a tackifier used is a hydrocarbon resin, which can improve the adhesion with the previous layers and the flexibility of the filament [45]. Waxes, such as partially crystalline polyolefin wax, can be used to reduce the viscosity and improve the stiffness of filaments [103]. Finally, a low molecular weight polyolefin can be used as a plasticizer to reduce the viscosity of the feedstock [104]. Since finding a formulation that works is a complex task, many times the formulation is not clearly specified in published papers and patents to prevent competitors from using the exact binder systems [58,59,105]. Some formulations available in the literature are shown in Table 3. The binder systems used in MEAM-HP for pellets or powders (Table 4) are a lot more similar to the feedstocks used in PIM; in fact, the goal is to develop machines that can use the readily available PIM feedstocks to 3D print objects. These feedstocks go back to the idea of decreasing the viscosity of the feedstock, thus their main binder components are again waxes or PEG [48,106]. Most of the MEAM-HP methods that do not use filaments are still in the development phase and as such, the feedstock formulations might not be the final formulations that really can be shaped, debound, and sintered.

Table 3. Examples of binder system compositions used in material extrusion additive manufacturing with highly-filled polymers (MEAM-HP) with filaments.

Main Component (50–90 vol %)	Backbone (0–50 vol %)	Additives (0–10 vol %)	Ref.
Elastomer and wax	Polymer	Plasticizer, tackifier, oleyl alcohol	[60,61,100]
Amorphous polyolefin	Amorphous polyolefin	Tackifier, wax, plasticizer, surfactant	[45,107,108]
Microcrystalline wax	Ethylene Vinyl Acetate (EVA)	None	[103,109]
Thermoplastic elastomer (TPE)	Grafted polyolefin	Unspecified compatibilizer	[62,73,74,110–115]
4 hydroxybenzoic acid-behenylester solid, and 4 hydroxybenzoic acid-ethyhexylester	Co-polyamide (PA) 6/12	None	[101]
HDPE	None	Isopropyl tri(dioctyl)pyrophosphato titanate, tri(dioctyl)phosphato zirconate or mixtures thereof	[102]
POM	Polyolefin, and other polymer (polyether, polyurethane, polyepoxide, polyamide, etc)	None	[116]
PA	None	Undisclosed	[105]
Undisclosed	Undisclosed	Stearic acid	[58,59]
LDPE wax	LDPE	None	[117]
Polypropylene	Elastomer	Wax, tackifier, plasticizer	[118–120]

Table 4. Examples of binder system compositions used in MEAM-HP with pellets or powders.

Main Component (50–100 vol %)	Backbone (0–50 vol %)	Additives (0–10 vol %)	Ref.
PE wax, paraffin wax, PEG	PP	None	[48]
PEG	None	None	[106]
Paraffin wax	LDPE	SA	[117]

The effect of binder formulation on the properties of filaments have been investigated and linked to the ability of those filaments to be printed to build a 3D object. For example, Kukla et al. [110] studied the variation in the amount of three different polyolefins as a backbone in the feedstock containing thermoplastic elastomer (TPE, main binder component) and 316L steel powders. It was found that the modulus of the feedstock filaments could be increased with an increase in backbone content. The resulting feedstock properties with medium viscosity (~1000 Pa·s) and medium to high secant modulus (400 to 2700 MPa) were found to be printable in conventional FFF printers in a continuous manner.

Agarwala et al. [118] initially developed a binder system to be used for the production of parts of silicon nitride (Si_3N_4) via filaments. The filaments produced with the feedstocks were too brittle to be spooled. However, the high material stiffness enabled the production of parts by hand-feeding the equipment. For this reason, Agarwala et al. [61] later implemented a design of an experiment for optimizing the binder. Four polymers were employed: a backbone polymer, an elastomer for improving the flexibility, a wax for reducing the viscosity, and a tackifier for promoting the adhesion. A tradeoff had to be attained, since the wax reduced the flexibility, and the elastomer increased the viscosity. After the optimization of the system and the pre-treating of the powder with a dispersant for reducing the viscosity, flexible filaments with a diameter of 1.78 ± 0.05 mm could be spooled.

Bhat et al. [121] studied the use of polyethylene systems for the FFF of alumina feedstocks. The designed binders were composed of a polyethylene wax as a plasticizer and a linear low density polyethylene. The plasticizer content varied from 0 to 100 vol % in the binders and in the feedstocks containing 50 vol % of alumina powder. A higher plasticizer content resulted in a lower viscosity, but also in a lower compressive strength of the component. The best properties were attained with feedstocks containing 40 vol % of plasticizer, from which straight filaments could be used for MEAM.

These examples demonstrate that finding the right combination of polymers and their proportions in the binder is not a simple task and for this reason, most of the actual binder formulations that work are not described in detail in the literature. However, it is clear that the selection of the optimal binder systems can be directly based on the compatibility with the powder, resulting viscosity, and mechanical properties (modulus and flexibility) required for MEAM with filaments.

3.1.2. Powder Fillers

Many additive manufacturing technologies rely on the use of powder as their building material; MEAM-HP is not an exception. The powder material used in MEAM-HP is in principle the same metal or ceramic powder as that used in PIM; this represents an advantage compared to AM techniques that rely on powder bed technology, which require a very specific particle size distribution for the process to work properly. In general, MIM utilizes particles with an average size between 5 to 15 μm [34]. Typical mean particle sizes in ceramic CIM are 1 to 2 μm, but also submicron or nano sizes are used in advanced CIM [122]. Therefore, the typical particle sizes used in PIM are fine enough to prevent the plugging of the nozzle (diameter range 0.3 to 0.8 mm) of the MEAM machine.

Many metals and ceramics are available for PIM [34] and in principle they should also work with MEAM-HP; however, not all of them have been tested. Table 5 shows a list of the types of powders that have been successfully shaped by MEAM-HP, debound, and sintered as found in the scientific literature. Table 6 shows the types of materials that are claimed to be available by companies producing MEAM-HP equipment. All companies advertise many other powder fillers as being under development; these materials have been excluded from the list.

Table 5. Ceramics and metals investigated for use in MEAM-HP.

Metal or Ceramic	Type	Powder Content in Feedstock (vol %)	Ref.
Ceramic	Silicon nitrate (Si_3N_4)	55 and 60	[60,61,104]
Ceramic	Fused silica (SiO_2)	56, 60, and 65	[60,61,119]
Ceramic	Lead zirconium titanate	50 and 52.6	[60,61,123]
Ceramic	Zirconia	85	[106]
Ceramic	Yttria stabilized zirconia	47	[114]
Ceramic	Strontium ferrite ($SrFe_{12}O_{19}$)	53, 55, and 60	[112,124]
Ceramic	Alumina	50	[121]
Ceramic	Mullite + Alumina + MgO	47.93 + 6.85 + 0.69 = 55.47	[120]
Ceramic	Fused silica + MgO	53 + 3 = 56	[120]
Ceramic	Titanium dioxide + MgO	51 + 4 = 55	[120]
Metal	Stainless steel (17-4PH)	55 and 60	[58,59,61,62,73]
Metal	Stainless steel (316L)	50 and 55	[62,74,105,110–112,125]
Metal	Stainless steel (AISI 630)	79	[106]
Metal	Tungsten carbide-cobalt	50	[60,61]
Metal	Carbonyl iron	65	[48]
Metal	Titanium (Ti6Al4V)	55	[90]
Metal	Rare earth magnet (NdFeB)	55	[113]

Table 6. Feedstocks with powders currently offered by companies; powders in beta phase of development are excluded from the list.

Company	Powders	Ref.
Markforged Inc.	Stainless steels 316L and 17-4PH.	[35]
Desktop Metal Inc.	4140 (chrome moly), copper, Kovar F-15, Inconel 625, 316L (austenitic), 17-4 PH, and tool steel H13	[36]
AIM3D GmbH	Stainless steels 17-4PH, 316L, 410L, 430 and 440C, tool steel M2, and low-alloy steel 4340	[50]
EVO-tech GmbH	Stainless steel 316L	[126]

Changing the characteristics of the powder can drastically influence the mechanical and flow properties of the feedstock materials, as has been reported in the PIM and the particulate composite literature [127–129]. For feedstock materials to be used in FFF this is particularly critical because changing the characteristics of the powder can lead to filaments with low mechanical properties that are not printable.

One main factor that affects the properties of highly-filled filaments is the change in the physical and chemical properties of the metal or ceramic fillers. This change can be attributed to the different size, morphology, and chemical composition of the different metals and ceramics, which lead to a distinct interaction with the binder system. Examples of how changing the particles affects the properties of filaments have been presented in the AM literature. For example, Kukla et al. [112] presented the tensile properties of filaments filled with stainless steel (316L), titanium (Ti6Al4V), copper (Cu), rare earth alloy (NdFeB), aluminum (AlSi10Mg), strontium ferrite ($Fe_{12}O_{19}Sr$), and yttria stabilized zirconia (YSZ), see Figure 3a. As expected, the tensile properties of the filaments greatly varied as the filler particles were changed and it was observed that the processability of the different filaments varied. Feedstocks containing ceramic fillers ($Fe_{12}O_{19}Sr$ and YSZ) have the stiffest and less flexible filaments and they have to be fed into the printer head manually, i.e., one strand at a time. Among the metal-filled filaments, 316L was the stiffest, followed by the Cu, Ti6Al4V, NdFeB, and AlSi10Mg. All of the metallic filaments with the exception of NdFeB could be continuously printed on conventional MEAM machines having a feeding unit with counter-rotating wheels. The NdFeB-filled filaments had low stiffness and were fragile and they could only be printed on a MEAM machine with a feeding mechanism consisting of conveyor belts as opposed to rotating wheels. The conveyor belts provided more contact between the feeding mechanism and the filament and prevented the breakage of the filament during feeding into the liquefier unit [112]. It can be concluded that changing fillers will influence the processability of highly-filled filaments and thus different printing conditions or even printing setups might be required to process them by MEAM.

Keeping the same powder type but with different characteristics has also been investigated. For example, Wu et al. [58] studied two 17-4PH stainless steels powders (a spherical powder with an average particle size of 22 μm, and an irregular powder with an average particle size 10 μm). Small powders were preferred for the use of smaller nozzles. In another study, Kukla et al. [111] reported that increasing the average particle size (from 5.5. to 8.6 μm) of round steel particles (316L) used in the filament, while maintaining all parameters constant, can lead to unprintable filaments. The increase in the average particle size resulted in the decrease of apparent viscosity and secant modulus of the filaments by ca. 42% and elongation at break by ca. 35%. The decrease in secant modulus can be linked to a decrease in stiffness and thus it was responsible for the tendency to buckle at the feeding mechanism of conventional FFF machines, resulting in failure during printing. One possible solution to this problem could be to modify the proportion of backbone in the binder system, as discussed previously [111].

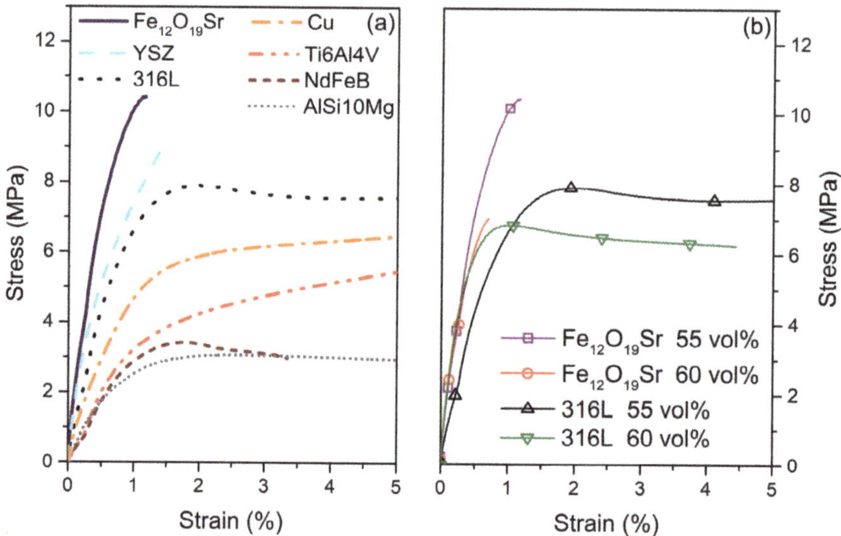

Figure 3. Tensile response of filaments with (**a**) different powders [112] and (**b**) different powder contents [124].

The powder content in the filament has also been investigated. Gonzalez-Gutierrez et al. [124] characterized feedstocks and filaments with different contents of Fe$_{12}$O$_{19}$Sr (Figure 5b). It was observed that increasing the content from 55 to 60 vol % made the filaments significantly less ductile (i.e., shorter strain-stress curve); however, the stiffness remained almost the same. The filaments with the highest powder content had to be fed manually and even the printed parts were easily broken if not handled properly during the removal from the building platform. Figure 3b also shows unpublished results from our group in which the 316L steel content in the filaments was increased from 55 to 60 vol %. In the case of steel, the 5 vol % powder increase led to a stiffer filament and, as in the case of $Fe_{12}O_{19}Sr$, a more brittle and fragile filament; but contrary to the filaments with $Fe_{12}O_{19}Sr$, both steel filaments could be fed continuously from the spool to the printer head. What can be concluded is that the maximum volume content that yields printable filaments is very much material-dependent since the mechanical properties are greatly dependent on the particle-matrix chemical interaction [124].

The processability of filaments by MEAM is not only dependent on the mechanical properties of the filament, but also on the flow properties of the feedstock as well as on the processing conditions, the geometry of the filament, and the design of the printing head. This was discussed by Venkataraman et al. [130] for ceramic and metallic feedstocks with different binder systems. According to these authors, the filament will buckle during the printing process when the extrusion pressure exceeds the critical buckling stress of the material, i.e., $\Delta P' > \sigma_{cr}$. The critical stress was considered to be approximately equal to the filament buckling stress by Euler's criterion. This criterion depends on the geometry of the filament, the elastic modulus (E), and the length between the feeding rollers and the liquefier unit (L). If the filament is a cylinder with radius R, then the critical stress (σ_{cr}) can be calculated as [131]:

$$\sigma_{cr} \approx \sigma_E = \frac{\pi^2 E}{4(L/R)^2} \tag{1}$$

For a non-Newtonian fluid with apparent viscosity η_a, the pressure drop (ΔP) in a capillary rheometer with radius r and length l for a given volumetric rate Q is defined as [132]:

$$\Delta P = \frac{8\eta_a Q l}{\pi r^4} \tag{2}$$

Venkataraman et al. [130] assumed that there is a linear scaling factor k correlating the pressure recorded in a capillary rheometer with the one during the printing process with the relation $\Delta P = k\Delta P'$. Combining this relation with Equations (1) and (2), it can be stated that the filaments will buckle when:

$$E/\eta_a < \frac{32 Q l (L/R)^2}{\pi^3 r^4 k} \tag{3}$$

As can be observed in Equation (3), the buckling of the filament will also depend on the radius (kr^4) and length (l/k) of the nozzle, as well as on the volumetric flow employed (Q/k) during printing. According to the results obtained for feedstocks with different binders and powders, the buckling of the filaments will occur when the E/η_a parameter is below the experimental critical range of 3×10^5 to 5×10^5 s^{-1} in the range of shear rates commonly employed in filament MEAM (100 to 200 s^{-1}) [130].

A similar work was conducted by Rangajaran et al. [133], who investigated the rheology and the mechanical properties of a feedstock containing 55 vol % of Si_3N_4. In this case, the parameter k defining the relationship between the pressure measured in the capillary rheometer and the pressure in the FFF nozzle was supposed to be proportional to the diameter difference in both devices and equal to 1.1. Using the same hypotheses as Venkataraman et al. [130], the buckling will occur when $1.1\Delta P > \sigma_{cr}$. The relation was experimentally validated [133], but further research should be conducted dealing with the relation of ΔP with the flow rates, the material properties, and the geometrical parameters of the nozzles.

3.2. Effect of Processing on Properties of Feedstocks and Filaments

Feedstocks for MEAM-HP can be produced in a similar manner as PIM feedstocks: the metal or ceramic powder is mixed with the molten binder constituents and the filler is dispersed in the binder. One of the main requirements is that the resulting compound has a homogeneous distribution of powder particles and binder components. This helps to minimize the segregation of components during the shaping process and later on to obtain isotropic shrinkage after debinding and sintering. Avoiding the segregation of components is crucial to prevent visual defects, excessive porosity, warpage, and cracks in the sintered part [34]. An example of a feedstock material with well-dispersed particles is shown in Figure 4 [62,73]. In the optical micrograph to the left, the shining spots are steel particles and in the SEM micrograph to the right, the particles are fully covered with the binder system, but they are still visible and no agglomerates can be seen.

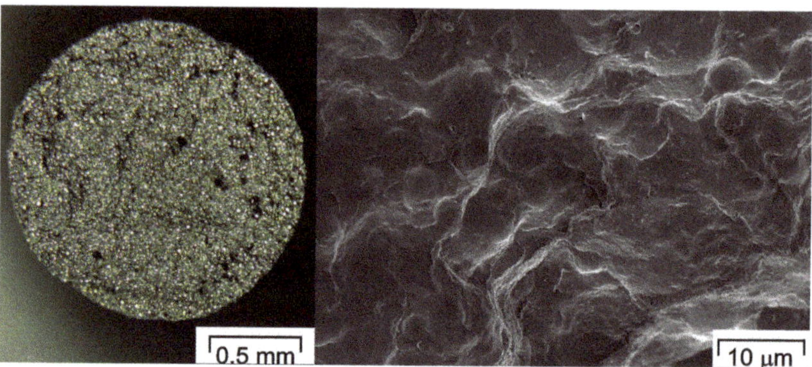

Figure 4. Optical microscopy and SEM images of a filament filled with steel particles [62,73].

Feedstocks can be prepared either in batch or continuous processes. Batch operations include roll mill kneaders and mixers, while continuous processing can be done in screw extruders and shear rolls [34]. In general, feedstocks for MEAM-HP have been produced by using a kneader with Z-blades [60,119,120,123], kneader with counter-rotating rollers [62,73], twin roll mill [134], sigma blade mixer [135], and co-rotating twin screw extruder [62,114]. For materials that have a tendency to agglomerate, co-rotating screws or shear rolls might be the best option, since the high shear achieved with the screw design helps to break down agglomerates and disperse the individual particles [136]. Moreover, for hard-to-disperse powders such as Si_3N_4, zirconia, and other ceramics, the powder is first coated with surfactants (e.g., oleyl alcohol and stearic acid) during a ball-milling step before blending with the rest of the binder components. After milling, the powder has to be sieved to further remove agglomerates and increase the homogeneity of the fillers in the feedstocks [104,107,133,137]. The temperature of mixing is dependent on the viscosity of the feedstock and the thermal stability of the components used in the binder system. The viscosity of the feedstock determines the shear stress generated during compounding, which in turn determines how well the powder agglomerates are broken and the individual particles are dispersed. Therefore, it is recommended to use higher temperatures for feedstocks with a higher viscosity and lower temperatures for feedstocks with a low viscosity [61].

Filaments are generally produced via extrusion. For example, in capillary rheometers [118,137] for small batches, and in single screw extruders [104,119,120,137] or twin screw extruders [105,137] for larger productions. When dealing with powders prone to agglomeration, extruders with breaker plates and screens can be used to reduce the agglomeration in the filaments; this was observed by Clancy et al. [137] when working with Si_3N_4 feedstocks.

It is very important to produce filaments with tight tolerances on the diameter, because the feeding mechanisms in a filament-based MEAM machine sets the feeding speed, and thus the flow rate, based on the assumption that the filament has a constant diameter. If the filament diameter is smaller than the specified value, the flow rate is lower than desirable and thus strands with smaller widths and thicknesses are deposited. This is referred to as underflow and it results in poor bonding between the deposited strands or creates voids between adjacent strands that will not be closed even after sintering.

If the diameter is larger than the specified value, the filament cannot be fed into the liquefier or it may lead to overflow. Overflow leads to poor definition of the fine features of the part. Also, the filaments should be as round as possible (i.e., with minimum ovality) to be properly gripped by the feeding mechanism and so ensure constant feed without slippage [61]. In order to produce filaments with constant dimensions and roundness, the extruder is coupled with a conveyor belt or a haul off unit that pulls away the filament, which is finally spooled in a winding unit [114]. Depending on the thermal conductivity of the filament, it may be necessary to cool down the filament by water or by air. Most highly-filled filaments have a high enough thermal conductivity that no water cooling is necessary. The filament's diameter and roundness need to be monitored with laser micrometers [137] or other optical sensors so that the different processing parameters can be adjusted. Such parameters include, for example, the extruder temperatures and rotational speed, the speed of the conveyor belt, the haul off unit, and the spooling device [114].

The effect of different compounding strategies on the properties of filaments has been observed in zirconia feedstocks [114]. It was observed that the filaments produced after a high shear compounding step in a co-rotating twin screw extruder had much better tensile properties than filaments produced after compounding in a contra-rotating roller mixer. The mechanical properties of both types of filaments are shown in Table 7 [114]. By looking at the values in Table 7, it can be seen that the filament produced after compounding in the twin screw extruder is more ductile and less stiff, since its ultimate tensile strength (UTS) and its elongation at UTS are larger, while its secant modulus is smaller than those of the filament produced after compounding in the roller mixer. Filaments made after twin screw extrusion were easier to print in MEAM machines.

Table 7. Tensile properties of filaments produced by different compounding strategies [114].

Compounding Method	Ultimate Tensile Strength—UTS (MPa)	Elongation at UTS (%)	Secant Modulus (MPa)
Twin screw extruder	13.7	3.07	1250
Roller mixer	10.6	1.28	1860

4. Building of Parts

The fabrication process is crucial to obtain good quality sintered parts. If there are large gaps between the deposited strands of material and other defects, these defects will remain in the sintered parts, affecting the mechanical performance and functionality of the final part. Therefore, optimizing the printing parameters to obtain a homogeneous part is very important.

The building of parts via MEAM-HP has been reported in the literature. Table 8 summarizes the details of different models of MEAM machines and some parameters used for different types of feedstocks. The setup parameters are highly dependent on the setup design, the binder composition, as well as the type of powder used; therefore, it is clear in Table 8 that all of the parameters are different. Not all of the parameters are described in the literature, so preliminary runs are recommended for new materials in different setups.

Table 8. Equipment and processing parameters used to process materials for MEAM-HP.

MEAM Model	Fillers in Feedstocks	Building Parameters Given	Refs.
Stratasys FDM^{TM} 1650 (Filament-based)	Mullite, fused silica, and titanium dioxide	Ext. Temp: 235–237 °C Envelope Temp: 48 °C Material flow rate: 130%	[120]
Hage3D-72L (Filament-based)	Stainless steel 316L at 55 vol %	Nozzle diameter: 0.5 mm Ext. Temp: 240 °C Deposition speed: 50 mm/s	[125]
Hage3D-72L (Filament-based)	Yttria stabilized zirconia at 47 vol % Modified stainless steel 17-4PH powder at 47 vol %	Ext. Temp: 220–240 °C Bed Temp: 20 °C Print speed: 10 mm/s Layer thickness: 0.25 mm	[114]
Wanhao Duplicator i3 v2 (Filament-based)	Stainless steel 316L and 17-4PH at 55 vol %	Printing surface: glass or PP Nozzle diameter: 0.6 mm Ext. Temp: 210–260 °C Bed Temp: 60 °C Flow rate: 100–200% Deposition speed: 40–80 mm/s Layer thickness: 0.15–0.20 mm	[62,73]
Ultimaker 2 (Filament-based)	Stainless steel 316L	Nozzle diameter: 0.8 mm Feed speeds: 0.5–7 mm/s Deposition speed: 14 mm/s Ext. Temp: 235–240 °C Built rate: 0.62–5 mm^3/s	[105]
Stratasys 3D Modeler (Filament based)	Si$_3$N$_4$ (Honeywell's GS44 grade)	Nozzle diameter: 0.25 mm Ext. Temp: 185 °C Envelope Temp: 37 °C Layer thickness: 0.254 mm	[104]
Mini-Extruder Deposition (MED) (Screw-based)	Piezoelectric ceramic ECG9/PZT [107]	Top liquefier temp.: 145–160 °C Lower liquefier temp.: 135–145 °C Nozzle diameter: 0.3 and 0.6 mm Pellet size: 1–5 mm	[46]
Fused Deposition of Metals (FDMe)(Screw-based)	Carbonyl iron at 57 to 59 vol %	Top liquefier temp.: 155–159 °C Lower liquefier temp.: 180–18 7 °C Nozzle diameter: 2 mm	[48]

Effect of Processing on Properties of Built Parts

Several studies have been performed to investigate the effect of the printing parameters on the properties of the built parts via MEAM [40,73,138]. It was observed that increasing the temperature of extrusion, bed, and/or envelope improves the adhesion between adjacent strands and as such the mechanical properties. This is due to the increased mobility of the polymeric chains, leading to greater inter-diffusion between the strands [40,138]. In general, decreasing the layer thickness improves the mechanical properties of a built part, but it was found that a certain threshold is needed to avoid over-compression of the deposited strand, which negatively affects the mechanical properties. This threshold is dependent on the material that is being deposited [138]. Finally, the extrusion or volumetric flow rate should be selected carefully depending on the material. For example, Allahverdi et al. [139] conducted the MEAM of advanced electroceramic components using a proprietary binder system with alumina or lead-zirconium-titanate (PZT). It was observed that increasing the volumetric flow rate increased the propensity of the filaments to buckle and therefore the extrusion process was stopped.

Other parameters that affect the quality of the final part are the improperly deposited strands (also referred to as roads) or the spacing between strands, which can lead to incomplete bonding between adjacent strands or between layers. In turn, all of these factors lead to a systematic variation in density and defect appearance in shaped parts [104]. For unfilled polymers or polymers filled with anisotropic fillers, the orientation of the deposited strands and the layering sequence can affect the mechanical properties of the shaped part. During the extrusion and deposition processes, anisotropy can be created due to shear gradients in the nozzle and due to the nozzle movement relative to the building platform. These shear gradients also align the anisotropic fillers in the direction of flow and deposition, and if severe enough, they can result in an orientation-dependent shrinkage and warping during debinding and sintering. Iyer et al. [104] performed systematic printing trials on ceramic-filled feedstocks with printing raster patterns of $0°$, $90°$, and $+45°/-45°$. They observed that raster patterns of $0°$ or $90°$ led to warping of the built parts, but warpage was avoided when a cross-hatched pattern $+45°/-45°$ was used.

5. Post-Shaping Operations

5.1. Surface Treatment

After building the part via MEAM, the deposited strands are very noticeable and the lines do not disappear after sintering, as it can be seen in Figure 5a,b for 17-4PH steel parts. When dealing with highly-filled polymers, the strand dimensions cannot be as small as those achieved with unfilled polymers since using very small nozzles results in blockage. The rigid particles inside the feedstock cannot be deformed like the molten binder system and hence they block the nozzle. One possible solution for obtaining a smoother surface is to make thermo-mechanical treatments at the surface. Burkhardt et al. [125] investigated grinding, sandblasting, and laser structuring of shaped parts with 316L steel particles. Grinding and sandblasting for 25 s reduced the depth of the channels at the inter-strand space, but they were not sufficient to fully eliminate the strand marks. Sandblasting also had a negative effect in that it increased any defects in the part related to bad contact between strands (i.e., the spaces between strands were augmented). Finally, the surface treatment with a low power laser for 25 s yielded the best results; the strands were no longer visible and small defects could be covered using this technique (Figure 5c) [125]. The reason for this is that the laser melts the binder system and redistributes it on the surface of the part, carrying with it the filler particles. Treating the surface of green parts can be beneficial when dealing with hard metals or ceramics, which are very difficult to be polished mechanically or will require high energy lasers to alter the surface of the sintered parts.

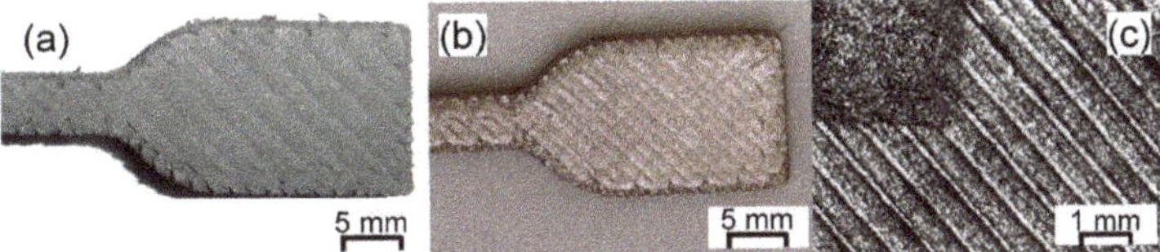

Figure 5. (**a**) 17-4PH steel printed part, (**b**) sintered part without surface treatment, and (**c**) 316L steel sintered part with laser surface treatment in the upper corner and without surface treatment [73,125]. © Carlo Burkhardt (OBE GmbH & Co. KG). Figure 5c first published by EPMA in the World PM2016 Proceedings [125].

Another possibility for smoothing the surface of shaped parts could be the use of solvent vapor. Kuo and Mao [140] developed an acetone-vapor system to smooth acrylonitrile butadiene styrene (ABS) parts fabricated by MEAM. A similar technique was also reported by Garg et al. [141] and by Takagishi and Umezu [142]. All groups observed that by partially dissolving the outer surface of the shaped parts the surface roughness was significantly reduced with minimal variations to the geometric accuracy. Similar processes have been described in the informal literature available on the internet, but no literature was found regarding parts fabricated with highly-filled polymers.

5.2. Debinding

The polymeric binder system must be removed without disrupting the MEAM-shaped part; this process is commonly referred to as debinding and it is also applicable for parts produced by PIM. Polymers have to be removed completely from the green part since carbon residues can influence the sintering process and affect the quality of the final product in a negative way. Moreover, binder removal is one of the most critical steps in the SDS process [62], since defects can be produced by inadequate debinding, examples of which include bloating, blistering, surface cracking, and large internal voids. There are three main debinding techniques are thermal, solvent, and catalytic methods [34].

The most common way to remove the binder from the shaped/printed parts is to heat the binder so it melts and flows out of the part and/or the binder is thermally degraded and diffuses out of the shaped part. The temperatures used to carry out thermal debinding depend on the formulation of the binders used. For example, Onagoruwa et al. [120] had a binder system consisting of PP/elastomer/tackifier/wax and they heated up their shaped parts in a setter bed of alumina powder. At temperatures lower than 200 °C, some of the binder components melted and they were removed via capillary action, and at temperatures higher than 200 °C, evaporation and thermal decomposition of the residual binders occurred. A similar strategy was employed by Wu et al. [59], but the parts were buried in a carbon powder with a very high surface area to aid the debinding through capillary suction at low temperatures, referred to as wicking. Iyer et al. [104] used a two-stage thermal debinding cycle to remove their binder from the Si_3N_4 particles. The first cycle was carried out in a nitrogen atmosphere, where the binder was wicked out of the material using a setter bed. The second cycle was performed in air, where the residual binder was burned off. The composition of the binder system can be modified to improve the thermal debinding rates, which tend to be very slow to prevent damage to the shaped part. This was shown in feedstocks containing 55 vol % piezoelectric powder, in which the proportions of base binder, tackifier, wax, and plasticizer were adjusted not only to improve the processability by MEAM, but also to exhibit good burnout properties to reduce the time required to thermally decompose the binder without any blistering or cracking in the finished samples [45].

Another way to remove the binder system is to use first a solvent extraction step in a special solvent debinding unit, followed by a thermal debinding step in the same furnace where sintering will take place. This strategy has been investigated in feedstocks consisting of a cyclohexane-soluble thermoplastic elastomer (TPE) and a polyolefin as an insoluble component [112,125]. The solvent extraction rate is a dissolution and diffusion process. Therefore, it is dependent on the temperature, the time, and on the particle characteristics such as the shape and size distribution of the particles. This is illustrated in Figure 6a, where the solvent extraction of feedstocks with 55 vol % 316L steel was performed at different temperatures (75, 60, and 40 °C). After 6 h at 75 °C, 90% of the soluble binder component was removed while only 85% was removed at 40 °C; this is because the dissolution and diffusion of substances increase as the temperature increases. Figure 6a also compares the debinding rate of 55 vol % strontium ferrite-filled feedstocks at 60 °C with that of steel with a similar powder content and temperature. The big difference between the debinding rates of strontium ferrite and steel feedstocks can be attributed to the different particle characteristics. In general, smaller irregular particles (e.g., $SrFe_{12}O_{19}$) have more surface area to which the binder system can adhere and smaller pores between the particles; these two factors can slow down the diffusion process first of the solvent and then of the dissolved polymer [112].

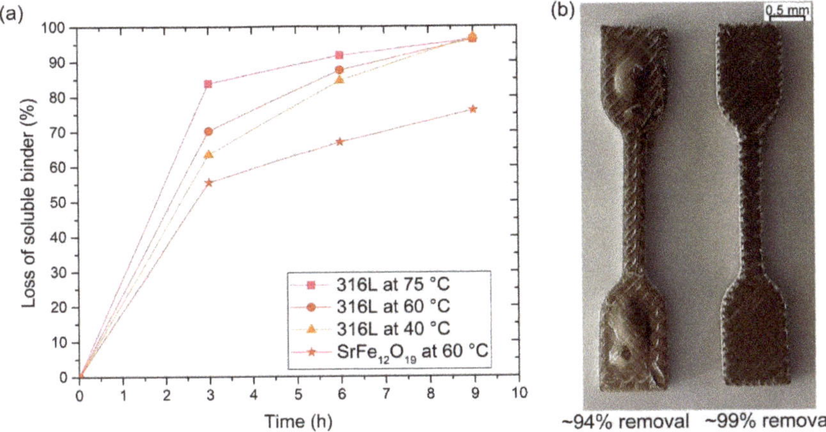

Figure 6. (**a**) Solvent extraction rate for MEAM feedstocks at different temperatures (75, 60, and 40 °C) and different fillers (316L steel and strontium ferrite); (**b**) 17-4PH sintered parts with different levels of solvent extraction [62].

When doing solvent debinding, it is important to remove a certain level of the soluble binder component. Gonzalez-Gutierrez et al. [62] showed what happens when the soluble component is not properly removed in Figure 6b. This figure shows two sintered parts, one of them had approximately 94% of the soluble binder removed (left), while the other one had approximately 99% (right). The part with the lower of soluble binder removal had bubbles at the center (left). Bloating of specimens due to incomplete debinding is a common defect in PIM [143]. The cause of this defect is the vapor formed during the degradation of the polymers. In the areas rich in binder, this vapor cannot be easily evacuated and when the partial pressure of the trapped vapor is higher than the atmospheric pressure the bubbles appear [143]. Therefore, the creation of a porous structure by the removal of the soluble component is critical for the following thermal debinding step [62].

The Austrian company EVO-tech GmbH in collaboration with BASF SE offers filaments for MEAM that can be catalytically debound [126]. Such technology is in the process of being patented by BASF SE [116]. The filaments are composed of a polyoxymethylene binder with an external proprietary coating to retain their flexibility. The coating can be observed in the scanning electron microscopy (SEM) image shown in Figure 7. The catalytic debinding process focuses on a solid-to-vapor catalytic degradation of the main binder component. Such catalytic degradation occurs for example when polyoxymethylene-based feedstocks are exposed to acid vapors, such as nitric acid. This results in a much faster binder removal when compared to thermal or solvent debinding [34].

It is worth noting that in solvent and catalytic debinding methods, a skeleton of insoluble or non-degraded polymer remains to impart adequate strength and shape retention up to the onset of sintering. This remaining backbone is thermally removed usually between 200 °C and 600 °C in a pre-sintering step, depending on its chemical composition [34].

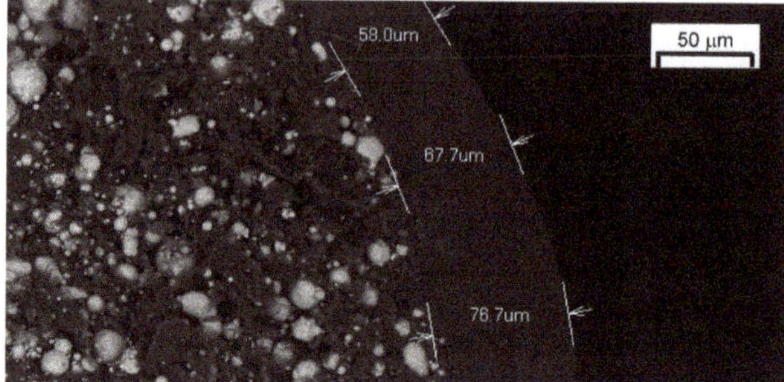

Figure 7. SEM cross-sectional view of POM-based filament produced by BASF SE where a coating layer is observed (courtesy of RHP-Technology GmbH, Seibersdorf, Austria).

5.3. Sintering

The last step of the process to obtain dense metal, ceramic, or metal/ceramic parts is sintering. Sintering is a thermal treatment that transforms metallic or ceramic powders into bulk materials. Sintering is performed at temperatures below the melting temperature of the major constituent in the powder, generally within 70 to 90% of the powder's melting point [34].

Rearrangement, particle movement, and mass transport happen during sintering. Before the sintering occurs, the parts have a highly porous structure formed by particles with a large free surface and therefore a high surface energy. As heat is applied and the temperature increased, the system tends to reduce the surface energy by the formation of solid bonds between the particles. When temperatures increase beyond one half to two-thirds of the melting temperature of the powder material, atomic diffusion and chemical changes on the surface of the particles occur and lead to the formation of solid bonds. These bonds, known as necks [144], continue growing, resulting in the reduction of the porosity and the densification of the part. At the final stage of sintering, the pores are isolated and the density increases up to 99% of the theoretical value [144]. Nevertheless, the increase of the grain growth also occurs in this stage, which hinders the process and the densification rate. At least six different mechanisms for mass transfer involved in sintering have been identified. These mechanisms include: surface diffusion, lattice diffusion, grain boundary diffusion, evaporation-condensation, viscous flow, and plastic flow [144]. All of these mechanisms contribute to the growth of necks between particles and their bonding, leading to an increase in the strength of the consolidated powders. Additionally, some of these mechanisms also lead to the reduction of the porosity and therefore the shrinkage and densification. Surface diffusion is the mechanism that produces surface smoothing, particle joining, and pore rounding, but not densification. If the material has a high vapor pressure, sublimation and vapor transport produce the same effects as surface diffusion. Diffusion along the grain boundaries and through the lattice produces both neck growth and densification. Bulk viscous flow plays an important role in densification, when a wetting liquid is present, while plastic deformation is important when mechanical pressure is applied [34].

Shrinkage of the shaped parts by MEAM is always observed after sintering due to the reduction of the porosity and the densification of the parts Examples of printed and sintered parts are shown in Figure 8. Shrinkage and density values obtained in the literature are shown in Table 9.

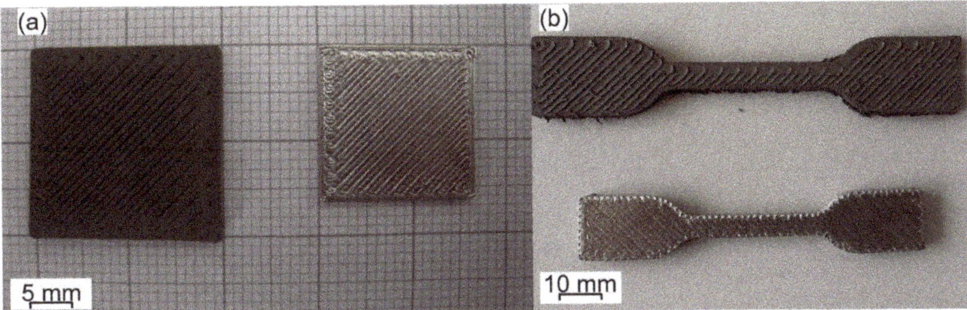

Figure 8. Comparison of printed and sintered parts: (**a**) 316L [125] and (**b**) 17-4PH steels [62].

It is important to mention that the shrinkage does not occur in the same amount in all dimensions, as observed by Lous et al. [145], Agarwala et al. [41], and Kukla et al. [113]. Anisotropic shrinkage has also been reported for parts produced by PIM [146]. The anisotropic shrinkage in PIM is the result of polymer orientation, which can be influenced by the injection molding parameters [147]. Besides polymer orientation, in MEAM-HP, shrinkage and density can be influenced by the presence of gaps between deposited strands. The more gaps, the larger the shrinkage and the lower the density of the sintered parts, since larger gaps cannot be closed during sintering [41]. The shrinkage can also be affected by the orientation of the filler particles, as discussed by Kukla et al. [113]; they observed a smaller shrinkage when the parts were printed on a polypropylene plate, while a larger shrinkage was produced when the parts were printed on top of a permanent magnet, which aligned the anisotropic NdFeB particles in the direction of the magnetic field. Since anisotropic shrinkage is inevitable, it should be included in the CAD design of the parts for the MEAM of ceramics and metals, and the printing strategy needs to be considered and optimized to prevent excessive variation in the shrinkage.

Table 9. Linear shrinkage and density after the sintering of parts produced by MEAM-HP.

Material	Linear Shrinkage (%)	Percentage Density from Theoretical (%)	Ref.
Fused silica	1–4	70	[60]
Mullite	10–12	N/A	[120]
Fused silica	8–12	N/A	[120]
316L stainless steel	19.2 ± 0.02	95	[125]
Piezoelectric ceramics	16–20	N/A	[145]
Silicon nitride	12–20	95–98	[41]
NdFeB	16–19	94–96	[113]

5.4. Co-Sintering

One advantage of using MEAM as a shaping technology is that multi-material parts can be easily fabricated compared to powder bed technologies or vat photopolymerization. The multi-material process merely requires a printing head that can work with two or more materials simultaneously. However, the main limitation for applying MEAM-HP for the fabrication of multi-material components is the co-sintering step. For successful co-sintering, the powders have to be sintered together at the same temperature and in the same atmosphere. In order to avoid excessive mechanical stress under cooling, the coefficient of thermal expansion and the shrinkage behavior of both materials should be similar [148].

One of the first demonstrations of the possibility of producing multi-material components was described by Jafari et al. [149]. Filaments with two different piezoelectric ceramics were prepared and parts with alternating layers of these two types of ceramics were produced by MEAM-HP. It was observed that the microstructure and the dielectric properties of the co-sintered parts had no sign of delamination; therefore, it was concluded that MEAM-HP could be used for the fabrication of multi-material transducers with improved performance.

More recently, the possibility of producing parts made of 17-4PH steel and zirconia via MEAM-HP has been demonstrated [114,150]. These two powder materials in their commercial state have very different sintering activity since their particle sizes are orders of magnitude different; zirconia has an average particle size around 0.6 μm, while this steel has an average size around 20 μm. In order to increase the sintering activity of the steel powder, it was re-shaped by attrition milling and ball milling. The milling produced particles with a higher specific surface area. Also, after the milling, the initially spherical steel particles became irregular and angularly shaped; thus, the packing of the steel powder was changed and the overall sintering behavior became comparable between zirconia and steel [150]. An example of a part shaped by MEAM with zirconia and steel is shown in Figure 9a,b; a co-sintered part with 17-4PH and zirconia is shown in Figure 9c.

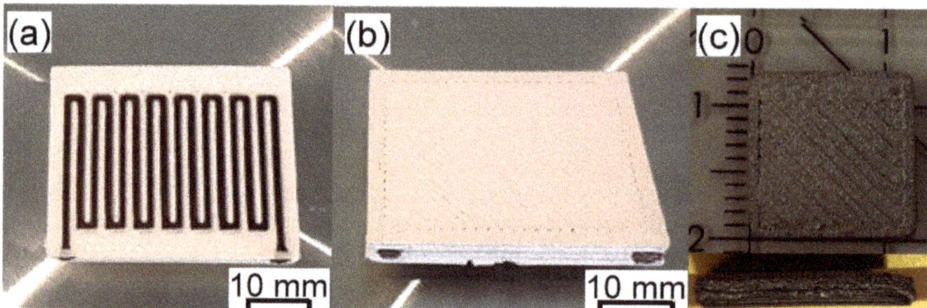

Figure 9. Printed part with a base of zirconia and an insert of 17-4PH steel: (**a**) middle of the printed job, and (**b**) printed job completed. (**c**) Co-sintered part of 17-4PH steel (top) and zirconia (bottom). Please notice that during sintering the zirconia lost its white color due to the reducing atmosphere [114].

5.5. Post-Sintering Operations

Post-sintering operations of the MEAM-printed parts such as the improvement of surface texture, accuracy, aesthetics, and a matte surface finish are often desired depending on the application of the printed part. A simple bead blasting of the surface can help even the surface texture, remove sharp corners from stair-stepping, and give an overall matte appearance. If a smooth or polished finish is desired, then wet or dry sanding and hand-polishing are performed; this is particularly important if no surface treatment was performed before debinding and sintering. In many cases, it is desirable to paint the surface (e.g., with cyanoacrylate or a sealant) prior to sanding or polishing. Painting the surface has the dual benefit of sealing porosity and, by viscous forces, smoothing the stair-step effect, thus making sanding and polishing easier and more effective [2].

Depending on the material used, AM parts can be effectively colored by simply dipping the part into a dye of the appropriate color. If painting is required, the part may need to be sealed prior to painting. Common automotive paints are quite effective in these instances. Another aesthetic enhancement (which also strengthens the part and improves wear resistance) is chrome plating. Ni, Cu, and other coatings can also be applied to the surface for aesthetic improvements. This is particularly important for materials that are sensitive to oxidation and for which the oxidation process causes a significant decay in the wanted properties, for example, the magnetic performance of NdFeB parts [113].

Another post-sintering operation is the infusion process for the production of multi-material parts. The production of metal-ceramic composites was investigated by Onagoruwa et al. [120] and Bandyopadhyay et al. [119]. Onagoruwa et al. [120] developed a PP-based binder which was used for the production of mullite and fused silica preforms with high porosity. The use of MEAM enabled the production of porous ceramic parts in which the pore geometries and connectivity could be controlled. Aluminum was infiltrated into the preforms, obtaining uniform as well as gradient composites. The process was further studied by the authors for the production of aluminum-alumina composites [119]. Fused silica preforms were fabricated by MEAM and sintered. Aluminum was infiltrated into the preforms at ambient atmosphere and the reaction of both components during the infiltration produced ceramic aluminum oxide (with higher stability compared to the fused silica). The displaced silicon moved to the melted metal, and after cooling metal-ceramic composite parts were obtained [119]. Another infiltration process was performed by Bandyopadhyay et al. [151] for the production of lead-zirconium-titanate (PZT)-epoxy composites by two routes. In the direct route, filaments containing 52 vol % of lead-zirconium-titanate (PZT) powder and 48 vol % of proprietary binder were used to produce via MEAM simple shapes with intricate internal structures, such as ladder structures. After debinding and sintering the MEAM-shaped parts, they were infiltrated with an acoustic epoxy resin to form piezocomposites. In the indirect route, molds having the negative structure were created by MEAM using a commercial polymer/wax material. Subsequently, a slurry containing 45 vol % of PZT was infiltrated in the molds and the binders and bonds were burnt out prior to sintering. The sintered parts were then also infiltrated with the acoustic epoxy [151]. Lous et al. [152] used polymer-infiltrated piezoelectric ceramic skeletons fabricated via MEAM-HP to develop transducers for medical imaging; they concluded that the such produced parts had a similar sensitivity to that of a commercial transducer, but the ringing was much longer due to the lack of optimization on the backing layers produced by MEAM-HP.

6. Comparison to Other Manufacturing Techniques

MEAM-HP is still in its early stages of development. Therefore, there is not a lot of literature comparing the properties of parts produced by this technique to parts produced by traditional manufacturing methods such as casting and PIM or to other AM techniques such as powder bed fusion, binder jetting, or vat photopolymerization. Just a few examples were found, and they are all presented in this section.

Bandyopadhyay et al. [119] tested the binder developed by Onagoruwa et al. [120] with piezoelectric lead-zirconium-titanate (PZT), comparing production by the direct and indirect methods [119,151]. As previously discussed in Section 5.4, parts were directly produced with filaments of feedstocks containing their binders. However, in the indirect method, molds were first produced using FFF common materials and ceramic slurry was then casted. The electromechanical properties of the thus-fabricated composites were superior to those of the composites solely prepared by casting, due to the control in the phase periodicity obtained in the AM process.

Agarwala et al. [60] prepared silica parts for investment casting with comparable properties to those silica parts produced by conventional core-making techniques by using MEAM. Traditional core-making techniques involve the machining or injection molding of positive patterns of the actual cores and shells in wax or other polymers. These positive molds are then sequentially dipped in a ceramic slurry. The traditional technique is labor-intensive and MEAM, with debinding and sintering, represents a viable alternative. In the same study, Agarwala et al. [60] also compared silica parts shaped by MEAM-HP to parts shaped by PIM. It was observed that the microstructure of sintered parts produced by MEAM-HP had no evidence of delamination or inter-strand debonding, thus the mechanical properties measured were within the acceptable limits of commercial silica parts.

Griffith and McMillin [135] prepared two types of feedstock materials for the production of alumina parts. They mixed a special binder system with the alumina powder and milled the mixture to prepare a powder that could be shaped by the selective laser sintering (SLS) of the binder, debound, and sintered to obtain a full alumina part. They also prepared feedstock materials as discontinuous filaments to shape parts using MEAM. The parts produced by MEAM were debound and sintered under the same conditions as the parts shaped by binder-SLS. After sintering, the density and porosity of the parts were measured. It was observed that the parts shaped by binder-SLS had a much lower density (53 to 65% of theoretical) and higher porosity (36 to 47%) than the parts shaped by MEAM (density 96% of theoretical). The low density obtained by the SLS process indicated that the same feedstock material should not be used for both technologies, but rather an optimized feedstock should be found for each of the technologies. Nevertheless, it was suggested that the parts shaped by binder-SLS could be used to produce ceramic cores for investment casting, while the parts shaped by MEAM could be used in structural applications [135].

7. Summary and Future Direction

Material extrusion additive manufacturing (MEAM) is a very popular way to shape three-dimensional objects with thermoplastic materials. However, when using highly-filled polymers (HP) combined with the post-shaping operations of debinding and sintering, metallic, ceramic, and metallic-ceramic parts can be produced.

MEAM can be performed in three main types of setups, which differ in the way the material is fed to the nozzle with either plungers, screws, or as a filament. The use of filaments is the most widespread. In order to carry out the shaping, debinding, and sintering successfully, appropriate feedstock materials have to be developed for each of the MEAM types. The feedstock materials are made by combining different ingredients in a binder system with powder of the desired material. The binder system has to provide enough flowability, when in the molten state, and enough mechanical strength for handling and processing the material before melting and after re-solidifying. The mechanical strength is particularly important when performing filament-based MEAM, since the filaments should not break or buckle during the feeding of the material into the nozzle. The binder system should also thermally decompose or dissolve without affecting the shape of the formed part during the debinding step, and the binder system should degrade and be fully eliminated before sintering of the particles takes place to prevent contamination of the produced part that could lead to poor properties. For this reason, it is clear that the development of such feedstocks is a complicated task and thus the full description of their recipe is not publicly available, since it represents a competitive advantage for their developers.

Once the feedstock material has been developed, the building job has to be optimized by varying the processes parameters and the building strategy (i.e., layer height, temperatures, direction of the deposition, and infill grade), since these factors and other factors will affect the properties and shape of the final part. The debinding of shaped parts by MEAM-HP can be performed in a similar manner as in powder injection molding (PIM), but the majority of results available in the literature are for thermal and solvent debinding. Finally, sintering is conducted the same way as in other powder technologies such as PIM. It is important to point out that there will be a significant anisotropic shrinkage similar to the one observed in powder injection molding; therefore, preliminary trials should be performed to optimize the final shape and performance of parts produced by MEAM-HP, debound, and sintered.

When comparing the results of MEAM-HP with other shaping technologies, it was observed that MEAM-HP produces objects with acceptable properties when the overall process is optimized. After optimization, dense parts of metal, ceramic, or metal-ceramic can be produced for structural applications. However, it is possible that even if the parts retain some porosity after this manufacturing technique, this could be an advantage for certain applications where a large surface area is beneficial; examples of this include filtration, medical applications, or catalytic applications [153,154]. In conclusion, the authors believe that MEAM-HP is a complementary technology to the many other manufacturing technologies available today for the production of metal, ceramic, and multi-material parts, so it is worthwhile to keep improving the materials and processes to make it commercially viable and useful.

Some of the work that could be done in the future to improve MEAM-HP includes:

i.   The development of improved simulation tools that can predict the properties of built parts by knowing the material properties, processing parameters, and building strategies. This could allow the optimization of the printing process before it is actually done.

ii.   The development of simulations tools for the debinding and sintering processes.

iii.   The development of automatic monitoring systems that supervise the building process that stop and adjust the processing parameters to reduce the amount of unwanted defects on the printed part.

iv.   The development of mechanisms to smooth the surface of the deposited strands as they are being deposited to reduce the surface roughness of the printed parts and to increase the mechanical properties of printed parts.

v.   The development of new binder systems for filaments and pellets, which are faster to remove and can use other solvents, such as water.

vi.   The development of new feedstock materials with different filler particles that can be co-sintered for the fabrication of new multi-material components with new functionalities.

vii.   The development of simulation tools that could speed up the development of new highly-filled polymers to be used in MEAM. Such simulation tools should take into account the compatibility and interaction of the different polymeric components, as well as the chemical and physical characteristics of powders.

viii.   The improvement of the reliability of screw-based MEAM machines to replace the filaments by pellets, since filaments are hard to make and limit the amount of powder that can be added to the feedstock.

ix.   The improvement of the properties of the interface of multi-material components so as to promote good adhesion between the different materials and obtain multi-material components with a long service life.

As it can be seen, there are areas in research and development in the field of MEAM-HP that should be improved in order to achieve a reliable process that can be used for the fabrication of unique products with multifunctional properties.

Author Contributions: All authors participated in the scientific discussion before the preparation of the manuscript. J.G.-G. and S.C. compiled and summarized the literature sources used in this review paper. S.C. and S.S. drew the schematic Figures 1 and 2. J.S. contributed to the conceptualization of the review paper and its realization with a structured outline. J.G. and J.S. wrote the first complete version of the paper. C.H., C.K., S.C., and S.S. provided comments for improving the manuscript. J.G.-G., S.C., and J.S. wrote the final version of the manuscript and checked the correctness of the references.

Acknowledgments: This work was performed as part of the European projects REProMag and CerAMfacturing and the Austria-China bilateral cooperation project FlexiFactory3Dp. Both European projects have received funding from the European Union's Horizon 2020 research and innovation program under Grant Agreement No. 636881 and No. 678503. FlexiFactory3Dp has received funding from the Austrian Research Promotion Agency under the program Production of the Future, Grant Agreement No. 860385.

Conflicts of Interest: The authors declare no conflict of interest.

References

1.   Singh, S.; Ramakrishna, S.; Singh, R. Material issues in additive manufacturing: A review. *J. Manuf. Process.* **2017**, *25*, 185–200. [CrossRef]

2.   Gibson, I.; Rosen, D.; Stucker, B. *Additive Manufacturing Technologies. 3D Printing, Rapid Prototyping, and Direct Digital Manufacturing*, 2nd ed.; Springer: New York, NY, USA, 2015.

3.   Bourell, D.L. Perspectives on Additive Manufacturing. *Annu. Rev. Mater. Res.* **2016**, *46*, 1–18. [CrossRef]

4.   F42 Committee. *Specification for Additive Manufacturing File Format (AMF) Version 1.2*; ASTM International: West Conshohocken, PA, USA, 2016.

5.   Chakravorty, D. STL File Format for 3D Printing—Simply Explained: Are There Any Alternatives to the STL File Format? Available online: https://all3dp.com/what-is-stl-file-format-extension-3d-printing/#pointnine (accessed on 8 May 2018).

6.   3MF Consortium. Enabling the Full Potential of 3D Printing. Available online: https://3mf.io/ (accessed on 8 May 2018).

7. Hegab, H.A. Design for additive manufacturing of composite materials and potential alloys: A review. *Manuf. Rev.* **2016**, *3*, 11. [CrossRef]

8. Giannatsis, J.; Dedoussis, V. Additive fabrication technologies applied to medicine and health care: A review. *Int. J. Adv. Manuf. Technol.* **2009**, *40*, 116–127. [CrossRef]

9. Goffard, R.; Sforza, T.; Clarinval, A.; Dormal, T.; Boilet, L.; Hocquet, S.; Cambier, F. Additive manufacturing of biocompatible ceramics. *Adv. Prod. Eng. Manag.* **2013**, *8*, 96–106. [CrossRef]

10. Murr, L.E.; Gaytan, S.M.; Medina, F.; Lopez, H.; Martinez, E.; Machado, B.I.; Hernandez, D.H.; Martinez, L.; Lopez, M.I.; Wicker, R.B.; et al. Next-generation biomedical implants using additive manufacturing of complex, cellular and functional mesh arrays. *Philos. Trans. Ser. A Math. Phys. Eng. Sci.* **2010**, *368*, 1999–2032. [CrossRef] [PubMed]

11. EN ISO/ASTM. *Additive Manufacturing—General Principles—Terminology;* (52900:2017-02); International Organization for Standardization (ISO): Geneve, Switzerland, 2017.

12. Gu, D.D.; Meiners, W.; Wissenbach, K.; Poprawe, R. Laser additive manufacturing of metallic components: Materials, processes and mechanisms. *Int. Mater. Rev.* **2013**, *57*, 133–164. [CrossRef]

13. Kruth, J.P.; Froyen, L.; van Vaerenbergh, J.; Mercelis, P.; Rombouts, M.; Lauwers, B. Selective laser melting of iron-based powder. *J. Mater. Process. Technol.* **2004**, *149*, 616–622. [CrossRef]

14. Santos, E.C.; Shiomi, M.; Osakada, K.; Laoui, T. Rapid manufacturing of metal components by laser forming. *Int. J. Mach. Tools Manuf.* **2006**, *46*, 1459–1468. [CrossRef]

15. Murr, L.E.; Gaytan, S.M.; Ramirez, D.A.; Martinez, E.; Hernandez, J.; Amato, K.N.; Shindo, P.W.; Medina, F.R.; Wicker, R.B. Metal Fabrication by Additive Manufacturing Using Laser and Electron Beam Melting Technologies. *J. Mater. Sci. Technol.* **2012**, *28*, 1–14. [CrossRef]

16. Melchels, F.P.W.; Feijen, J.; Grijpma, D.W. A review on stereolithography and its applications in biomedical engineering. *Biomaterials* **2010**, *31*, 6121–6130. [CrossRef] [PubMed]

17. Majewski, C.E.; Oduye, D.; Thomas, H.R.; Hopkinson, N. Effect of infra-red power level on the sintering behaviour in the high speed sintering process. *Rapid Prototyp. J.* **2008**, *14*, 155–160. [CrossRef]

18. Ellis, A.; Noble, C.J.; Hopkinson, N. High Speed Sintering: Assessing the influence of print density on microstructure and mechanical properties of nylon parts. *Addit. Manuf.* **2014**, *1–4*, 48–51. [CrossRef]

19. Kumar, S.; Kruth, J.-P. Composites by rapid prototyping technology. *Mater. Des.* **2010**, *31*, 850–856. [CrossRef]

20. Zein, I.; Hutmacher, D.W.; Tan, K.C.; Teoh, S.H. Fused deposition modeling of novel scaffold architectures for tissue engineering applications. *Biomaterials* **2002**, *23*, 1169–1185. [CrossRef]

21. Malinauskas, M.; Rekstyt, S.; Lukosevicius, L.; Butkus, S.; Balciunas, E.; Peciukaityt, M.; Baltriukien, D.; Bukelskien, V.; Butkevicius, A.; Kucevicius, P.; et al. 3D Microporous Scaffolds Manufactured via Combination of Fused Filament Fabrication and Direct Laser Writing Ablation. *Micromachines* **2014**, *5*, 839–858. [CrossRef]

22. Turner, B.N.; Gold, S.A. A review of melt extrusion additive manufacturing processes: II. Materials, dimensional accuracy, and surface roughness. *Rapid Prototyp. J.* **2015**, *21*, 250–261. [CrossRef]

23. Turner, B.N.; Strong, R.; Gold, S.A. A review of melt extrusion additive manufacturing processes: I. Process design and modeling. *Rapid Prototyp. J.* **2014**, *20*, 192–204. [CrossRef]

24. Caffrey, T.; Wohlers, T.; Campbell, I. (Eds.) *Wohlers Report 2016. 3D Printing and Additive Manufacturing State of the Industry: Annual Worldwide Progress Report;* Annual Worldwide Progress Report; Wohlers Associates: Fort Collins, CO, USA, 2016.

25. Guo, N.; Leu, M.C. Additive manufacturing: Technology, applications and research needs. *Front. Mech. Eng.* **2013**, *8*, 215–243. [CrossRef]

26. Gao, W.; Zhang, Y.; Ramanujan, D.; Ramani, K.; Chen, Y.; Williams, C.B.; Wang, C.C.; Shin, Y.C.; Zhang, S.; Zavattieri, P.D. The status, challenges, and future of additive manufacturing in engineering. *Comput.-Aided Design* **2015**, *69*, 65–89. [CrossRef]

27. Mireles, J.; Espalin, D.; Roberson, D.; Zinniel, B.; Medina, F.; Wicker, R. Fused Deposition Modeling of Metals. In Proceedings of the Solid Freeform Fabrication Symposium, Austin, TX, USA, 6–8 August 2012; pp. 836–845.

28. Valkenaers, H.; Vogeler, F.; Ferraris, E.; Voet, A.; Kruth, J.P. A Novel Approach to Additive Manufacturing: Screw Extrusion 3D-Printing. In Proceedings of the 10th International Conference on Multi-Material Micro Manufacture, San Sebastian, Spain, 8–10 October 2013; Azcarate, S., Dimov, S., Eds.; Research Publishing: San Sebastian, Spain, 2013; pp. 235–238.

29. Wohlers, T. Desktop Metal: A Rising Star of Metal AM Targets Speed, Cost and High-Volume Production. Metal AM [Online]. 2017, pp. 89–92. Available online: http://www.metal-am.com/wp-content/uploads/sites/4/2017/06/MAGAZINE-Metal-AM-Summer-2017-PDF-sp.pdf (accessed on 11 July 2017).

30. Campbell, I.; Wohlers, T. Markforged: Taking a Different Approach to Metal Additive Manufacturing. Metal AM [Online]. 2017, pp. 113–115. Available online: http://www.metal-am.com/wp-content/uploads/sites/4/2017/06/MAGAZINE-Metal-AM-Summer-2017-PDF-sp.pdf (accessed on 11 July 2017).

31. Schuh, C.A.; Myerberg, J.S.; Fulop, R.; Chiang, Y.-M.; Hart, A.J.; Schroers, J.; Vereminski, M.D.; Mykulowycz, N.; Shim, J.J.; Fontana, R.R.; et al. Methods and Systems for Additive Manufacturing. International Patent PCT/US2016/067378, 16 December 2016.

32. Martínez-Vázquez, F.J.; Pajares, A.; Miranda, P. A simple graphite-based support material for robocasting of ceramic parts. *J. Eur. Ceram. Soc.* **2018**, *38*, 2247–2250. [CrossRef]

33. Cesarano, J. A Review of Robocasting Technology. *MRS Proc.* **1998**, *542*, 125. [CrossRef]

34. Gonzalez-Gutierrez, J.; Stringari, G.; Emri, I. Powder Injection Molding of Metal and Ceramic Parts. In *Some Critical Issues for Injection Molding*; Wang, J., Ed.; InTech: Rijeka, Croatia, 2012; pp. 65–86.

35. Markforged Inc. Complete Metal Solution. Available online: https://markforged.com/metal-x/ (accessed on 7 June 2017).

36. Desktop Metal Inc. Prototype and Mass Produce with the Same Alloys. Available online: https://www.desktopmetal.com/products/materials/ (accessed on 29 March 2018).

37. Crump, S. Apparatus and Method for Creating Three-Dimensional Objects. US Patent 07/429,012, 30 October 1989.

38. Arburg GmbH & Co. KG. Freeformer System: No Mold Required for a Fully Functional Part. Available online: https://www.arburg.com/us/us/products-and-services/additive-manufacturing/freeformer-system/ (accessed on 19 July 2017).

39. Sells, E.; Bailard, S.; Smith, Z.; Bowyer, A.; Olliver, V. RepRap: The Replicating Rapid Prototyper: Maximizing Customizability by Breeding the Means of Production. *Handb. Res. Mass Cust. Personal.* **2009**, *2*, 568–580. [CrossRef]

40. Spoerk, M.; Gonzalez-Gutierrez, J.; Sapkota, J.; Schuschnigg, S.; Holzer, C. Effect of the printing bed temperature on the adhesion of parts produced by fused filament fabrication. *Plast. Rubber Compos.* **2018**, *47*, 17–24. [CrossRef]

41. Agarwala, M.K.; Jamalabad, V.R.; Langrana, N.A.; Safari, A.; Whalen, P.J.; Danforth, S.C. Structural quality of parts processed by fused deposition. *Rapid Prototyp. J.* **1996**, *2*, 4–19. [CrossRef]

42. Spoerk, M.; Gonzalez-Gutierrez, J.; Lichal, C.; Cajner, H.; Berger, G.; Schuschnigg, S.; Cardon, L.; Holzer, C. Optimisation of the Adhesion of Polypropylene-Based Materials during Extrusion-Based Additive Manufacturing. *Polymers* **2018**, *10*, 490. [CrossRef]

43. Elkins, K.; Nordby, H.; Janak, C.; Gray, R.W.; Bøhn, H.H.; Baird, D.G. Soft Elastomers for Fused Deposition Modeling. In Proceedings of the 8th Solid Freeform Fabrication Symposium, Austin, TX, USA, 11–13 August 1997; pp. 441–448.

44. Wang, X.; Jiang, M.; Zhou, Z.; Gou, J.; Hui, D. 3D printing of polymer matrix composites: A review and prospective. *Compos. Part B Eng.* **2016**. [CrossRef]

45. McNulty, T.F.; Mohammadi, F.; Bandyopadhyay, A.; Shanefield, D.J.; Danforth, S.C.; Safari, A. Development of a binder formulation for fused deposition of ceramics. *Rapid Prototyp. J.* **1998**, *4*, 144–150. [CrossRef]

46. Bellini, A.; Shor, L.; Guceri, S.I. New developments in fused deposition modeling of ceramics. *Rapid Prototyp. J.* **2005**, *11*, 214–220. [CrossRef]

47. Bellini, A. Fused Deposition of Ceramics: A Comprehensive Experimental, Analytical and Computational Study of Material Behavior, Fabrication Process and Equipment Design. Ph.D. Thesis, Drexen University, Philadelpia, PA, USA, 2002.

48. Cruz, N.; Santos, L.; Vasco, J.; Barreiros, F.M. Binder System for Fused Deposition of Metals. In Proceedings of the Euro PM2013, Congress & Exhibition, Gothenburg, Sweden, 15–18 September 2013; EPMA: Athens, Greece, 2013; pp. 79–84.

49. Drescher, P.; Lieberwirth, C.; Seitz, H. Process and Installation for Manufacturing the Additive of Amorphous Crystalline and/or Semi-Crystalline Metal Components—Selective Amorphous Metal Extrusion (SAME). DE Patent 201410018080, 6 December 2014.

50. AIM3D GmbH. Edelstahl. Available online: http://www.aim3d.de/materialien/edelstaehle/ (accessed on 7 July 2017).

51. Pollen AM Inc. Meet PAM: Pellet Additive Manufacturing. Available online: https://www.pollen.am (accessed on 22 August 2017).

52. Koslow, T. Pollen Introduces Pam: Their New Professional-Grade Multi-Material 3D Printer. Available online: https://3dprint.com/140595/pollen-pam-multi-material/ (accessed on 22 August 2017).

53. Li, L.; Tirado, A.; Nlebedim, I.C.; Rios, O.; Post, B.; Kunc, V.; Lowden, R.R.; Lara-Curzio, E.; Fredette, R.; Ormerod, J.; et al. Big Area Additive Manufacturing of High Performance Bonded NdFeB Magnets. *Sci. Rep.* **2016**, *6*, 36212. [CrossRef] [PubMed]

54. Cincinnati Inc. BAAM: Fact Sheet. Available online: http://wwwassets.e-ci.com/PDF/Products/baam-fact-sheet.pdf (accessed on 7 June 2017).

55. Gonzalez-Gutierrez, J.; Duretek, I.; Kukla, C.; Poljšak, A.; Bek, M.; Emri, I.; Holzer, C. Models to Predict the Viscosity of Metal Injection Molding Feedstock Materials as Function of Their Formulation. *Metals* **2016**, *6*, 129. [CrossRef]

56. German, R.M. (Ed.) *Powder Metallurgy Science*, 2nd ed.; Metal Powder Industries Federation: Princeton, NJ, USA, 1994.

57. German, R.M. Metal powder injection molding (MIM): Key trends and markets. In *Handbook of Metal Injection Molding*; Heaney, D.F., Ed.; Woodhead Publishing: Cambridge, UK; Philadelphia, PA, USA, 2012; pp. 1–25.

58. Wu, G.; Langrana, N.A.; Rangarajan, S.; McCuiston, R.; Sadanji, R.; Danforth, S.C.; Safari, A. Fabrication of Metal Components using FDMet: Fused Deposition of Metals. In Proceedings of the Solid Freeform Fabrication Symposium, Austin, TX, USA, 9–11 August 1999; pp. 775–782.

59. Wu, G.; Langrana, N.A.; Sadanji, R.; Danforth, S.C. Solid freeform fabrication of metal components using fused deposition of metals. *Mater. Des.* **2002**, *23*, 97–105. [CrossRef]

60. Agarwala, M.K.; van Weeren, R.; Bandyopadhyay, A.; Whalen, P.J.; Safari, A.; Danforth, S.C. Fused Deposition of Ceramics and Metals: An Overview. In Proceedings of the Solid Freeform Fabrication Symposium, Austin, TX, USA, 12–14 August 1996.

61. Agarwala, M.K.; van Weeren, R.; Bandyopadhyay, A.; Safari, A.; Danforth, S.C.; Priedeman, W.R. Filament Feed Materials for Fused Deposition Processing of Ceramics and Metals. In Proceedings of the Solid Freeform Fabrication Symposium, Austin, TX, USA, 12–14 August 1996.

62. Gonzalez-Gutierrez, J.; Godec, D.; Kukla, C.; Schlauf, T.; Burkhardt, C.; Holzer, C. Shaping, Debinding and Sintering of Steel Components via Fused Filament Fabrication. In *CIM 2017 Computer Integrated Manufacturing and High Speed Machining, Proceedings of the 16th International Scientific Conference on Production Engineering, Zadar, Croatia, 8–10 June 2017*; Abele, E., Udijak, T., Ciglar, D., Eds.; Croatian Association of Production Engineering: Zagreb, Croatia, 2017; pp. 99–104.

63. Shahzad, K.; Deckers, J.; Kruth, J.-P.; Vleugels, J. Additive manufacturing of alumina parts by indirect selective laser sintering and post processing. *J. Mater. Process. Technol.* **2013**, *213*, 1484–1494. [CrossRef]

64. Shahzad, K.; Deckers, J.; Zhang, Z.; Kruth, J.-P.; Vleugels, J. Additive manufacturing of zirconia parts by indirect selective laser sintering. *J. Eur. Ceram. Soc.* **2014**, *34*, 81–89. [CrossRef]

65. Bai, Y.; Williams, C.B. An exploration of binder jetting of copper. *Rapid Prototyp. J.* **2015**, *21*, 177–185. [CrossRef]

66. Nandwana, P.; Elliott, A.M.; Siddel, D.; Merriman, A.; Peter, W.H.; Babu, S.S. Powder bed binder jet 3D printing of Inconel 718: Densification, microstructural evolution and challenges☆ *Curr. Opin. Solid State Mater. Sci.* **2017**, *21*, 207–218. [CrossRef]

67. Gonzalez, J.A.; Mireles, J.; Lin, Y.; Wicker, R.B. Characterization of ceramic components fabricated using binder jetting additive manufacturing technology. *Ceram. Int.* **2016**, *42*, 10559–10564. [CrossRef]

68. Homa, J.; Schwentenwein, M. A Novel Additive Manufacturing Technology for High-Performance Ceramics. In *Advanced Processing and Manufacturing Technologies for Nanostructured and Multifunctional Materials*; John Wiley & Sons, Inc: New York, NY, USA, 2014; pp. 33–40.

69. Schwentenwein, M.; Homa, J. Additive Manufacturing of Dense Alumina Ceramics. *Int. J. Appl. Ceram. Technol.* **2015**, *12*, 1–7. [CrossRef]

70. Scheithauer, U.; Schwarzer, E.; Richter, H.-J.; Moritz, T. Thermoplastic 3D Printing-An Additive Manufacturing Method for Producing Dense Ceramics. *Int. J. Appl. Ceram. Technol.* **2015**, *12*, 26–31. [CrossRef]

71. Scheithauer, U.; Poetschke, J.; Weingarten, S.; Schwarzer, E.; Vornberger, A.; Moritz, T.; Michaelis, A. Droplet-based additive manufacturing of hard metal components by thermoplastic 3D Printing (T3DP). *J. Ceram. Sci. Technol.* **2017**, *8*, 155–158.

72. Schwarzer, E.; Götz, M.; Markova, D.; Stafford, D.; Scheithauer, U.; Moritz, T. Lithography-based ceramic manufacturing (LCM)—Viscosity and cleaning as two quality influencing steps in the process chain of printing green parts. *J. Eur. Ceram. Soc.* **2017**. [CrossRef]

73. Gonzalez-Gutierrez, J.; Godec, D.; Guran, R.; Spoerk, M.; Kukla, C.; Holzer, C. 3D printing conditions determination for feedstock used in fused filament fabrication (FFF) of 17-4PH stainless steel parts. *Metalurgija* **2018**, *57*, 117–120.

74. Kukla, C.; Gonzalez-Gutierrez, J.; Cano, S.; Burkhardt, C.; Hampel, S.; Moritz, T.; Holzer, C. Fused Filament Fabrication for the production of metal and/or ceramic parts and feedstocks therefore. In Proceedings of the 19th Plansee Seminar, Reutte, Austria, 29 May–2 June 2017; Plansee Group Service GmbH: Reutte, Austria, 2017.

75. German, R.M. Markets applications, and financial aspects of global metal powder injection moulding (MIM) technologies. *Met. Powder Rep.* **2012**, *67*, 18–26. [CrossRef]

76. Bloemacher, M.; Weinand, D. CatamoldTM-A new direction for powder injection molding. *J. Mater. Process. Technol.* **1997**, *63*, 918–922. [CrossRef]

77. Farrow, G.; Conciatori, A.B. Polyacetal Binders for Injection Moulding of Ceramics. DE3477767 D1, 24 May 1989.

78. Liu, L.; Ni, X.L.; Yin, H.Q.; Qu, X.H. Mouldability of various zirconia micro gears in micro powder injection moulding. *J. Eur. Ceram. Soc.* **2015**, *35*, 171–177. [CrossRef]

79. Gonzalez-Gutierrez, J.; Oblak, P.; Megen, Z.M.; Emri, I. Viscosity and creep compliance of polyoxymethylene copolymers of various average molecular weights. *Polimery* **2015**, *61*, 620–627. [CrossRef]

80. Wiech, R.E. Manufacture of Parts from Particulate Material. U.S. Patent 497,698, 12 April 1976.

81. Kong, X.; Barriere, T.; Gelin, J.C. Determination of critical and optimal powder loadings for 316 L fine stainless steel feedstocks for micro-powder injection molding. *J. Mater. Process. Technol.* **2012**, *212*, 2173–2182. [CrossRef]

82. Thomas-Vielma, P.; Cervera, A.; Levenfeld, B.; Várez, A. Production of alumina parts by powder injection molding with a binder system based on high density polyethylene. *J. Eur. Ceram. Soc.* **2008**, *28*, 763–771. [CrossRef]

83. Sotomayor, M.E.; Ospina, L.M.; Levenfeld, B.; Várez, A. Characterization of 430 L porous supports obtained by powder extrusion moulding for their application in solid oxide fuel cells. *Mater. Charact.* **2013**, *86*, 108–115. [CrossRef]

84. Aggarwal, G.; Smid, I.; Park, S.J.; German, R.M. Development of niobium powder injection molding. Part II: Debinding and sintering. *Int. J. Refract. Met. Hard Mater.* **2007**, *25*, 226–236. [CrossRef]

85. Park, S.-J.; Wu, Y.; Heaney, D.F.; Zou, X.; Gai, G.; German, R.M. Rheological and Thermal Debinding Behaviors in Titanium Powder Injection Molding. *Met. Mat. Trans. A* **2009**, *40*, 215–222. [CrossRef]

86. Martin, R.; Vick, M.; Kelly, M.; Souza, J.P.d.; Enneti, R.K.; Atre, S.V. Powder injection molding of a mullite–zirconia composite. *J. Mater. Res. Technol.* **2013**, *2*, 263–268. [CrossRef]

87. Ma, J.; Qin, M.; Zhang, L.; Tian, L.; Li, R.; Chen, P.; Qu, X. Effect of ball milling on the rheology and particle characteristics of Fe–50%Ni powder injection molding feedstock. *J. Alloy. Compd.* **2014**, *590*, 41–45. [CrossRef]

88. Heng, S.Y.; Raza, M.R.; Muhamad, N.; Sulong, A.B.; Fayyaz, A. Micro-powder injection molding (μPIM) of tungsten carbide. *Int. J. Refract. Met. Hard Mater.* **2014**, *45*, 189–195. [CrossRef]

89. Chuankrerkkul, N.; Davies, H.A.; Messer, P.F. Application of PEG/PMMA Binder for Powder Injection Moulding of Hardmetals. *MSF* **2007**, *561–565*, 953–956. [CrossRef]

90. Yang, S.; Zhang, R.; Qu, X. X-ray tomographic analysis of powder-binder separation in SiC green body. *J. Eur. Ceram. Soc.* **2013**, *33*, 2935–2941. [CrossRef]

91. Chakartnarodom, P.; Chuankrerkkul, N. Statistical Analysis of Binder Behavior during Debinding Step in Powder Injection Molding (PIM). *AMR* **2014**, *970*, 172–176. [CrossRef]

92. Krauss, V.A.; Oliveira, A.; Klein, A.N.; Al-Qureshi, H.A.; Fredel, M.C. A model for PEG removal from alumina injection moulded parts by solvent debinding. *J. Mater. Process. Technol.* **2007**, *182*, 268–273. [CrossRef]

93. Yang, W.-W.; Yang, K.-Y.; Hon, M.-H. Effects of PEG molecular weights on rheological behavior of alumina injection molding feedstocks. *Mater. Chem. Phys.* **2003**, *78*, 416–424. [CrossRef]

94. Yang, W.-W.; Yang, K.-Y.; Wang, M.-C.; Hon, M.-H. Solvent debinding mechanism for alumina injection molded compacts with water-soluble binders. *Ceram. Int.* **2003**, *29*, 745–756. [CrossRef]

95. Hauf, G. Verfahren zur Herstellung von Formkörpern aus Keramischen und/Oder Metallischen Pulvern und Dabei Verwendetes Bindersystem. DE Patent 103 47747.0, 14 October 2003.

96. Wohlfromm, H.; Assmann, J.; ter Maat, J.; Mehdizadeh, A. Thermoplastic Masses Containing Binding Agents for the Production of Metallic Molds. International Patent PCT/EP2007/056857, 5 July 2007.

97. Sterzel, H.-J.; ter Maat, J.; Ebenhoech, J.; Meyer, M. Thermoplastic Materials for the Production of Ceramic Moldings. U.S. Patent 823,607, 17 January 1992.

98. LaSalle, J.C.; Behi, M.; Glandz, G.A.; Burlew, J.V. Aqueous Nonferrous Feedstock Material for Injection Molding. U.S. Patent 09/768,411, 24 January 2001.

99. Yang, X.; Petcavich, R.J. Powder and Binder Systems for Use in Metal and Ceramic Powder Injection Molding. U.S. Patent 6008281, 13 January 1999.

100. Danforth, S.C.; Agarwala, M.K.; Bandyopadhyay, A.; Langrana, N.; Jamalabad, V.R.; Safari, A.; van Weeren, R. Solid Freeform Fabrication Methods. U.S. Patent 5900207, 4 May 1999.

101. Daute, P.; Jaeckel, M.; Waldmann, J. Sinterable Feedstock for Use in 3d Printing Devices. International Patent PCT/EP2014/064646, 8 July 2014.

102. Heikkila, K.E. Surface Modified Particulate and Sintered Extruded Products. WIPO Patent WO/2015/006697, 15 January 2015.

103. Pekin, S.; Zangvil, A.; Ellingson, W. Binder Formulation in EVA-wax system for Fused Deposition of Ceramics. In Proceedings of the Solid Freeform Fabrication Symposium, Austin, TX, USA, 10–12 August 1998.

104. Iyer, S.; McIntosh, J.; Bandyopadhyay, A.; Langrana, N.; Safari, A.; Danforth, S.C.; Clancy, R.B.; Gasdaska, C.; Whalen, P.J. Microstructural Characterization and Mechanical Properties of Si_3N_4 Formed by Fused Deposition of Ceramics. *Int. J. Appl. Ceram. Technol.* **2008**, *5*, 127–137. [CrossRef]

105. Riecker, S.; Clouse, J.; Studnitzky, T.; Andersen, O.; Kieback, B. Fused Deposition Modeling-Opportunities for cheap metal AM. In Proceedings of the World PM2016 Congress & Exhibition, Hamburg, Germany, 9–13 October 2016.

106. Giberti, H.; Strano, M.; Annoni, M.; Yuan, Y.; Menon, L.; Xu, X. An innovative machine for Fused Deposition Modeling of metals and advanced ceramics. *MATEC Web Conf.* **2016**, *43*, 3003. [CrossRef]

107. McNulty, T.F.; Shanefield, D.J.; Danforth, S.C.; Safari, A. Dispersion of Lead Zirconate Titanate for Fused Deposition of Ceramics. *J. Am. Ceram. Soc.* **1999**, *82*, 1757–1760. [CrossRef]

108. Pistor, C.M. Thermal Properties of Green Parts for Fused Deposition of Ceramics (FDC). *Adv. Eng. Mater.* **2001**, *3*, 418–423. [CrossRef]

109. Pekin, S.; Bukowski, J.; Zangvil, A. A Study on Weight Loss Rate Controlled Binder Removal From Parts Produced by FDC. In Proceedings of the Solid Freeform Fabrication Symposium, Austin, TX, USA, 10–12 August 1998.

110. Kukla, C.; Duretek, I.; Schuschnigg, S.; Gonzalez-Gutierrez, J.; Holzer, C. Properties for PIM Feedstocks Used in Fused Filament Fabrication. In Proceedings of the World PM2016 Congress & Exhibition, Hamburg, Germany, 9–13 October 2016.

111. Kukla, C.; Gonzalez-Gutierrez, J.; Duretek, I.; Schuschnigg, S.; Holzer, C. Effect of Particle Size on the Properties of Highly-Filled Polymers for Fused Filament Fabrication. In Proceedings of the 32nd PPS International Conference, Lyon, France, 25–29 July 2016; pp. 274–277.

112. Kukla, C.; Gonzalez-Gutierrez, J.; Cano, S.; Hampel, S.; Burkhardt, C.; Moritz, T.; Holzer, C. Fused Filament Fabrication (FFF) of PIM Feedstocks. In *Proceedings of VI Congreso Nacional de Pulvimetalurgia y I Congreso Iberoamericano de Pulvimetalurgia*; Ciudad Real, Spain, 7–9 June 2017.

113. Kukla, C.; Gonzalez-Gutierrez, J.; Burkhardt, C.; Weber, O.; Holzer, C. The Production of Magnets by FFF-Fused Filament Fabrication. In Proceedings of the Euro PM2017 Congress & Exhibition, Milan, Italy, 1–5 October 2017; EPMA: Shrewsbury, UK, 2017; pp. 1–6.

114. Abel, J.; Scheithauer, U.; Hampel, S.; Cano, S.; Müller-Köhn, A.; Günther, A.; Kukla, C.; Moritz, T. Fused Filament Fabrication (FFF) of metal-ceramic components. *J. Vis. Exp.* under revision.

115. Kukla, C.; Cano, S.; Kaylani, D.; Schuschnigg, S.; Holzer, C.; Gonzalez-Gutierrez, J. Debinding behaviour of feedstock for material extrusion additive manufacturing of zirconia. *Powder Metall.* **2017**. submitted.

116. Nestle, N.; Hermant, M.; Schimdt, K. Mixture for Use in a Fused Filament Fabrication Process. International Patent PCT/EP2015/066728, 22 July 2015.

117. Ren, L.; Zhou, X.; Song, Z.; Zhao, C.; Liu, Q.; Xue, J.; Li, X. Process Parameter Optimization of Extrusion-Based 3D Metal Printing Utilizing PW–LDPE–SA Binder System. *Materials* **2017**, *10*, 305. [CrossRef] [PubMed]

118. Agarwala, M.K.; van Weeren, R.; Vaidyanathan, R.; Bandyopadhyay, A.; Carrasquillo, G.; Jamalabad, V.R.; Langrana, N.; Safari, A.; Garofalini, S.H.; Danforth, S.C.; et al. Structural Ceramics by Fused Deposition of Ceramics. In Proceedings of the Solid Freeform Fabrication Symposium, Austin, TX, USA, 7–9 August 1995.

119. Bandyopadhyay, A.; Das, K.; Marusich, J.; Onagoruwa, S. Application of fused deposition in controlled microstructure metal-ceramic composites. *Rapid Prototyp. J.* **2006**, *12*, 121–128. [CrossRef]

120. Onagoruwa, S.; Bose, S.; Bandyopadhyay, A. Fused Deposition of Ceramics (FDC) and Composites. In Proceedings of the Solid Freeform Fabrication Symposium, Austin, TX, USA, 6–8 August 2001.

121. Bhat, V.V.; Geetha, K.; Das, R.N.; Gurumoorthy, B.; Umarji, A.M. Characterization of Polyolefin-Aluminacompounded mix for FDC processing. In Proceedings of the Solid Freeform Fabrication Symposium, Austin, TX, USA, 8–10 August 2000.

122. Stanimirovic, Z.; Stanimirovic, I. Ceramic Injection Molding. In *Some Critical Issues for Injection Molding*; Wang, J., Ed.; InTech: Rijeka, Croatia, 2012; pp. 131–148.

123. Venkataraman, N.; Rangarajan, S.; Matthewson, M.J.; Safari, A.; Danforth, S.C.; Yardimci, A.; Guceri, S.I. Mechanical and Rheological Properties of Feedstock Material for Fused Deposition of Ceramics and Metals (FDC and FDMet) and their Relationship to Process Performance. In Proceedings of the Solid Freeform Fabrication Symposium, Austin, TX, USA, 9–11 August 1999.

124. Gonzalez-Gutierrez, J.; Duretek, I.; Holzer, C.; Arbeiter, F.; Kukla, C. Filler Content and Properties of Highly Filled Filaments for Fused Filament Fabrication of Magnets. In Proceedings of the ANTEC 2017, Anaheim, CA, USA, 8–10 May 2017; Society of Plastics Engineers, Ed.; Society of Plastics Engineers: Bethel, CT, USA, 2017; pp. 1–4.

125. Burkhardt, C.; Freigassner, P.; Weber, O.; Imgrund, P.; Hampel, S. Fused Filament Fabrication (FFF) of 316L Green Parts for the MIM process. In Proceedings of the World PM2016 Congress & Exhibition, Hamburg, Germany, 9–13 October 2016; EPMA: Shrewsbury, UK, 2016; pp. 1–7.

126. EVO-Tech GmbH. FMP: Filament Metal Printing. Available online: http://evo-tech.eu/de/Filament-Metal-Printing (accessed on 22 August 2017).

127. Hausnerova, B.; Kitano, T.; Kuritka, I.; Prindis, J.; Marcanikova, L. The Role of Powder Particle Size Distribution in the Processability of Powder Injection Molding Compounds. *Int. J. Polym. Anal. Charact.* **2011**, *16*, 141–151. [CrossRef]

128. Brostow, W.; Buchman, A.; Buchman, E.; Olea-Mejia, O. Microhybrids of metal powder incorporated in polymeric matrices: Friction, mechanical behavior, and microstructure. *Polym. Eng. Sci.* **2008**, *48*, 1977–1981. [CrossRef]

129. Uhm, Y.R.; Kim, J.; Son, K.J.; Kim, C.S. Effect of particle size, dispersion, and particle–matrix adhesion on W reinforced polymer composites. *Res. Chem. Intermed.* **2014**, *40*, 2145–2153. [CrossRef]

130. Venkataraman, N.; Rangarajan, S.; Matthewson, M.J.; Harper, B.; Safari, A.; Danforth, S.C.; Wu, G.; Langrana, N.; Guceri, S.I.; Yardimci, A. Feedstock material property-process relationships in fused deposition of ceramics (FDC). *Rapid Prototyp. J.* **2000**, *6*, 244–253. [CrossRef]

131. Beer, F.P.; Johnston, E.R.; DeWolf, J.T. *Mechanics of Materials*, 2nd ed.; McGraw-Hill: London, UK, 1992.

132. Barnes, H.A.; Hutton, J.F.; Walters, K. *An Introduction to Rheology*; Elsevier Science Pub. Co.: Amsterdam, The Netherlands; New York, NY, USA, 1989.

133. Rangarajan, S.; Qi, G.; Venkataraman, N.; Safari, A.; Danforth, S.C. Powder Processing, Rheology, and Mechanical Properties of Feedstock for Fused Deposition of Si_3N_4 Ceramics. *J. Am. Ceram. Soc.* **2000**, *83*, 1663–1669. [CrossRef]

134. Grida, I.; Evans, J.R. Extrusion freeforming of ceramics through fine nozzles. *J. Eur. Ceram. Soc.* **2003**, *23*, 629–635. [CrossRef]

135. Griffin, E.A.; McMillin, S. Selective Laser Sintering and Fused Deposition Modeling Processes for Functional Ceramic Parts. In Proceedings of the Solid Freeform Fabrication Symposium, Austin, TX, USA, 7–9 August 1995.

136. Edirisinghe, M.J.; Evans, J.R.G. Compounding Ceramic Powders Prior to Injection Moulding. *Br. Ceram. Proc.* **1986**, *38*, 67–81.

137. Clancy, R.; Jamalabad, V.R.; Whalen, P.; Bhargava, P.; Dai, C.; Rangarajan, S.; Wu, S.; Danforth, S.C.; Langrana, N.; Safari, A. Fused Deposition of Ceramics: Progress Torwards a Robust and Controlled Process for Commercialization. In Proceedings of the Solid Freeform Fabrication Symposium, Austin, TX, USA, 11–13 August 1997.

138. Spoerk, M.; Arbeiter, F.; Cajner, H.; Sapkota, J.; Holzer, C. Parametric optimization of intra- and inter-layer strengths in parts produced by extrusion-based additive manufacturing of poly(lactic acid). *J. Appl. Polym. Sci.* **2017**, *134*, 45401. [CrossRef]

139. Allahverdi, M.; Danforth, S.C.; Jafari, M.A.; Safari, A. Processing of advanced electroceramic components by fused deposition technique. *J. Eur. Ceram. Soc.* **2001**, *21*, 1485–1490. [CrossRef]

140. Kuo, C.-C.; Mao, R.-C. Development of a Precision Surface Polishing System for Parts Fabricated by Fused Deposition Modeling. *Mater. Manuf. Process.* **2015**, *31*, 1113–1118. [CrossRef]

141. Garg, A.; Bhattacharya, A.; Batish, A. On Surface Finish and Dimensional Accuracy of FDM Parts after Cold Vapor Treatment. *Mater. Manuf. Process.* **2015**, *31*, 522–529. [CrossRef]

142. Takagishi, K.; Umezu, S. Development of the Improving Process for the 3D Printed Structure. *Sci. Rep.* **2017**, *7*, 39852. [CrossRef] [PubMed]

143. Enneti, R.K.; Park, S.J.; German, R.M.; Atre, S.V. Review: Thermal Debinding Process in Particulate Materials Processing. *Mater. Manuf. Process.* **2012**, *27*, 103–118. [CrossRef]

144. Banerjee, S.; Joens, C.J. Debinding and sintering of metal injection molding (MIM) components. In *Handbook of Metal Injection Molding*; Heaney, D.F., Ed.; Woodhead Publishing: Cambridge, UK; Philadelphia, PA, USA, 2012; pp. 133–180.

145. Lous, G.M.; Cornejo, I.A.; McNulty, T.F.; Safari, A.; Danforth, S.C. Fabrication of curved ceramic/polymer composite transducer for ultrasonic imaging applications by fused deposition of ceramics. In Proceedings of the Eleventh IEEE International Symposium on Applications of Ferroelectrics, Montreux, Switzerland, 24–27 August 1998; pp. 239–242.

146. Krug, S.; Evans, J.; ter Maat, J. Differential sintering in ceramic injection moulding: Particle orientation effects. *J. Eur. Ceram. Soc.* **2002**, *22*, 173–181. [CrossRef]

147. Huang, M.-S.; Hsu, H.-C. Influence of injection moulding and sintering parameters on properties of 316L MIM compact. *Powder Met.* **2013**, *54*, 299–307. [CrossRef]

148. Scheithauer, U.; Slawik, T.; Schwarzer, E.; Richter, H.-J.; Moritz, T.; Michaelis, A. Additive manufacturing of metallic and ceramic multimaterial components using thermoplastic 3D printing (T3DP). *J. Ceram. Sci. Technol.* **2015**, *6*, 125–131.

149. Jafari, M.A.; Han, W.; Mohammadi, F.; Safari, A.; Danforth, S.C.; Langrana, N. A novel system for fused deposition of advanced multiple ceramics. *Rapid Prototyp. J.* **2000**, *6*, 161–175. [CrossRef]

150. Scheithauer, U.; Johne, R.; Weingarten, S.; Schwarzer, E.; Abel, J.; Müller-Köhn, A.; Günther, A.; Moritz, T. CerAMfacturing: Development of ceramic and multi material components by additive manufacturing methods for personalized medical products. *J. 3D Print. Med.* **2018**, *2*, 15–25. [CrossRef]

151. Bandyopadhyay, A.; Panda, R.K.; Janas, V.F.; Danforth, S.C.; Safari, A. Processing of Piezocomposites via Solid Freeform Fabrication (SFF) Techniques. In Proceedings of the Solid Freeform Fabrication Symposium, Austin, TX, USA, 12–14 Augus 1996.

152. Lous, G.M.; Cornejo, I.A.; McNulty, T.F.; Safari, A.; Danforth, S.C. Fabrication of Piezoelectric CeramicPolymer Composite Transducers Using Fused Deposition of Ceramics. *J. Am. Ceram. Soc.* **2000**, *83*, 124–128. [CrossRef]

153. Atisivan, R.; Bose, S.; Bandyopadhyay, A. Porous Mullite Preforms via Fused Deposition. *J. Am. Ceram. Soc.* **2001**, *84*, 221–223. [CrossRef]

154. Zhou, X.; Liu, C.-J. Three-dimensional Printing for Catalytic Applications: Current Status and Perspectives. *Adv. Funct. Mater.* **2017**, *27*, 1701134. [CrossRef]

MDPI

St. Alban-Anlage 66

4052 Basel

Switzerland

Tel. +41 61 683 77 34

Fax +41 61 302 89 18

www.mdpi.com

Materials Editorial Office

E-mail: materials@mdpi.com

www.mdpi.com/journal/materials